G. Brunnett
H. Bieri
G. Farin (eds.)

Geometric Modelling

Dagstuhl 1999

Computing
Supplement 14

SpringerWienNewYork

Professor Dr. Guido Brunnett
Fakultät für Informatik, TU Chemnitz,
Chemnitz, Germany

Dr. Hanspeter Bieri
Institut für Informatik und angewandte Mathematik,
Universität Bern,
Bern, Switzerland

Professor Dr. Gerald Farin
Department of Computer Science and Engineering,
Arizona State University
Tempe, AZ, USA

Printed in Austria

Typesetting: Scientific Publishing Services (P) Ltd., Madras
Printing and Binding: Manz Crossmedia, 1051 Wien

Printed on acid-free and chlorine-free bleached paper

SPIN: 10794546

With 204 Figures

CIP-data applied for

ISSN 0344-8029
ISBN 3-211-83603-9 Springer-Verlag Wien New York

Preface

The fourth Dagstuhl seminar on Geometric Modelling took place in May 1999 and was organized by Hanspeter Bieri (University Bern), Guido Brunnett (Technical University Chemnitz) and Gerald Farin (Arizona State University). This workshop brought together experts from the fields of Computer Aided Geometric Design and Computational Geometry to discuss the state-of-the-art and current trends of Geometric Modelling. 56 participants from Austria, Canada, Croatia, England, France, Germany, Greece, Hungary, Israel, Korea, Netherlands, Norway, Spain, Swiss and USA were present.

Participation in the Dagstuhl workshops is by invitation only, thus ensuring a high level of expertise among the attendees. In addition, all papers for this book underwent a careful refereeing process. We would like to thank the referees for their efforts.

The topics discussed on the workshop included classical surface and solid modelling as well as geometric foundations of CAGD. However, the focus of this workshop was on new developments as surface reconstruction, mesh generation and multiresolution models. Taken together these topics show that Geometric Modelling still is a lively field that provides fundamental methods to different application areas as CAD/CAM, Computer Graphics, Medical Imaging and Scientific Visualization.

As a special highlight of the workshop two prominent researchers Prof. Michael J. Pratt and Prof. Larry L. Schumaker have been awarded the John Gregory Memorial Award for their fundamental contributions to Geometric Modelling and their enduring influence on this field.

March, 2001

Guido Brunnett
Hanspeter Bieri
Gerald Farin

Contents

(*Listed in Current Contents*)

Computing [Suppl] 14, 1–18 (2001)

Computing

Converting Orthogonal Polyhedra from Extreme Vertices Model to B-Rep and to Alternating Sum of Volumes

A. Aguilera, Puebla, and **D. Ayala,** Barcelona

Abstract

In recent published papers we presented the Extreme Vertices Model (EVM), a concise and complete model for representing orthogonal polyhedra and pseudopolyhedra (OPP). This model exploits the simplicity of its domain by allowing robust and simple algorithms for set-membership classification and Boolean operations that do not need to perform floating-point operations.

Several applications of this model have also been published, including the suitability of OPP as geometric bounds in Constructive Solid Geometry (CSG).

In this paper, we present an algorithm which converts from this model into a B-Rep model. We also develop the application of the Alternating Sum of Volumes decomposition to this particular type of polyhedra by taking advantage of the simplicity of the EVM. Finally we outline our future work, which deals with the suitability of the EVM in the field of digital images processing.

AMS Subject Classifications: I. 3 Computer graphics; I. 3.1 Computational geometry and object modeling.

Key Words: Solid modeling, boundary representation, orthogonal polyhedra, alternating sum of volumes, extreme vertices model.

1. Introduction

In previous papers we presented a specific model for OPP, the Extreme Vertices Model, (EVM). This model is very concise, and although it only needs to store some of the OPP vertices, it has been proved to be complete. In [2] we presented the EVM for OP, a Boolean operations algorithm and an application consisting in using OP as geometric bounds in CSG. In [3] the domain was extended to OPP and we proved the completeness of the model and all the remaining formal properties. We also analyzed set-membership classification algorithms in the EVM. The problems of point and plane classification were extensively detailed in [4].

In this paper we present two contributions related with the model. We first present an algorithm which converts from EVM into a B-Rep. Then, we develop the application of the Alternating Sum of Volumes decomposition to the particular type of OPP by taking advantage of the simplicity of the EVM.

The paper is arranged as follows. The section below includes a brief review on the EVM, focusing particularly on those concepts and properties which are needed in the following sections. Section 3 explains the EVM to B-Rep conversion algorithm. Section 4 introduces the ASV decomposition and Section 5 develops the application of this technique to OPP. Finally, the last section outlines future work which is oriented to the study of the suitability of the EVM in the field of digital images processing.

2. The Extreme Vertices Model

2.1. Orthogonal Polyhedra (OP) and Pseudo-Polyhedra (OPP)

An OP is a polyhedron with all of its edges and faces oriented in three orthogonal directions. A pseudo-polyhedron [26] is a regular polyhedron with a non-manifold boundary. An OPP is an orthogonal pseudo-polyhedron. The class of OPP involves a drastic restriction with respect to general polyhedra, concerning geometry, as follows from its definition. However, with respect to topology, OPP do not involve any restriction at all, and they can have any genus and any number of shells.

In order to classify vertices for our purposes, we have done an exhaustive analysis of the neighborhood of an orthogonal vertex [1]. Vertices in an OPP follow the same pattern as do nodes in the marching cubes algorithm [20]. There are $2^8 = 256$ combinations which are grouped, by applying rotational symmetries, into 22 cases [25] and, by grouping complementary cases into the 14 basic patterns [20]. Figure 1 shows the 22 cases (from *a* to *v*). The 14 basic patterns are those from *a* to *n*.

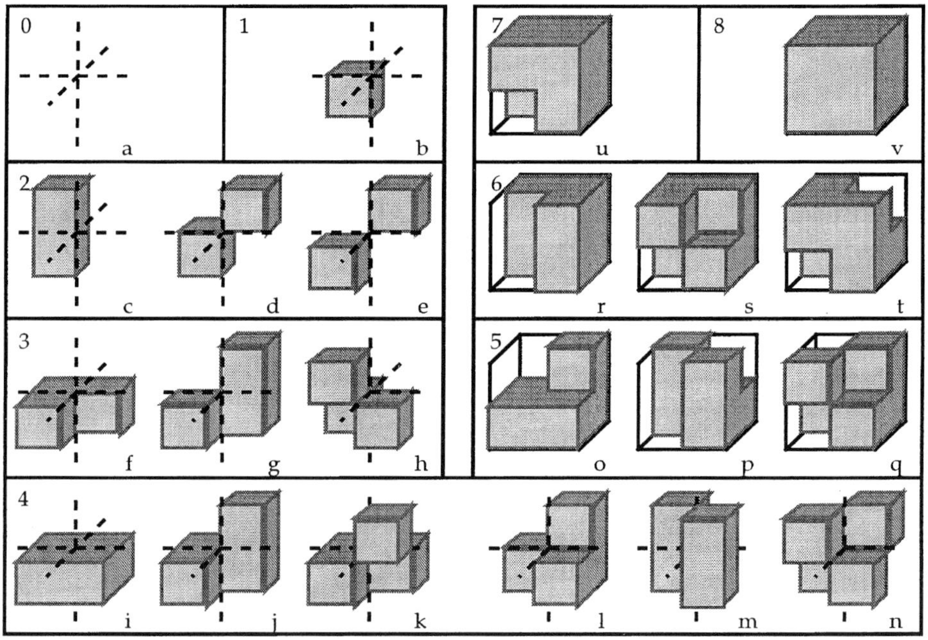

Figure 1. The 22 possible cases

The 14 basic patterns have finally been grouped into 8 classes depending on the number of manifold and non-manifold incident edges. The name of the vertex indicates the total number of incident edges, if it is a non-manifold vertex (N) and, in this case, the number of non-manifold incident edges. See Fig. 2 and the following table.

Vertex name	V3	V4	V4N1	V4N2	V5	V6	V6N1	V6N2
Patterns	b, f	j	g	k	d	e, l	h	n

2.2. *EVM Definitions*

Let P be an orthogonal pseudo-polyhedron (OPP).

A *brink* is the longest uninterrupted segment built out of a sequence of collinear and contiguous two-manifold edges of P. See Fig. 3.

In a brink each ending vertex is V3, V4N1 or V6N3 and the remaining (interior) vertices are V4, V4N2, V5N1 or V6. Vertices V6N6 do not belong to any brink. Vertices V3, V4N1, and V6N3 share the property of having exactly three incident two-manifold and linearly independent edges, regardless of the number of incident non-manifold edges, and are called *extreme vertices, EV*.

The *Extreme Vertices Model, EVM*, is a representation scheme for OPP in which any OPP is represented by its set of EV. From a theoretical point of view, vertices in the EVM can be without any order. However, for implementation purposes, they will be considered in an ordered way so that brinks parallel to axis X, Y or Z will appear consecutively. There are six different possible sortings: XYZ, XZY, YXZ, YZX, ZXY, and ZYX. For instance, when extreme vertices are sorted in the XZY or ZXY way, brinks parallel to the Y axis appear directly as pairs of consecutive EV (see Fig. 4).

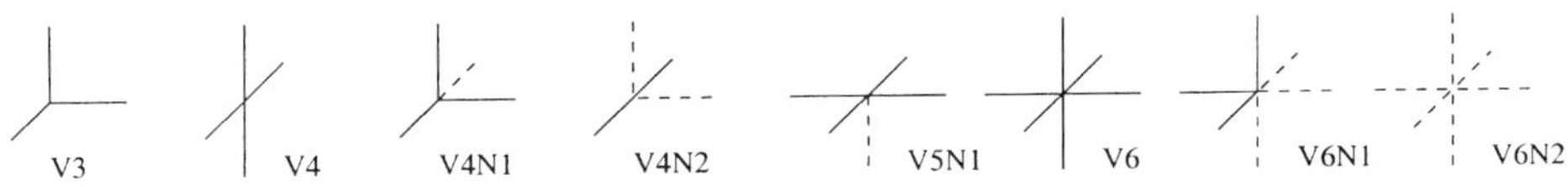

Figure 2. Vertex classification

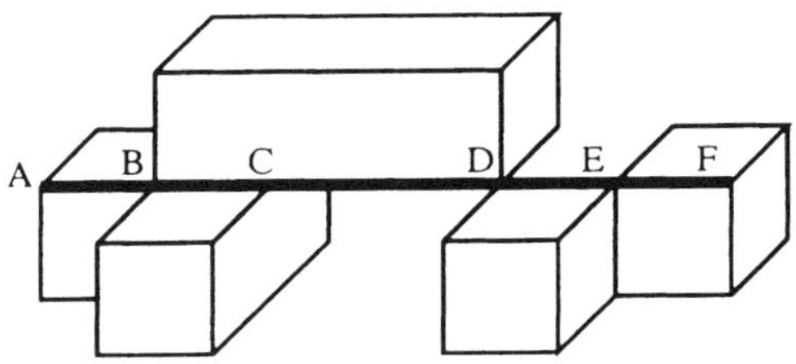

Figure 3. An OPP with a brink having five edges and six vertices

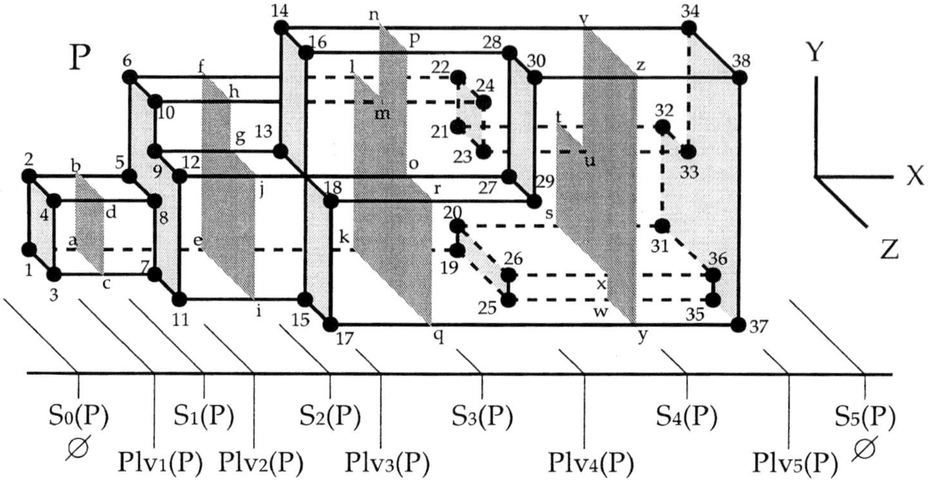

Figure 4. An OP, P, with five planes of vertices (light regions) and four slices with their corresponding sections (dark regions). Extreme vertices are numbered in the same way as they appear in a XZY-sorted EVM

A *plane of vertices* is the set of vertices lying on a plane perpendicular to a main axis of P. A *slice* is the region between two consecutive planes of vertices. A *section* is the resulting polygon from the intersection between P and an orthogonal plane. If P has n planes of vertices, $plv_i(P), i = 1 \ldots n$, it will have $n - 1$ slices and it can be expressed as $P = \bigcup_{k=1}^{k=n-1} slice_k(P)$. Each slice $slice_k(P)$ has its representing section $S_k(P)$ and there are two more empty sections, the initial and final sections, $S_0(P)$ and $S_n(P)$ (see Fig. 4). Planes of vertices and sections are 2D-OPP embedded in 3D space, and we will sometimes need to work with their 2D projection, so we will denote by $\bar{P}$ the projection of (d − 1)-dimensional OPP, P, which is embedded in E^d, onto a main plane parallel to P. In order to obtain such a projection we only need not to consider the first coordinate of all P vertices.

All these definitions can be extended to any dimension [10]. In this paper we are concerned with dimension ≤ 3.

2.3. Properties of the EVM

All the following properties are formally demonstrated in [1].

The first property concerning EVM is that coordinate values of non-extreme vertices may be obtained from EV coordinates. Some non-extreme vertices correspond to the intersection of two or three perpendicular brinks and, therefore, their coordinates come directly from the EV coordinates of these brinks. Coordinates of the remaining non-extreme vertices are obtained from EV and from the non-extreme vertices first obtained.

The two next properties shown below relate to sections and planes of vertices of P. We can compute sections from planes of vertices:

$$S_0(P) = \emptyset$$

$$\overline{S_i(P)} = \overline{S_{i-1}(P)} \otimes^* \overline{plv_i(P)}, \quad \forall i \in [1, np]$$

And conversely, we can compute planes of vertices from sections:

$$\overline{plv_i(P)} = \overline{S_{i-1}(P)} \otimes^* \overline{S_i(P)}, \quad \forall i \in [1, np]$$

where $\otimes^*$ is the regularized XOR operation.

Applying the definition of the $\otimes$ operation, this last equation can be expressed as: $\overline{plv_i(P)} = \overline{S_{i-1}(P)} \otimes^* \overline{S_i(P)} = (\overline{S_{i-1}(P)} -^* \overline{S_i(P)}) \cup (\overline{S_i(P)} -^* \overline{S_{i-1}(P)}), \quad \forall i \in [1, np]$, and, thus, we can decompose any plane of vertices into two terms that we will call *forward difference* ($\overline{FD_i(P)} = (\overline{S_{i-1}(P)} -^* \overline{S_i(P)})$ and *backward difference* ($\overline{BD_i(P)} = (\overline{S_i(P)} -^* \overline{S_{i-1}(P)})$).

The following property guarantees that the correct orientation of all faces of P can be obtained from its EVM. $FD_i(P)$ is the set of faces on $plv_i(P)$ whose normal vector points to one side of the main axis perpendicular to $plv_i(P)$, while $BD_i(P)$ is the set of faces whose normal vector points to the other side (see Fig. 5).

Concerning the so-called formal properties [23] we have proved that EVM is complete (non-ambiguous) and unique. The domain is limited to OPP and there is a validity condition for a finite point set to be a valid EVM for some OPP [1].

2.4. Boolean Operations

The EVM behaves nicely with Boolean operations. The XOR operation applied to EVM represented OPP fulfills the following nice property:

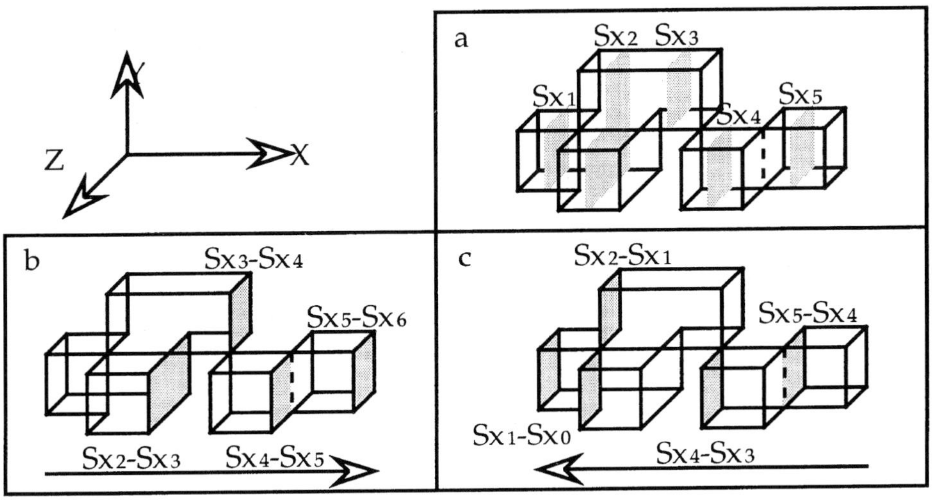

Figure 5. An OPP with its **a** sections, **b** forward differences and **c** backward differences, perpendicular to X

Theorem 1. *Let P and Q be two d-D* ($d \leq 3$) *OPP, having* $EVM(P)$ *and* $EVM(Q)$ *as their respective models, then* $EVM(P \otimes^* Q) = EVM(P) \otimes EVM(Q)$.

This theorem is formally proved in [1]. It is proved by induction over the dimension and the basis of the induction (case 1D) is proved exhaustively. The property means that XOR between two OPP, which are infinite sets of points, can be carried out by applying the operation XOR to their EVM models, which are finite sets of EV.

The following two properties are corollaries of the previous one and are used in the application presented in Section 5.

Corollary 1. *If* $Q \subseteq P$ *then* $EVM(P -^* Q) = EVM(P) \otimes EVM(Q)$

Corollary 2. *If P and Q are quasi-disjoint, then* $EVM(P \cup^* Q) = EVM(P) \otimes EVM(Q)$

General Boolean operations between two OPP can be carried out by applying the same operation over the corresponding OPP sections. The corresponding algorithm is presented in [2] and consists in a geometric merger between the EVM of both operands.

2.5. The Splitting Operation

Set-membership classification tests were given a general analysis in [3] and the particular cases of point and plane classification were detailed in [4]. However, here we review the splitting operation as it is needed in the application presented in Section 5.

The classification of an OPP, *P*, against a splitting plane, *SP*, perpendicular to a main axis produces two polyhedra, *Q* and *R*, in the IN and OUT half-spaces of *SP* respectively. EVM(Q) and EVM(R) will be subsets of EVM(P) except for some possible new vertices that will be created and that will lie on *SP*. If *SP* is perpendicular to the *C* axis, only brinks parallel to this axis, C-brinks, have to be considered and these brinks appear as consecutive pairs of vertices in a conveniently ordered EVM (see Fig. 6).

Let $V_b = V_{2k-1}$ and $V_e = V_{2k}$ be the beginning and ending vertices of the *kth* C-brink. The classification of this brink with respect to *SP* gives the following two cases:

- V_b and V_e lie in the same half-space of *SP* or one of them is ON. Then both of them will be assigned either to *Q* or to *R*.
- Each vertex, V_b and V_e, belongs to a different half-space. A new vertex V_i is computed as the intersection between the C-brink and *SP*. The brink is split into two brinks: the brink from V_b to V_i goes to *Q* and the brink from V_i to V_e goes to *R*. V_i is obtained without any floating-point computation. Consider an ZYX-sorted EVM and let $x = x_p$ be the *SP* equation, $V_b = (x_1, y, z)$ and $V_e = (x_2, y, z)$, then $V_i = (x_p, y, z)$.

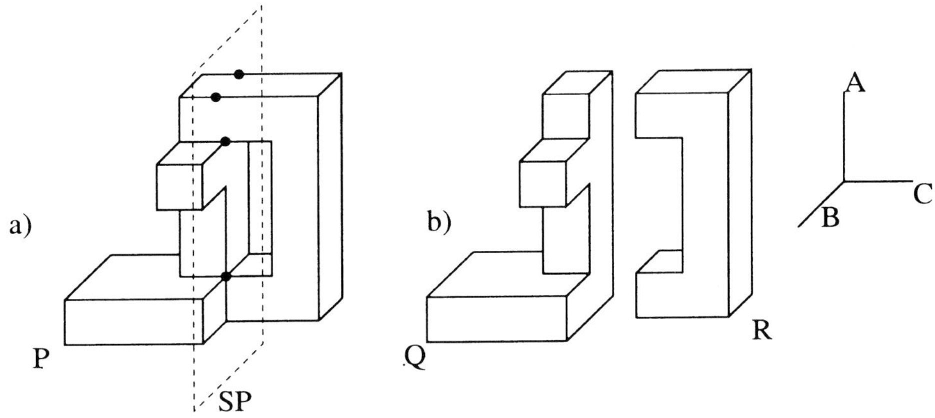

Figure 6. Splitting operation. **a** Object P and splitting plane SP. Dots show new vertices created. **b** Objects Q and R

2.6. Applications

There are a number of published papers dealing with OP. In [13], [14] the problem of converting a B-Rep into a Peterson-style CSG is studied for OP.

In [7] a method is presented for simplifying OP. This method has been extended to general polyhedra but it uses OP in its process [6], [5].

In [10] a representation scheme for OPP in any dimension is presented and operations such as face detection and Boolean operations are studied. This representation is very similar to ours but it includes all the vertices with assigned colors. The authors work in the field of dynamical systems and restrict the state-space to being OPP [11].

Concerning EVM-represented OPP, in [2] the suitability of OPP as geometric bounds in CSG is discussed and the use of OPP as geometric approximations of general polyhedra is presented in [1].

The restricted class of convex and orthogonal polyhedra, i.e., orthogonal boxes, has been widely used in many applications [22], [12], [24].

3. EVM to B-Rep Conversion Algorithm

The EVM is complete, which means that all the geometry and topology can be obtained from it. In this section we present a conversion algorithm from EVM to a B-Rep.

The input of the algorithm is the set of points constituting the EVM of an OPP and the output is the set of faces of a B-Rep model each one with its corresponding normal vector and with its associated set of edges. Edges are oriented according to the normal vector of the face to which they belong. Therefore the output consists in the geometrical information corresponding to the normal vectors of each face and the coordinates of each vertex and the topological relations $f : \{e\}$ and $e : \{v\}$.

The algorithm does not provide edges ordered in the traveling order around faces and does not make distinction between edges belonging to the external boundary and to the possible internal boundaries (holes) of a face. If such order and distinction are required then a well-known postprocess is needed [12] which applies a domino-like procedure to obtain contours, and several point polygon containment tests in order to classify contours as external or as holes. An outline of the algorithm is showed below:

```
procedure EVM to BRep(input p : EVM, output q : BRep)
  var
    dim : integer   {current dimension}
    dir : boolean   {orientation for faces and edges}
  endvar
  q := IniBRep(); dim := 3; dir := true
  Sort(p, XYZ); ProcessDifferences(p, dim, dir, q)
  Sort(p, YZX); ProcessDifferences(p, dim, dir, q)
  Sort(p, ZXY); ProcessDifferences(p, dim, dir, q)
endprocedure
procedure ProcessDifferences(input p : EVM, input dim : integer,
  input dir : boolean, inputoutput q : BRep)
  var
    Si, Sj : EVM    {two consecutive sections}
    plv : EVM       {the current plane or line of vertices}
    ForwardDif, BackwardDif : EVM
  endvar
  dim := dim − 1; Si := ∅; plv := GetPlv(p, dim)
  while ¬EndEVM(p) do
        Sj := ComputeSeccion(Si, plv, dim)
        ForwardDif := OpBool(Si, Sj, dim, '−')
        BackwardDif := OpBool(Sj, Si, dim, '−')
        if dim = 2 → ProcessDifferences(ForwardDif, dim, dir, q)
                     Sort(ForwardDif, ACB)
                     ProcessDifferences(ForwardDif, dim, dir, q)
                     ProcessDifferences(BackwardDif, dim, ¬dir, q)
                     Sort(BackwardDif, ACB)
                     ProcessDifferences(BackwardDif, dim, ¬dir, q)
        □ dim = 1 → AddEdgeBRep(ForwardDif, dir, q)
```

$AddEdgeBRep(BackwardDif, \neg dir, q)$

endif

$Si := Sj;\ plv := GetPlv(p, dim)$

endwhile

endprocedure

The algorithm works first for dimension 3 (3D) and then for dimension 2 (2D). In 3D, the set of EV of the EVM is sorted in three ways thus making it possible to obtain the faces parallel to each coordinate plane. Moreover the property concerning forward and backward differences (FD, BD), showed in Section 2.3, allows to determine which of these faces have the normal vector pointing to the interior of the solid and which of them have it pointing out the solid. Then, in 2D, the sets of EV corresponding to FD and to BD are sorted in two orderings which enables us to obtain the edges parallel to each coordinate axis, also correctly oriented thanks to the mentioned property. FD and BD are initially already sorted in one way (the sorting which comes from the algorithm when it works in 3D, say ABC) and so we only need to sort them in the other possible way (ACB).

Planes and lines of vertices, sections, and FD and BD are EVM-represented 2D or 1D orthogonal objects. Planes of vertices come directly from the EVM. Sections are computed by means of XOR operations and FD and BD computation involve Boolean differences. The variable *dir* is used to assign the correct orientation to each face and edge: *dir* = *TRUE* indicates that the FD normal vector points say to the solid interior while *dir* = *FALSE* indicates that the BD normal vector points to the solid exterior.

When computing FD and BD (2D and 1D) not only the correct orientation of faces and edges is obtained but also vertices that did not appear in the EVM come up.

Figure 7 shows how this algorithm works in two examples corresponding to the planes of vertices plv_2 and plv_4 of Fig. 5. In both cases a V6 vertex, V, appears which was not in the EVM. For plv_2 when the algorithm works in 3D (sorting *XBC*), the whole plane of vertices belongs to BD, then this BD is processed in 2D (sorting *XZY*). When the 1D BD $S_{xz2} - S_{xz1}$ is computed, both vertex V and edge (V, 4) appear and when the 1D FD $S_{xz1} - S_{xz2}$ is computed, vertex V and edge (3, V) both appear. For plv_4 when the algorithm works in 3D this plane of vertices is split into two faces which correspond to the 2D FD and BD and the vertex V is then obtained.

In [1] the worst case and experimental complexities of this algorithm and of all the processes in which it is based (computing sections from planes of vertices and Boolean operations) are widely analyzed.

The first issue to remark is that the basic operation of all the processes involved in this algorithm is the *XOR* operation between finite sets of points. Therefore the algorithm is robust because does not perform any floating point operation.

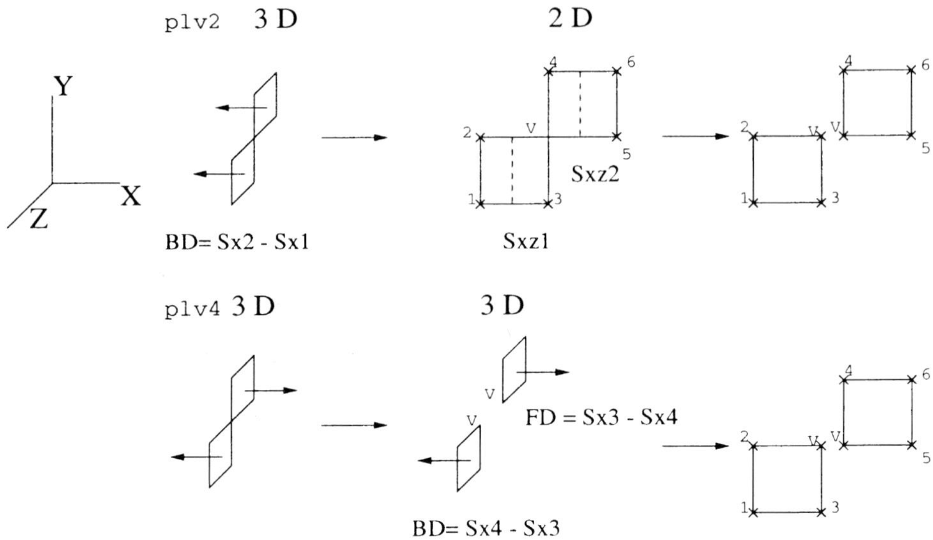

Figure 7. Working example for the EVM to B-Rep conversion algorithm

As in most algorithms concerning EVM the bottle neck process is the computation of all the sections of the object from the EVM (i.e. from its planes of vertices) and it is this process that gives the worst case complexity to the conversion algorithm. The worst case complexity of computing all sections is $O(n \times np)$, n being the number of extreme vertices and np being the number of planes of vertices. As np ranges from 2 to n, the worst case complexity is quadratic. However, experimental results show that the average experimental complexity is far less than quadratic but slightly greater than linear. Performing a numerical regression to the data used in these experimental results of the form $y = ax^b$ the coefficient b obtained was $b = 1.221$.

Finally, it has to be noted that also as in most algorithms concerning EVM a preprocess is needed to sort extreme vertices and so there is a preprocess of $O(nlgn)$ complexity.

4. The Alternating Sum of Volumes Decomposition

A great amount of work has been done in the field of form feature recognition and there are several approaches in the literature. A survey of feature recognition can be found in [21].

As this section focuses on the Alternating Sum of Volumes decomposition, only this method will be reviewed. The earliest representations of form feature recognition [17], [29] propose a convex decomposition method which uses convex hulls and Boolean differences. Reflecting the nature of alternating volume contribution, this technique was called *Alternating Sum of Volumes* (ASV) [29] and form features can be automatically obtained by a manipulation of the resulting expression [26], [28].

4.1. *Alternating Sum of Volumes*

Let $CH(P)$ be the convex hull of a polyhedron P and $CHD^*(P)$ the regularized convex hull difference, also called the deficiency, of P, $CHD^*(P) = CH(P) -^* P$. The ASV decomposition of a polyhedron P, $ASV(P)$, is defined by means of the following recursive expression [15]:

$$ASV(D_k) = \begin{cases} D_k & \text{if } D_k \text{ is convex} \\ H_{k+1} -^* ASV(D_{k+1}) & \text{otherwise} \end{cases}$$

where $D_0 = P$, $H_k = CH(D_{k-1})$ and $D_k = H_k -^* D_{k-1}$ (D stands for deficiency).

The ASV decomposition allows P to be expressed as

$P = (H_1 -^* (H_2 -^* (H_3 -^* (H_4 -^* (\ldots)))))$ or

$P = H_1 -^* H_2 +^* H_3 -^* H_4 +^* \cdots$

$+^*$ being the quasi-disjoint union operator. The name ASV for this decomposition comes from this last expression.

In an ASV decomposition any convex hull H_k completely encloses all the subsequent convex hulls, i.e., $H_j \subseteq H_k, j > k$. This makes it possible to distinguish between terminating and non-terminating ASV series. ASV terminates when a deficiency D_n is found to be convex for some n. In this case $H_{n+1} = D_n, D_{n+1} = \emptyset$ and the relation $H_k \subset H_{k-1}$ holds for all $k \leq n$. Conversely, when two consecutive convex hulls coincide, $H_k = H_{k-1}$, the corresponding deficiency D_k becomes equal to a previous one D_{k-2} and the process becomes cyclic. This deficiency is said to be *ASV-irreducible* or *non-convergent* [15] [26].

Through the manipulation of an ASV series, form features of a given object can be extracted automatically. The following expression

$$P = H_1 -^* H_2 +^* H_3 -^* H_4 +^* \cdots = H_1 -^* (H_2 -^* H_3) -^* (H_4 -^* H_5) -^* \cdots$$

expresses P as the result of a series of volumes to be removed from an initial convex raw material H_1.

4.2. *Non-Convergence of ASV and its Remedy*

When a deficiency becomes ASV-irreducible, the ASV decomposition is non-convergent. This problem can be solved by decomposing the non-convergent deficiency into subsets that are themselves convergent and finding the ASV series of each subset. This method is called Alternating Sum of Volumes with Partitioning (ASVP) [16] and consists in splitting the irreducible deficiency by a plane passing through vertices with two or more non-collinear concave incident edges.

4.3. *Non-Extremal Faces Reduction*

A face of a polyhedron P is an *extremal face* if the corresponding plane is a supporting plane of P, i.e., P is on one side of the closed half-space determined by the plane. Otherwise it is a *non-extremal face* [16].

In the ASV decomposition, the boundary of H_k consists of the set of extremal faces of D_{k-1} plus a new set of fictitious hull faces, while the boundary of D_k consists of the set of non-extremal faces of D_{k-1} plus the same set of fictitious hull faces. Therefore, the set of non-extremal faces of D_k is a subset of the set of non-extremal faces of D_{k-1}.

The convergence condition may be expressed in terms of non-extremal faces reducibility. A D_k with any non-extremal face is a convex deficiency, and then the ASV decomposition converges. So the remedy for non-convergence is to partition non-extremal faces of the irreducible deficiency in such a way that the resulting sets of non-extremal faces can be reduced.

5. Extracting Orthogonal Form Features using the ASV Technique

In this section we apply the ASV method to EVM-represented OPP by taking advantage of its simplicity. We will call this derived method ASOV (O stands for orthogonal).

Theorem 2. *Let P be an OPP, $CH(P)$ its convex hull and $OH(P)$ its orthogonal hull (minimum bounding box). Let A be the set of faces of P lying on the boundary of $OH(P)$ and B the set of faces of P lying on the boundary of $CH(P)$. Then $A = B$.*

This theorem is proved in [14]. It follows from it that computing deficiency sets with respect to $OH(P)$ is equivalent to computing them with respect to $CH(P)$. Therefore, we can use orthogonal hulls instead of convex hulls and, as the initial polyhedron is an OPP, we are guaranteed that all the objects in the ASOV decomposition will be OPP. And the EVM will be used to handle all the necessary operations.

Let P be an OPP and $D_0 = P$, $H_k = OH(D_{k-1})$, $D_k = H_k -^* D_{k-1}$, the same recursive expression as in the general case, holds:

$$ASOV(D_k) = \begin{cases} D_k & \text{if } D_k \text{ is a box} \\ H_{k+1} -^* ASOV(D_{k+1}) & \text{otherwise} \end{cases}$$

Figure 8 shows an example.

In order to compute the ASV decomposition of P we need to compute the EVM of both the bounding box and the deficiency of an OPP.

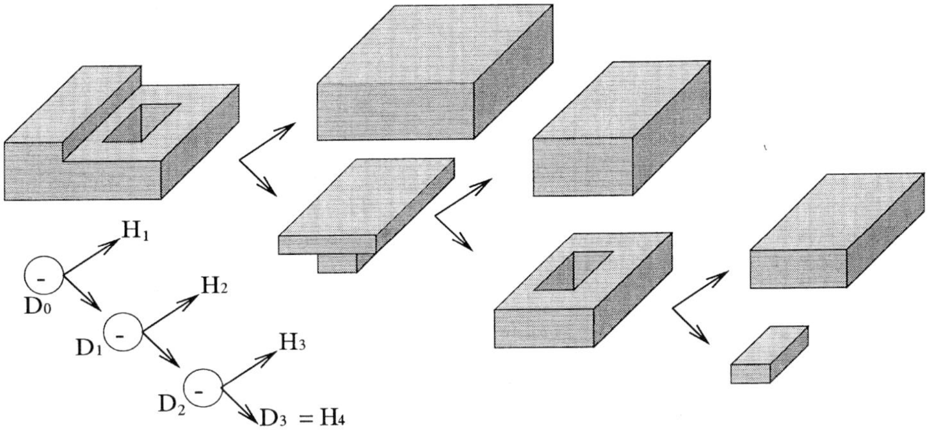

Figure 8. Example of Alternating Sum of Volumes

Theorem 3. $EVM(H_k) = MaxMin(EVM(D_{k-1}))$.

Proof: $H_k = OH(D_{k-1})$. As H_k is a box, $EVM(H_k)$ contains all the vertices of H_k. Moreover, MaxMin (Vertices(D_{k-1})) = MaxMin ($EVM(D_{k-1})$) because as the first property of the EVM stands, coordinate values of non-extreme vertices can be obtained from coordinate values of EV. □

Theorem 4. $EVM(D_k) = EVM(H_k) \otimes EVM(D_{k-1})$.

Proof: $EVM(D_k) = EVM(H_k -^* D_{k-1})$. Since $D_{k-1} \subseteq H_k$, by corollary 1 $EVM(H_k -^* D_{k-1}) = EVM(H_k) \otimes EVM(D_{k-1})$. □

Conversely, we can obtain P from its ASOV decomposition:

Theorem 5. *Let $H_i, \forall i \in [1, n]$ be the resulting boxes of the ASOV decomposition of an OPP, P, then* $EVM(P) = (\otimes^*)_{i=1}^{i=n} EVM(H_i)$.

Proof: The proof follows from the fact that $D_{k-1} \subseteq H_k, \forall k \in [1, n]$ and from corollary 1. □

Detecting and solving non-convergence is also derived from the general method.

Definition 1. *A full extremal face (FEF) of an OPP, P, is a face of P which coincides completely with a face of OH(P).*

Now we will demonstrate that the ASOV decomposition convergence is related with the existence of full extremal faces.

Lemma 1. *$H_k \subset H_{k-1} \Leftrightarrow D_{k-2}$ has at least one FEF.*

Proof: $\Leftarrow$ Let F be a FEF of D_{k-2}. Since $H_{k-1} = OH(D_{k-2})$ then F is also a FEF (i.e., a face) of H_{k-1}. Thus, neither D_{k-1} nor $H_k = OH(D_{k-1})$ contain F (remember that $D_{k-1} = H_{k-1} -^* D_{k-2} = H_{k-1} \otimes D_{k-2}$). This implies that $H_{k-1} \neq H_k$. Moreover, as in the general case $H_k \subseteq H_{k-1}$. Then, $H_k \subset H_{k-1}$.

$\Rightarrow H_k = OH(D_{k-1})$ and $H_{k-1} = OH(D_{k-2})$. $H_k \subset H_{k-1}$ means that H_k and H_{k-1} differ at least in one face (for instance, faces on planes $y = y^k_{MAX}$ of H_k and $y = y^{k-1}_{MAX}$ of H_{k-1} had to be such that $y^k_{MAX} < y^{k-1}_{MAX}$). And this implies that a part of H_{k-1} has to be removed (the part from y^{k-1}_{MAX} to y^k_{MAX}) and this is only possible if the corresponding face of H_{k-1} fully coincides with a face of D_{k-2}, i.e., D_{k-2} has at least this FEF. □

Figure 9 shows two 2D examples of an ASOV decomposition by the deficiency series. FEF are marked with a cross in their middle. Dashed lines in D_k correspond to the orthogonal hull H_{k+2}. The example in Fig. 9a converges: all the deficiencies have at least one FEF and we can see how the deficiencies shrink in a direction perpendicular to the FEF. The example in Fig. 9b do not converge: there is no FEF in deficiency D_1 and, therefore, $H_2 = H_1$, which is the non-convergence condition.

Theorem 6. *ASOV converges if there is at least one FEF.*

Proof: As in the general case, ASOV converges when $H_k \subset H_{k-1}$. Then the proof follows from this fact and from Lemma 1. □

In order to remedy the problem of non-convergence, as in the general case, the irreducible deficiency is split by a plane (see Section 2.5).

Definition 2. *The splitting vertex, SV, is the first extreme vertex of an OPP, P, which does not coincide with a corner of OH(P).*

SV belongs to the first plane of vertices of P and therefore to an extremal face. Our approach chooses the splitting plane to be a plane through SV. There are three orthogonal planes passing through SV. One of these planes is a supporting plane (the plane that contains the first plane of vertices) and cannot be a splitting plane, so in this case we have to choose one of the remaining two planes.

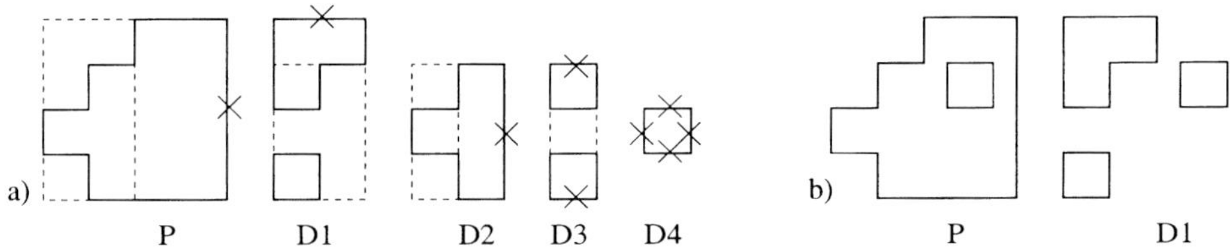

Figure 9. 2D examples of an ASOV decomposition

Moreover, if SV belongs to the first line of vertices, then the two planes intersecting in this line are supporting planes and so, in this case, there is only one possible splitting plane. Generally, it is appropriate to select as the splitting plane the plane through SV which is perpendicular to the lines of vertices of the OPP.

As SV belongs to an extremal face, it will coincide with a corner of the orthogonal hull of at least one of the two split parts and this leads to the conversion of the split extremal face into fully extremal faces, thus enabling convergence.

Then the ASOV with partitioning (ASOVP) is defined by the following recursive expression:

$$ASOVP(D_k^j) = \begin{cases} D_k^j & \text{if } D_k^j \text{ is convex, i.e., a box} \\ H_{k+1}^j -^* ASOV(D_{k+1}^j) & \text{if } D_k^j \text{ has at least one FEF} \\ ASOVP(D_0^{2j+1}) +^* ASOVP(D_0^{2j+2}) & \text{otherwise} \end{cases}$$

where $D_0^0 = P$, $H_k^j = OH(D_{k-1}^j)$, $D_k^j = H_k^j -^* D_{k-1}^j$.
D_0^{2j+1} and D_0^{2j+2} are the (disjoint) split parts of D_k^j, i.e., $D_k^j = D_0^{2j+1} +^* D_0^{2j+2}$.

Figure 10 shows an example of ASOV with partitioning.

We can obtain P from its ASOVP decomposition:

Theorem 7. *Let H_i^j be the resulting boxes of the ASOVP decomposition of an OPP, P, then $EVM(P)$ can be expressed as the regularized XOR between all the H_i^j.*

Proof: An ASOVP is a tree in which the operations are Boolean differences or quasi-disjoint unions.

1. If P has at least one FEF then $P = H -^* D$ and $D \subset H$. Then, by corollary 1, $EVM(P) = EVM(H) \otimes EVM(D)$.

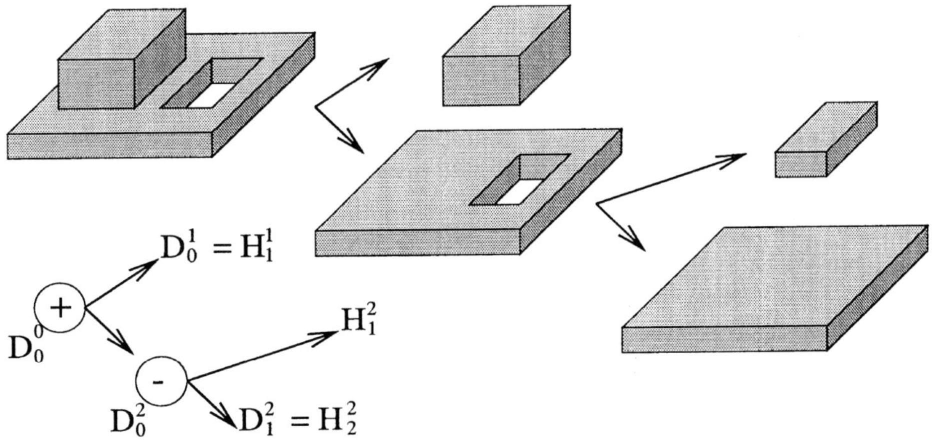

Figure 10. Example of ASOV with partitioning

2. If P has no FEF then $P = Q +^* R$ and Q and R are quasi-disjoint. Then, by corollary 2, $EVM(P) = EVM(Q) \otimes EVM(R)$. $\square$

In the example in Fig. 10, $D_0^0 = H_1^1 +^* H_1^2 -^* H_2^2$ and $EVM(D_0^0) = EVM(H_1^1) \otimes EVM(H_1^2) \otimes EVM(H_2^2)$.

6. Conclusions and Future Work

In this paper we have presented two contributions related with the EVM model.

The first one is a conversion algorithm from EVM to a B-Rep. The description of this algorithm can be understood as an informal proof of the completeness (non-ambiguity) of the EVM. This algorithm performs XOR operations between vertices and Boolean differences. The algorithm is robust because does not perform any floating-point operation: all the information that is not explicitly represented in the EVM is obtained by merging the extreme vertices coordinates.

The second contribution is the application of the Alternating Sum of Volumes decomposition to the particular type of OPP by taking advantage of the simplicity of the EVM. The method uses orthogonal hulls instead of convex hulls, replaces both Boolean difference and quasi-disjoint union by the XOR operation and uses a remedial process for non-convergence based on the potential of the EVM.

The restricted domain of the EVM reduces its applicability in the field of CAD. Applications of it are always directed to its use as an approximation of more complex solids. In Computer Aided Architectural Design (CADD) its applicability is bigger because a great number of modern apartment blocs are 3D objects belonging to the OP domain. We think that the applicability of our model can be more exploited in the field of digital images processing.

So, our future work is oriented to the study of suitability of the EVM in the field of digital images processing. We have a few initial conclusions. A digital image is an OPP and thus it can be represented by the EVM. Recently a definition of a well-composed picture has appeared in the literature [19], [18], and pictures with this property behave better than those without it when performing the most common operations. In fact, well-composed pictures are manifold sets. i.e., OP. Therefore an open problem is to determine whether a picture is well-composed and, if it is not, to process it in order for it to become well-composed. Among the problems that appear in this field are how to improve operations such as thinning and boundary extraction. Several approaches for representing a digital picture have been published that take these operations into account. We only cite two papers among those that we have begun to study. Bieri has proposed bintrees in [8] and hyperimages in [9] and Udupa has proposed a model called a shell in [27]. In the immediate future we will compare the EVM with these and other representations and study the suitability of the EVM for the most demanding image operations.

Acknowledgements

This work has been partially supported by a CICYT grant TIC99-1230-C02-02. The authors are very grateful to the referees whose comments and suggestions have aid to greatly improve the paper.

References

[1] Aguilera, A.: Orthogonal polyhedra: study and application. PhD thesis, LSI-Universitat Politècnica de Catalunya, (1998).

[2] Aguilera, A., Ayala, D.: Orthogonal polyhedra as geometric bounds in constructive solid geometry. In: ACM SM'97 (Hoffmann, C., Bronsvort, W., eds.), pp. 56–67. Atlanta, 1997.

[3] Aguilera, A., Ayala, D.: Domain extension for the extreme vertices model (EVM) and set-membership classification. In: CSG'98. Ammerdown (UK), pp. 33–47. Information Geometers Ltd., 1998.

[4] Aguilera, A., Ayala, D.: Solving point and plane vs. orthogonal polyhedra using the extreme vertices model (EVM). In: WSCG'98. The Sixth Int. Conf. in Central Europe on Computer Graphics and Visualization'98 (Skala, V., ed.), pp. 11–18. University of West Bohemia. Plzen (Czech Republic), 1998.

[5] Andújar, C., Ayala, D., Brunet, P.: Validity-preserving simplification of very complex polyhedral solids. In: Virtual Environments'99 (Gervautz, M., Hildebrand, A., Schmalstieg, D., eds.), pp. 1–10. Wien New York: Springer, 1999.

[6] Andújar, C., Ayala, D., Brunet, P., Joan-Arinyo, R., Solé, J.: Automatic generation of multiresolution boundary representations. Comput. Graphics Forum *15*, C87–C96 (1996).

[7] Ayala, D., Andújar, C., Brunet, P.: Automatic simplification of orthogonal polyhedra. In: Modeling, virtual worlds, distributed graphics: proceedings of the international MVD'96 workshop (Fellner, D., ed.), pp. 137–147. Infix, 1995.

[8] Bieri, H.: Computing the Euler characteristic and related additive functionals of digital objects from their bintree representation. Comput. Vision Graphics Image Proc. *40*, 115–126 (1987).

[9] Bieri, H.: Hyperimages – an alternative to the conventional digital images. In: EUROGRAPHICS'90 (Vandoni, C. E., Duce, D. A., eds.), pp. 341–352. Amsterdam: North-Holland, 1990.

[10] Bournez, O., Maler, O., Pnueli, A.: Orthogonal polyhedra: representation and computation. In: Hybrid systems: computation and control, pp. 46–60. Berlin Heidelberg New York Tokyo: Springer, 1999 (Lecture Notes in Computer Science 1569).

[11] Dang, T., Maler, O.: Reachability analysis via face lifting. In: Hybrid systems: computation and control (Henzinger, T. A., Sastry, S., eds.), pp. 96–109. Berlin Heidelberg New York Tokyo: Springer, 1998 (Lecture Notes in Computer Science 1386).

[12] Hoffmann, C. M.: Geometric and solid modeling. New York: Morgan Kauffmann, 1989.

[13] Juan-Arinyo, R.: On boundary to CSG and extended octrees to CSG conversions. In: Theory and practice of geometric modeling (Strasser, W., ed.), pp. 349–367. Berlin Heidelberg New York Tokyo: Springer, 1989.

[14] Juan-Arinyo, R.: Domain extension of isothetic polyhedra with minimal CSG representation. Comput. Graphics Forum *5*, 281–293 (1995).

[15] Kim, Y. S.: Recognition of form features using convex decomposition. Comput. Aided Des. *24*, 461–476 (1992).

[16] Kim, Y. S., Wilde, D.: A convergent convex decomposition of polyhedral objects. In: SIAM Conf. Geometric Design, (1989).

[17] Kyprianou, L. K.: Shape classification in computer-aided design. PhD thesis, University of Cambridge, 1980.

[18] Latecki, L.: 3D well-composed pictures. Graph. Models Image Proc. *59*, 164–172 (1997).

[19] Latecki, L., Eckhardt, U., Rosenfeld, A.: Well-composed sets. Comput. Vision Image Understand. *61*, 70–83 (1995).

[20] Lorensen, W., Cline, H.: Marching cubes: A high resolution 3D surfaces construction algorithm. Comput. Graphics *21*, 163–169 (1987).

[21] Pratt, M. J.: Towards optimality in automated feature recognition. Computing [Suppl] *10*, 253–274 (1995).

[22] Preparata, F. P., Shamos, M. I.: Computational geometry: an introduction. Berlin Heidelberg New York: Springer, 1985.

[23] Requicha, A.: Representations for rigid solids: Theory, methods, and systems. Comput. Surv. ACM *12*, 437–464 (1980).

[24] Samet, H.: The design and analysis of spatial data structures. Reading: Addison-Wesley, 1989.

[25] Srihari, S. N.: Representation of three-dimensional digital images. ACM Comput. Surv. *13*, 399–424 (1981).
[26] Tang, K., Woo, T.: Algorithmic aspects of alternating sum of volumes. Part 1: Data structure and difference operation. CAD *23*, 357–366 (1991).
[27] Udupa, J. K., Odhner, D.: Shell rendering. IEEE Comput. Graphics Appl. *13*, 58–67 (1993).
[28] Waco, D. L., Kim, Y. S.: Geometric reasoning for machining features using convex decomposition. CAD *26*, 477–489 (1994).
[29] Woo, T.: Feature extraction by volume decomposition. In: CAD/CAM Technology in Mechanical Engineering, (1982).

A. Aguilera
Universidad de las Américas-Puebla
Puebla, México
e-mail: aguilera@mail.udlap.mx

D. Ayala
Universitat Politècnica de Catalunya
Barcelona, Spain
e-mail: dolorsa@lsi.upc.es

Computing [Suppl] 14, 19–35 (2001)

Computing

Smooth Shell Construction with Mixed Prism Fat Surfaces

C. L. Bajaj*, Austin, and **G. Xu****, Beijing

Abstract

Several naturally occurring as well as manufactured objects have shell like structures, that is their boundaries consist of surfaces with thickness. In an earlier paper, we have provided a reconstruction algorithm for such shell structures using smooth fat surfaces within three-sided prisms. In this paper, we extend the approach to a scaffolding consisting of three and four-sided prisms. Within each prism the constructed function is converted to a spline representation. In addition to the adaptive feature of our earlier scheme, the new scheme has the following extensions: (a) four sided fat patches are employed; (b) the size of individual fat patches are bigger; (c) fairing techniques are combined to obtain nicely shaped fat surfaces.

AMS Subject Classification: 65D17.

Key Words: Shell, geometric modeling, curves and surfaces, splines.

1. Introduction

Many human manufactured and several naturally occurring objects have shell like structures, that is the object bodies consist of surfaces with thickness. Such surfaces are called *fat surfaces* in [2]. The problem of constructing smooth approximations to fat surface objects arises in creating geometric model such as airfoils, tin cans, shell canisters, engineering castings, sea shells, the earth's outer crust, the human skin, and so forth.

Problem Description. *As input we are given a matched triangulation pair* $\mathcal{T} = (\mathcal{T}^{(0)}, \mathcal{T}^{(1)})$ *(also called a fat triangulation) with attached normals at each vertex which presents a linearization of the inner and outer boundary surfaces of a shell domain. The goal is to reconstruct smooth fat surface whose bounding surfaces provide approximations of* $\mathcal{T}^{(0)}$ *and* $\mathcal{T}^{(1)}$*, respectively. Additionally mid-surfaces between the boundary surfaces are also provided.*

* Research supported in part by NSF grants CCR 9732306, KDI-DMS-9873326 and ACI-9982297.
** Project 19671081 supported by NSFC.

The matched pair of surface triangulation with normals could be obtained via several inputs, such as nearby iso-contours of volume data, point clouds, single surfaces (see methods in [2]).

Needless to say, one could solve this geometric modeling problem by classical or existing methods (see, e.g. [7–9]) of surface splines construction to construct individual boundary surfaces as well as mid-surfaces of the fat boundaries. However, besides the added space complexity of individually modeling the primary bounding surfaces and mid-surfaces, post local and/or global interactive surface modification would require extremely cumbersome surface–surface interference checks to be performed to preserve geometric model consistency.

The implicit method was shown effective for solving such a problem which was proposed in [2], in which the fat surface is defined by the contours of a single trivariate function F. The function is piecewise, and defined on a collection of triangular prisms in $\mathbb{R}^3$, such that it is C^1 and its contour $F(x, y, z) = \alpha$ for any $\alpha \in (-1, 1)$ provides a smooth mid-surface with $F(x, y, z) = -1$ and $F(x, y, z) = 1$ as the inner and outer boundaries of the shell structure. It should be pointed out that the simplicial hull scheme for constructing A-patches on tetrahedra (see [1, 5]) cannot serve to our purpose, since the simplicial hull, over which a trivariate function F is defined, has no thickness at each vertex.

In this paper, we extend the construction of the function F in [2] by incorporating quadrilateral patches, spline functions and fairing techniques, so that the size of several individual fat surface patches is bigger, the number of patches is fewer, and the "shape" of the fat surfaces is better.

2. Algorithm and Notations

This section gives the algorithm outline (see Fig. 1). Notations used are also introduced here.

2.1. Outline of the Algorithm

Step 1. *Decimation.* This step reduces the number of fat triangles and maintains features. We use the curvature adaptive decimation scheme of [2].

Step 2. *Merge triangles into quadrilaterals.* Merge certain adjacent triangles into quadrilaterals to further reduce the number of patches. Details of this step appear in Section 3.

Step 3. *Construct C^1 trivariate function approximations.* Construct a C^1 piecewise trivariate function $F^{(\sigma)}$, over a collection of 3-prisms and 4-prisms defined on the fat triangles and quadrilaterals, so that $S_\alpha^{(\sigma)} = \{p : F^{(\sigma)}(p) = \alpha, \alpha \in [-1, 1]\}$ are smooth surfaces and $S_{-1}^{(\sigma)}$ and $S_1^{(\sigma)}$ are approximation of $\mathscr{T}^{(0)}$ and $\mathscr{T}^{(1)}$, respectively. Here σ is a given integer related to the freedom of the spline function used. This step is detailed in Section 4.

Step 4. *Fairing*. Fairing by spline functions. Details are again in Section 4.

Step 5. (optional). *Capturing Sharp Features* is detailed in Section 5.

Step 6. *Display the fat surface*. Details are given in Section 6.

2.2. Notations

Our trivariate function $F^{(\sigma)}$ is piecewise defined on a collection of 3-prisms and 4-prisms. To define these prisms, we denote the i-th fat vertex (vertex pair)

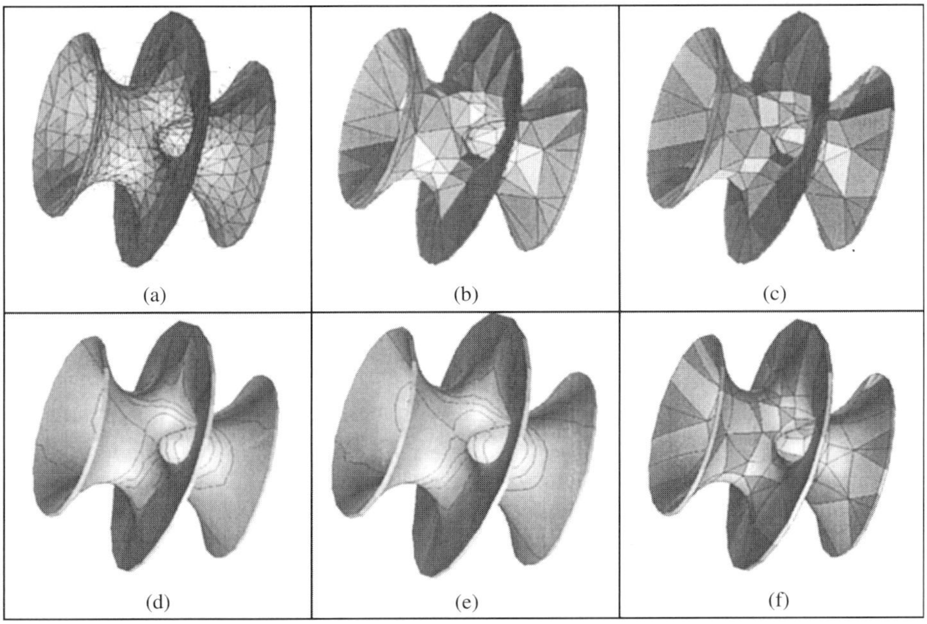

Figure 1. The algorithm steps: **a** is the input triangulation pair (917 fat triangles) with normals at vertices. **b** is the decimated result (265 fat triangles). **c** is the output (119 fat triangles and 73 fat quadrilaterals) of the merging step. **d** is a C^1 function construction without using splines. **e** is the fairing result using splines. The curves on the surfaces **d** and **e** are isophote lines. **f** is a display showing the mixed patch nature

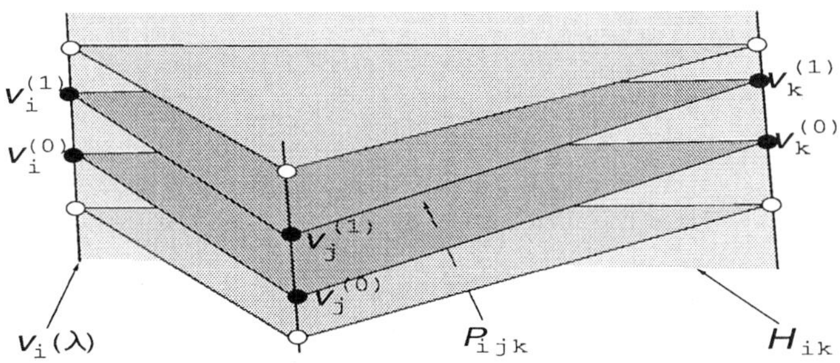

Figure 2. The volume prism cell D_{ijk} and a face $H_{jk}(t, \lambda)$ defined by a fat triangle $[V_i V_j V_k]$

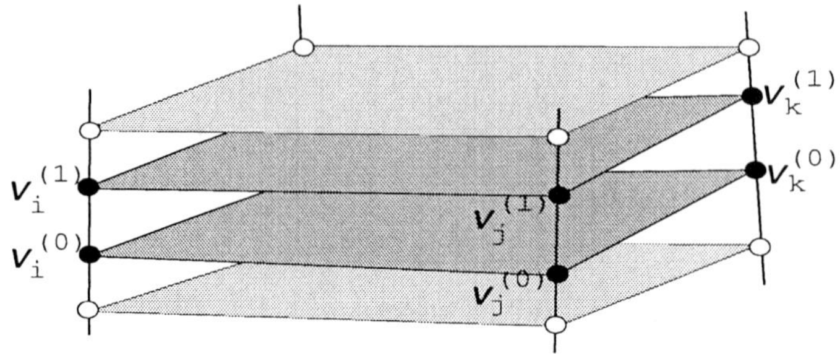

Figure 3. The volume prism cell D_{ijkl} defined by a fat quadrilateral $[V_iV_jV_kV_l]$

as $V_i = (V_i^{(0)}, V_i^{(1)}) \in \mathbb{R}^6$. Let $[V_iV_jV_k]$ be a fat triangle. Then the 3-prism D_{ijk} is a volume in $\mathbb{R}^3$ enclosed by the surfaces H_{ij}, H_{jk} and H_{ki} (see Fig. 2), where H_{lm} is a ruled surface defined by V_l and V_m:

$$H_{lm} = \{p : p = b_1 v_l(\lambda) + b_2 v_m(\lambda),\ b_1 + b_2 = 1,\ \lambda \in \mathbb{R}\}$$

with $v_i(\lambda) = V_i^{(0)} + \lambda N_i,\ N_i = V_i^{(1)} - V_i^{(0)}$. For any point $p = b_1 v_l(\lambda) + b_2 v_m(\lambda)$ with $b_1 + b_2 = 1$, (b_1, b_2, λ) will be called H_{lm}-coordinate of p. The 3-prism D_{ijk}, for $[V_iV_jV_k]$, is a volume which is represented explicitly as

$$D_{ijk} = \{p : p = b_1 v_i(\lambda) + b_2 v_j(\lambda) + b_3 v_k(\lambda),\ b_1 + b_2 + b_3 = 1,\ b_l \geq 0,\ \lambda \in \mathbb{R}\}.$$

We call (b_1, b_2, b_3, λ) as the D_{ijk}-coordinate of p. For each λ, $P_{ijk}(\lambda) := \{p : p = b_1 v_i(\lambda) + b_2 v_j(\lambda) + b_3 v_k(\lambda),\ b_1 + b_2 + b_3 = 1,\ b_l \geq 0\}$ defines a triangle. Let G_{ijk} be the point set that is the grouping of points into prism D_{ijk} in the decimation step.

Let $[V_iV_jV_kV_l]$ be a fat quadrilateral. The 4-prism D_{ijkl} for $[V_iV_jV_kV_l]$ is defined by (see Fig. 4)

$$\begin{aligned} D_{ijkl} = \{p : p = {} & B_{00}(u,v)v_i(\lambda) + B_{10}(u,v)v_j(\lambda) \\ & + B_{01}(u,v)v_l(\lambda) + B_{11}(u,v)v_k(\lambda), \quad u, v \in [0,1], \lambda \in \mathbb{R}\}, \end{aligned}$$

where $B_{00} = (1-u)(1-v)$, $B_{10} = u(1-v)$, $B_{01} = (1-u)v$, $B_{11} = uv$. We shall call (u, v, λ) as the D_{ijkl}-coordinate of p. The equation

$$p = B_{00}(u,v)v_i(\lambda) + B_{10}(u,v)v_j(\lambda) + B_{01}(u,v)v_l(\lambda) + B_{11}(u,v)v_k(\lambda) \tag{2.1}$$

defines a transform between (u, v, λ) and (x, y, z).

3. Merging Fat Triangles

Let $[V_iV_jV_k]$ and $[V_jV_kV_l]$ be two adjacent fat triangles of the decimated mesh. They could be merged to form a quadrilateral if the following condition is satisfied:

$$N_s^T[B_{00}(u,v)N_i + B_{01}(u,v)N_j + B_{10}(u,v)N_l + B_{11}N_k] > 0 \quad \text{for } p_s \in G_{ijkl}, \tag{3.1}$$

where $G_{ijkl} = G_{ijk} \cup G_{jkl}$, (u, v, λ) is the D_{ijkl}-coordinate of p_s, N_s is the normal at p_s and the term in the square brackets is the average of the normals at four vertices. Condition (3.1) implies that the angle between N_s and the averaging normal is less than $\pi/2$. We only need to consider the merging of one of $\mathscr{T}^{(0)}$ and $\mathscr{T}^{(1)}$, the other is correspondingly merged.

In [6], M. Eck and H. Hoppe also merge triangles into quadrilaterals where they attempt to pair up all the triangles by a graph matching. Since we allow a hybrid of triangular and rectangular patches (e.g., to keep sharp features (see § 5), some of the edges are not removable), and since our implementation and the tests show that the shape of quadrilateral surface patches become bad if the quadrilateral is too narrow, we do not seek to merge all the triangles into quadrilaterals. Instead, we grade each edge by the deviation from a rectangle of the quadrilateral formed by merging the two adjacent triangles. An edge is removed (that is its two adjacent triangles are merged) if condition (3.1) is satisfied and if the grade of this edge is less than a given threshold value and is less than its four neighbor edge grades. To grade an edge, for each vertex of the quadrilateral that is formed by merging the two adjacent triangles of the edge, compute the absolute value of the difference of the angle formed by the two incident edges and $\pi/2$, then choose the maximal value of the four absolute values, for the four vertices, as the grade of the edge. If a quadrilateral is a rectangle, then its grade is zero. The worst case is where its grade is close to $3\pi/2$, in which the angle at one vertex is close to 2π.

We notice that most of the CAGD models or some parts of the models come from curvilinear partition of objects. The triangulation is then formed by subdividing quadrilaterals, obtained from the curve partition, into triangles. Our triangle merging policy has the property that it recovers the original curve partition in most of the cases. Figure 4 shows such an example for a teapot.

4. Construct C^1 Trivariate Function Approximations

After step 2, we have a mesh consisting of fat triangles and fat quadrilaterals. For each triangle $[V_iV_jV_k]$ and quadrilateral $[V_iV_jV_kV_l]$ we have volumes D_{ijk} and D_{ijkl}

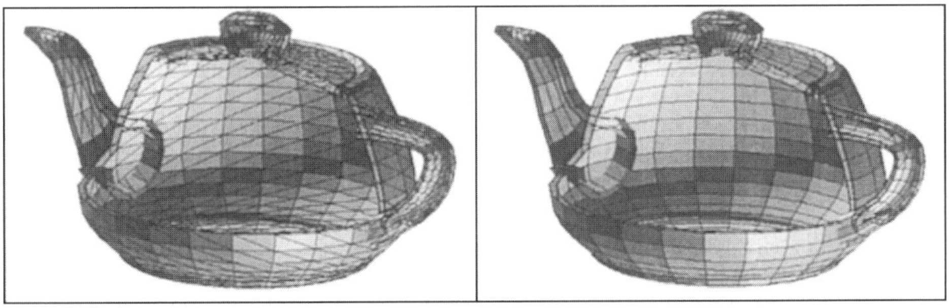

Figure 4. Left: the input triangulation pair approximate of a teapot that has 1428 fat triangles. Right: the merging result that has 294 fat triangles and 567 fat quadrilaterals. The threshold value that controls the merging is taken as $\pi/4$

with the grouped point sets G_{ijk} and G_{ijkl}, respectively. In this section, we construct a C^1 trivariate piecewise function $F = F^{(\sigma)} (\sigma \geq 0$ fixed) over the collection of these volumes, so that it is the required approximation. This function is constructed stepwise. First, the function is defined on the edges of the volumes (see §4.2), then on the faces (see §4.3) and finally in the volumes (see §4.4–§4.5).

4.1. Spline Functions

To achieve better approximation and better shape, spline functions defined on triangles and rectangles are utilized in the construction of F. On a triangular domain with a regular partition (see Fig. 5), C^1 cubic splines defined in BB form were given by Sabin, 1976 (see [10]). Figure 6 gives the BB-form coefficients of a typical base function defined on 13 sub-triangles. Note that these splines in general are not linearly independent (see Böhm, Farin and Kahmann [4]). But the collection we use are indeed linearly independent. For a regular partition of a triangle, say T, we shall associate a base function to each sub-triangle of the partition. To give proper indices for these bases, we label the sub-triangles as T_{ijk} for $(i,j,k) \in J^\sigma = J_1^\sigma \cup J_2^\sigma$, J_1^σ and J_2^σ are defined as follows:

$$J_1^\sigma = \{(i,j,k) : i,j,k \in \{1,2,3,\ldots,2^\sigma\}; \quad i+j+k = 2^\sigma + 2\},$$
$$J_2^\sigma = \{(i,j,k) : i,j,k \in \{1,2,3,\ldots,2^\sigma - 1\}; \quad i+j+k = 2^\sigma + 1\},$$

where 2^σ is the resolution of the partition. Figure 7 gives J_1 and J_2 for $\sigma = 2$. Now we denote the base function defined by Fig. 6 with center triangle T_{ijk} as N_{ijk}^σ.

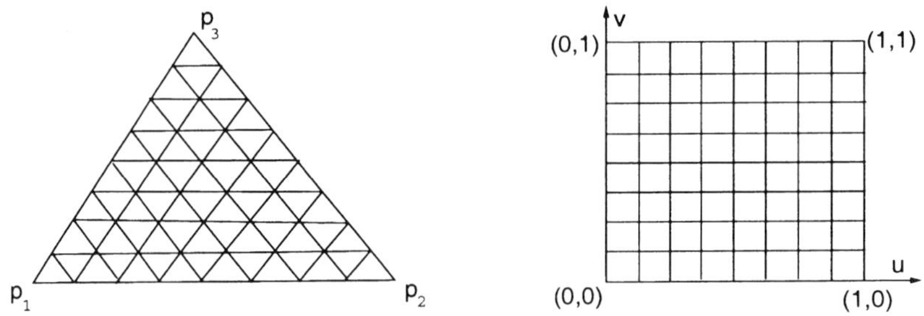

Figure 5. Regular partition of triangular and rectangular domains with resolution 2^σ for $\sigma = 3$

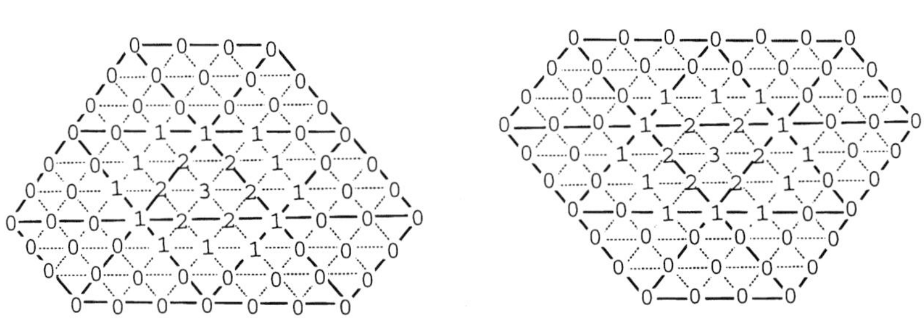

Figure 6. Bezier coefficients for two C^1 cubic spline basis functions. Each is defined on the union of 13 sub-triangles, which forms the support of the function

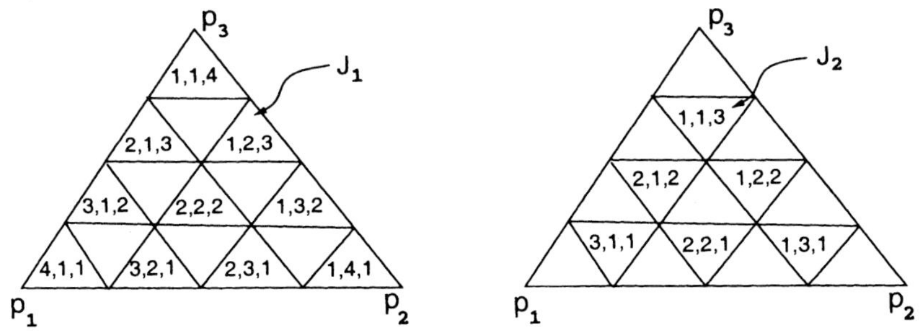

Figure 7. For the regular partition of a triangle with resolution 2^σ, the index set J^σ of the sub-triangles are divided into J_1^σ and J_2^σ. This figure shows them for $\sigma = 2$

Using N_{ijk}^σ, a C^1 cubic spline function on a regularly partitioned triangle is expressed as $\sum_{(i,j,k)\in J^\sigma} b_{ijk} N_{ijk}^\sigma$. On a rectangle, we use tensor-product B-splines $\sum_i \sum_j b_{ij} N_{i3}^\sigma(u) N_{j3}^\sigma(v)$, where $\{N_{i3}^\sigma(t)\}_{i=-1}^{2^\sigma+1}$ are C^2 cubic B-spline bases defined on the uniform knots $t_i = \frac{i}{2^\sigma}$, $i = 0, 1, \ldots, 2^\sigma$. Here we have shifted N_{i3}^σ so that t_i is the center of the support $\mathrm{supp}(N_{i3}^\sigma) = ((i-2)/2^\sigma, (i+2)/2^\sigma)$.

4.2. *F and ∇F on the Edge of the Volume*

The function value F and the gradient ∇F on the edge $v_i(\lambda)$ of a volume are defined by

$$F(v_i(\lambda)) = 2\lambda - 1, \quad \nabla F(v_i(\lambda)) = (1-\lambda)N_i^{(0)} + \lambda N_i^{(1)}.$$

We normalize the normals such that $N_i^T N_i^{(0)} = N_i^T N_i^{(1)} = 2$, and so that the definitions of $F(v_i(\lambda))$ and $\nabla F(v_i(\lambda))$ remain consistent. Here N_i^T is the transpose of $N_i = V_i^{(1)} - V_i^{(0)}$. Note $D_{N_i} F = N_i^T \nabla F$ holds on the edge, where $D_{N_i} F$ denotes the directional derivative of F in the direction N_i.

4.3. *F and ∇F on the Face of the Volume*

The function and gradient on a face of a volume determine the position and tangent of the constructed surface at the face. Since the C^1 construction of F in the volume requires the function and gradient to be C^2 and C^1 on the face, respectively (see (4.6) and (4.8)), we construct them in the following steps:

a. *Construct a C^1 function and a C^0 gradient by averaging pre-constructed volume functions.*

b. *Using the result of step* **a**, *construct the C^2 function and the C^1 gradient.*

4.3.1. C^1 Function and C^0 Gradient on the Face

Let H_{ij} be the face for the edge $[V_i V_j]$. The C^1 function $\tilde{F}_{ij}$ and the C^0 gradient $\nabla \tilde{F}_{ij}$ on H_{ij} are defined by averaging the C^1 functions and C^0 gradients on the two adjacent volumes. Hence the tasks of this sub-section are to construct the volume functions and then do averaging. The volume function constructed here is not our final result of F, since they are not smoothly and even not continuously join at the boundaries of the volumes. However, their average on the common face (regarded as a 2D function) are C^1 and C^0, respectively.

Let $[V_iV_jV_k]$ and $[V_iV_jV_l]$ be the two neighbor fat triangles of the edge $[V_iV_j]$. The case of one or two neighbors are quadrilaterals is similar. On the volume D_{ijk}, we construct a function of the following form

$$F_{ijk} = B_{ijk} + S^{\sigma}_{ijk}$$

where B_{ijk} is of cubic BB-form and S^{σ}_{ijk} is of spline form:

$$B_{ijk}(b_1, b_2, b_3, \lambda) = \sum_{i_1+i_2+i_3=3} b_{i_1i_2i_3}(\lambda) B^3_{i_1i_2i_3}(b_1, b_2, b_3),$$
$$S^{\sigma}_{ijk}(b_1, b_2, b_3, \lambda) = \sum_{(i_1+i_2+i_3)\in\tilde{J}^{\sigma}} (a_{i_1i_2i_3} + w_{i_1i_2i_3}\lambda) N^{\sigma}_{i_1i_2i_3}(b_1, b_2, b_3),$$

where $\tilde{J}^{\sigma} = J^{\sigma} - \{(2^{\sigma},1,1),(1,2^{\sigma},1),(1,1,2^{\sigma})\}$ and $B^3_{i_1i_2i_3}(b_1,b_2,b_3) = \frac{3!}{i_1!i_2!i_3!} b_1^{i_1} b_2^{i_2} b_3^{i_3}$. The BB-form part B_{ijk} is used to interpolate function values and gradients on the three edges of the volume. The spline part S^{σ}_{ijk} is a modification of B_{ijk} for achieving a better approximation in the volume.

The coefficients of B_{ijk} are defined by interpolating the data on the edge of the volume:

$$b_{300}(\lambda) = F(v_i(\lambda)), \quad b_{030}(\lambda) = F(v_j(\lambda)), \quad b_{003}(\lambda) = F(v_k(\lambda)),$$
$$b_{210}(\lambda) = F(v_i(\lambda)) + \tfrac{1}{3}[v_j(\lambda) - v_i(\lambda)]^T \nabla F(v_i(\lambda)),$$

$b_{201}, b_{120}, b_{021}, b_{102}$ and b_{012} are similarly defined. Also, b_{111} is defined by making the cubic B_{ijk} approximate a quadratic: $b_{111} = \frac{1}{4}(b_{210}+b_{120} + b_{021} + b_{012} + b_{102} + b_{201}) - \frac{1}{6}(b_{300} + b_{030} + b_{003})$.

S^{σ}_{ijk} is determined by fitting the points inside the volume D_{ijk} and fairing. Let $\{q_1^{(0)}, \ldots, q_{\mu 0}^{(0)}\} \subset \mathscr{T}^{(0)} \cap G_{ijk}$ be the vertex list. Similarly, let $\{q_1^{(1)}, \ldots, q_{\mu 1}^{(1)}\} \subset \mathscr{T}^{(1)} \cap G_{ijk}$. We compute the coefficients of the splines by the following equations:

$$\begin{cases} w_0\left[F_{ijk}(b_{1s}^{(t)}, b_{2s}^{(t)}, b_{3s}^{(t)}, \lambda_s^{(t)}) - (-1)^{t+1}\right] = 0, & s = 1,2,\ldots,\mu_t, \ t = 0,1, \\ w_1 n_{si}^T \dfrac{\partial S(v_{si})}{\partial b_1} = 0, & s = 0,1,2; \ i = 1,\ldots,2^{\sigma}-1, \\ w_1 n_{si}^T \dfrac{\partial S(v_{si})}{\partial b_2} = 0, & s = 0,1,2; \ i = 1,\ldots,2^{\sigma}-1, \\ \displaystyle\int\!\!\int \left(\frac{\partial^2 \tilde{S}}{\partial x^2}\right)^2 + 2\mu \frac{\partial^2 \tilde{S}}{\partial x^2}\frac{\partial^2 \tilde{S}}{\partial y^2} + \left(\frac{\partial^2 \tilde{S}}{\partial y^2}\right)^2 + 2(1-\mu)\left(\frac{\partial^2 \tilde{S}}{\partial x \partial y}\right)^2 = min. \end{cases} \tag{4.1}$$

where $(b_{1s}^{(t)}, b_{2s}^{(t)}, b_{3s}^{(t)}, \lambda_s^{(t)})$ are the D_{ijk}-coordinates of $q_s^{(t)}, n_{si}, s = 0, 1, 2; i = 1, \cdots, 2^\sigma - 1$, are the given normals on the three boundaries of the mid-surface. These normals are computed by averaging the mid-surface normals defined by $B_{ijk} = 0$. v_{si} are points on boundaries of mid-surface. $S(b_1, b_2)$ is the mid-surface defined by $F_{ijk} = 0$. $\tilde{S}(x, y) = S(b_1(x, y), b_2(x, y))$ with (x, y) to be a local Descartes coordinate. We choose $\frac{1}{2}(V_k^{(0)} + V_k^{(1)})$ as the origin of this system, $\frac{1}{2}[(V_i^{(0)} + V_i^{(1)}) - (V_k^{(0)} + V_k^{(1)})]$ as the x-direction. The y-direction is chosen to be perpendicular to x-direction and point to the side on which $\frac{1}{2}(V_j^{(0)} + V_j^{(1)})$ lies. Note that we do not use a (b_1, b_2) coordinate system directly, because the energy defined in this system is not rotation invariant. System (4.1) is solved in the least square sense. The first set of equations forces the surface interpolating the points in the volume. The second and third sets of equations force the mid-surface to have the given normals on the boundaries. The last minimization forces the surface to have minimal strain energy. Also w_0 and w_1 are weights balancing the three sets of constrains. The minimization leads to a nonlinear system of equations. The integrations in the system are computed by a 6-point numerical quadrature rule (see [3], page 35) on each sub-triangle. We solve the entire system by Newton iteration. Since the system behaves linearly, it converges fast. It needs, in general, 2 or 3 iterations to achieve a single word-length precision.

After F_{ijk} and F_{ijl} have been defined, then we are ready to define

$$\tilde{F}_{ij} = \frac{1}{2}\left(F_{ijk}|_{H_{ij}} + F_{ijl}|_{H_{ij}}\right), \quad \nabla\tilde{F}_{ij} = \frac{1}{2}\left(\nabla F_{ijk}|_{H_{ij}} + \nabla F_{ijl}|_{H_{ij}}\right).$$

If a four-sided polygon, say $[V_i V_j V_k V_l]$, is a neighbor of $[V_i V_j]$, then we define

$$F_{ijkl} = B_{ijkl} + S_{ijkl}^\sigma$$

with

$$B_{ijkl}(u, v, \lambda) = \sum_{i_1=0}^{3}\sum_{i_2=0}^{3} b_{i_1 i_2}(\lambda) B_{i_1}^3(u) B_{i_2}^3(v),$$

$$S_{ijkl}^\sigma(u, v, \lambda) = \sum_{i_1=0}^{2^\sigma}\sum_{i_2=0}^{2^\sigma} (a_{i_1 i_2} + w_{i_1 i_2}\lambda) N_{i_1 3}^\sigma(u) N_{i_2 3}^\sigma(v).$$

The coefficients $B_{ijkl}(u, v, \lambda)$ are determined as follows:

$$b_{00} = F(v_i(\lambda)), \quad b_{30} = F(v_j(\lambda)), \quad b_{33} = F(v_k(\lambda)), \quad b_{03} = F(v_l(\lambda)),$$
$$b_{10} = v_i(\lambda) + \tfrac{1}{3}(v_j(\lambda) - v_i(\lambda))^T \nabla F(v_i(\lambda)).$$

The other coefficients on the edge are similarly defined. Define

$$b_{11} = \tfrac{1}{3}(b_{10} + b_{01}) + \tfrac{1}{6}(b_{31} + b_{13}), \quad b_{22} = \tfrac{1}{3}(b_{32} + b_{23}) + \tfrac{1}{6}(b_{20} + b_{02}),$$
$$b_{21} = \tfrac{1}{3}(b_{20} + b_{31}) + \tfrac{1}{6}(b_{23} + b_{01}), \quad b_{12} = \tfrac{1}{3}(b_{02} + b_{13}) + \tfrac{1}{6}(b_{32} + b_{10}).$$

The coefficients of S^{σ}_{ijkl} are determined as that of S^{σ}_{ijk} by fitting the data inside the volume D_{ijkl} and fairing.

4.3.2. C^2 Function and C^1 Gradient on the Face

Now let us define the C^2 function F_{lm} and C^1 gradient ∇F_{lm}. Let

$$\begin{aligned} F_{lm}(t,\lambda) = {} & F(v_l(\lambda))H_0^3(t) + [v_m(\lambda) - v_l(\lambda)]^T \nabla F(v_l(\lambda))H_1^3(t) \\ & + F(v_m(\lambda))H_2^3(t) + [v_m(\lambda) - v_l(\lambda)]^T \nabla F(v_m(\lambda))H_3^3(t) \\ & + \phi^{\sigma}_{lm}(t) + \psi^{\sigma}_{lm}(t)\lambda, \end{aligned} \tag{4.2}$$

with

$$\begin{aligned} & H_0^3(t) = 1 - 3t^2 + 2t^3, && H_1^3(t) = t - 2t^2 + t^3, \\ & H_2^3(t) = 3t^2 - 2t^3, && H_3^3(t) = -t^2 + t^3, \\ & \phi^{\sigma}_{lm}(t) = \sum_{i=2}^{2^{\sigma}-2} \phi_i N^{\sigma}_{i3}(t), && \psi^{\sigma}_{lm}(t) = \sum_{i=2}^{2^{\sigma}-2} \psi_i N^{\sigma}_{i3}(t). \end{aligned}$$

From the construction of $\tilde{F}_{lm}$, we know that it has the same form as F_{lm} defined by (4.2) but with different ϕ^{σ}_{lm} and ψ^{σ}_{lm} that are C^1 cubic splines. We denote them as $\tilde{\phi}^{\sigma}_{lm}$ and $\tilde{\psi}^{\sigma}_{lm}$. Now we determine ϕ^{σ}_{lm} and ψ^{σ}_{lm} by approximating $\tilde{\phi}^{\sigma}_{lm}$ and $\tilde{\psi}^{\sigma}_{lm}$ in the least square sense:

$$\int_0^1 \left[\phi^{\sigma}_{lm}(t) - \tilde{\phi}^{\sigma}_{lm}(t)\right]^2 dt = min, \quad \int_0^1 \left[\psi^{\sigma}_{lm}(t) - \tilde{\psi}^{\sigma}_{lm}(t)\right]^2 dt = min. \tag{4.3}$$

Each of them leads to a system of linear equations. The integrations in these systems are computed by Gauss-Legendre quadrature rule on each sub-interval and then summed up. Let

$$d_1(\lambda) = v_m(\lambda) - v_l(\lambda), \quad d_2(t) = (1-t)N_l + tN_m, \quad d_3(t,\lambda) = d_1 \times d_2.$$

Then we define $\nabla F_{lm}(t,\lambda)$ by the following conditions:

$$\begin{cases} d_1^T \nabla F_{lm}(t,\lambda) = \dfrac{\partial F_{lm}(t,\lambda)}{\partial t}, \\ d_2^T \nabla F_{lm}(t,\lambda) = \dfrac{\partial F_{lm}(t,\lambda)}{\partial \lambda}, \\ d_3^T \nabla F_{lm}(t,\lambda) = d_3^T \nabla \breve{F}_{lm}(t,\lambda), \end{cases} \tag{4.4}$$

where $\nabla \breve{F}_{lm}(t,\lambda)$ is a C^1 approximation of $\nabla \tilde{F}_{lm}(t,\lambda)$:

$$\nabla \breve{F}_{lm}(t,\lambda) = (1-t)\nabla F(v_l(\lambda)) + t\nabla F(v_m(\lambda)) + \breve{\phi}^{\sigma}_{lm}(t) + \breve{\psi}^{\sigma}_{lm}(t)\lambda,$$

with $\breve{\phi}^{\sigma}_{lm}(t) = \sum_{i=2}^{2^\sigma-2} \phi_i N^{\sigma}_{i3}(t), \breve{\psi}^{\sigma}_{lm}(t) = \sum_{i=2}^{2^\sigma-2} \psi_i N^{\sigma}_{i3}(t)$. $\breve{\phi}^{\sigma}_{lm}$ and $\breve{\psi}^{\sigma}_{lm}$ are determined by

$$\int_0^1 ||\nabla \breve{F}_{lm}(t, \lambda_i) - \nabla \tilde{F}_{lm}(t, \lambda_i)||^2 \, dt = min, \quad \text{for } \lambda_0 = 0, \ \lambda_1 = 1.$$

It might be necessary to point out now that $\nabla \breve{F}_{lm}$ cannot be used as ∇F_{lm} even though it is C^1, since it may not satisfy the first two conditions of (4.4). It is clear that these two conditions must be satisfied because F_{lm} is previously defined. Though the right-handed side of the third equation of (4.4), which is a directional derivative, can be any value, it is reasonable to choose this value by approximating the existing information about ∇F_{lm}. Hence we use $\nabla \breve{F}_{lm}$ to compute this directional derivative.

Since $||d_3||^2[d_1, d_2, d_3]^{-1} = \left[d_1||d_2||^2 - d_2(d_1^T d_2), d_2||d_1||^2 - d_1(d_1^T d_2), d_3\right]^T$, (4.4) implies

$$\nabla F_{lm}(t, \lambda) = \frac{1}{||d_3||^2}\left\{\left[d_1||d_2||^2 - d_2(d_1^T d_2)\right]P + \left[d_2||d_1||^2 - d_1(d_1^T d_2)\right]Q + d_3 R\right\}, \tag{4.5}$$

where $P(t, \lambda) = \frac{\partial F_{lm}(t,\lambda)}{\partial t}$, $Q(t, \lambda) = \frac{\partial F_{lm}(t,\lambda)}{\partial \lambda}$, $R(t, \lambda) = d_3^T \nabla \breve{F}_{lm}(t, \lambda)$.

4.4. *F on the Volume* D_{ijkl}

Now we are ready to define F within the volumes. Let $[V_1 V_2 V_3 V_4]$ be a typical fat quadrilateral. Let F_u and F_v be defined by the cubic Hermite interpolation in the u and v directions, respectively:

$$F_u(u, v, \lambda) = H_0^3(u)F_{14}(v, \lambda) + H_1^3(u)d_u(v, \lambda)^T \nabla F_{14}(v, \lambda) + H_2^3(u)F_{23}(v, \lambda) + H_3^3(u)d_u(v, \lambda)^T \nabla F_{23}(v, \lambda),$$

$$F_v(u, v, \lambda) = H_0^3(v)F_{12}(u, \lambda) + H_1^3(v)d_v(u, \lambda)^T \nabla F_{12}(u, \lambda) + H_2^3(u)F_{43}(u, \lambda) + H_3^3(v)d_v(u, \lambda)^T \nabla F_{43}(u, \lambda),$$

where $d_u(v, \lambda) = H_{23}(v, \lambda) - H_{14}(v, \lambda)$, $d_v(u, \lambda) = H_{43}(u, \lambda) - H_{12}(u, \lambda)$. Then we define

$$F^{(\sigma)}(u, v, \lambda) = \frac{w_u F_u(u, v, \lambda) + w_v F_v(u, v, \lambda)}{w_u + w_v} + R^{\sigma}(u, v, \lambda) \tag{4.6}$$

with

$$R^{\sigma}(u, v, \lambda) = \sum_{i_1=2}^{2^\sigma-2} \sum_{i_2=2}^{2^\sigma-2} (a_{i_1 i_2} + w_{i_1 i_2}\lambda) N^{\sigma}_{i_1 3}(u) N^{\sigma}_{i_2 3}(v),$$

where $w_u = [(1-v)v]^2$, $w_v = [(1-u)u]^2$. The last term $R^\sigma(u, v, \lambda)$ in (4.6) is referred as *correction term*, which is used to fit the data in the volume and it does not change the surface on the face of the volume. Let $\{V_s^{(\tau)}\} \subset G_{1234} \cap \mathscr{T}^{(\tau)} (\tau = 0$ or $1)$, and $(u_s^{(\tau)}, v_s^{(\tau)}, \lambda_s^{(\tau)})$ be the D_{1234}-coordinate of $V_s^{(\tau)}$. Thus $a_{i_1 i_2}$ and $w_{i_1 i_2}$ are defined by

$$\begin{cases} w\left[F^{(\sigma)}(u_s^{(\tau)}, v_s^{(\tau)}, \lambda_s^{(\tau)}) - (-1)^{\tau+1}\right] = 0, \\ \displaystyle\int\int \left(\frac{\partial^2 S}{\partial u^2}\right)^2 + 2\mu \frac{\partial^2 S}{\partial u^2}\frac{\partial^2 S}{\partial v^2} + \left(\frac{\partial^2 S}{\partial v^2}\right)^2 + 2(1-\mu)\left(\frac{\partial^2 S}{\partial u \partial v}\right)^2 = min, \end{cases} \tag{4.7}$$

where $S(u, v)$ is the mid-surface defined by $F^{(\sigma)}(u, v, \lambda) = 0$. The first equality is in the least square sense. The weight w is put to address the importance of interpolating the points in the volume.

4.5. *F on Volume D_{ijk}*

Let $[V_1 V_2 V_3]$ be a typical fat triangle. We define

$$F^{(\sigma)}(b_1, b_2, b_3, \lambda) = \sum_{i=1}^{3} w_i D_i(b_1, b_2, b_3, \lambda) + T^\sigma(b_1, b_2, b_3, \lambda) \tag{4.8}$$

with

$$T^\sigma(b_1, b_2, b_3, \lambda) = \sum_{(i_1, i_2, i_3) \in J_3^\sigma} (a_{i_1 i_2 i_3} + w_{i_1 i_2 i_3}\lambda) N_{i_1 i_2 i_3}^\sigma (b_1, b_2, b_3)$$

where $w_i = \dfrac{\prod_{j \neq i} b_j^2}{\sum_{k=1}^{3} \prod_{j \neq k} b_j^2}$, $\quad J_3^\sigma = \{(i, j, k) \in J^\sigma : i > 1, j > 1, k > 1\}$,

$$\begin{aligned} D_i(b_1, b_2, b_3, \lambda) = {} & F(v_i(\lambda)) H_2^3(b_i) + F(p_i(b_1, b_2, b_3, \lambda)) H_0^3(b_i) \\ & + d_i(b_1, b_2, b_3, \lambda)^T \nabla F(v_i(\lambda)) H_3^3(b_i) \\ & + d_i(b_1, b_2, b_3, \lambda)^T \nabla F(p_i(b_1, b_2, b_3, \lambda)) H_1^3(b_i), \end{aligned}$$

$$p_i(b_1, b_2, b_3, \lambda) := \frac{b_j}{1 - b_i} v_j(\lambda) + \frac{b_k}{1 - b_i} v_k(\lambda),$$

$$d_i(b_1, b_2, b_3, \lambda) := -\frac{b_j}{1 - b_i} e_{ji}(\lambda) - \frac{b_k}{1 - b_i} e_{ki}(\lambda),$$

$$e_{ji}(\lambda) = v_j(\lambda) - v_i(\lambda), \quad e_{ki}(\lambda) = v_k(\lambda) - v_i(\lambda),$$

and $(i, j, k) \in \{(1, 2, 3), (2, 3, 1), (3, 1, 2)\}$. Again, the last term in (4.8) is called the *correction term*. The parameters $a_{i_1 i_2 i_3}$ and $w_{i_1 i_2 i_3}$ are similarly defined as $a_{i_1 i_2}$ and $w_{i_1 i_2}$ in (4.6) by fitting and fairing.

4.6. Basic Results

Theorem 4.1. *The function $F^{(\sigma)}$ constructed is C^1 on $\sum D_{ijk} \cup \sum D_{ijkl}$.*

Proof: First note that the function F^σ is C^1 within each of the volumes, since the gradient on the faces of the volumes is C^1 and the correction terms is C^1 in the volume. Second note that the function values and gradients of the correction term R^σ in (4.6) and the term T^σ in (4.8) vanish on the boundary of the corresponding volume. Hence these terms do not influence the continuity of the function F^σ. On each edge of the volumes, the C^1 continuity of F^σ can be similarly proved as Theorem 4.1 of [2]. Hence, a fact that needs to be proved is that the function values and gradients of F^σ on the boundary of the volumes coincide with function values and gradients defined in Section 4.3. This fact will guarantee that the function is C^1 on the boundary faces. For the 3-prisms, this fact could be proved similar to the proof of Theorem 4.1 of [2]. Hence the remaining is to prove the fact for the 4-prisms. Consider the function value and gradient of F^σ on the edge $u = 0$ for a typical fat quadrilateral $[V_1 V_2 V_3 V_4]$. It follows from (4.6) that

$$F^\sigma(0, v, \lambda) = F_u(0, v, \lambda) = F_{14}(v, \lambda).$$

Hence the function value is what we require. Computing partial derivatives of F^σ and using the relation (4.4), we have

$$\frac{\partial F^\sigma}{\partial u} = d_u(v, \lambda)^T \nabla F_{14}(v, \lambda),$$

$$\frac{\partial F^\sigma}{\partial v} = \frac{\partial F_{14}}{\partial v} = d_1(\lambda)^T \nabla F_{14}(v, \lambda),$$

$$\frac{\partial F^\sigma}{\partial \lambda} = \frac{\partial F_{14}}{\partial \lambda} = d_2(v)^T \nabla F_{14}(v, \lambda).$$

Differentiating (2.1) about x, y and z, we have

$$\begin{bmatrix} \frac{\partial u}{\partial x} & \frac{\partial v}{\partial x} & \frac{\partial \lambda}{\partial x} \\ \frac{\partial u}{\partial y} & \frac{\partial v}{\partial y} & \frac{\partial \lambda}{\partial y} \\ \frac{\partial u}{\partial z} & \frac{\partial v}{\partial z} & \frac{\partial \lambda}{\partial z} \end{bmatrix} = \left\{ [d_u(v, \lambda), d_1(\lambda), d_2(v)]^T \right\}^{-1}.$$

Computing partial derivatives of F^σ with respect to x, y and z and combining the two sets of equations above, we obtain $\nabla F^\sigma = \nabla F_{14}(v, \lambda)$. Hence, F^σ is C^1. □

To show the smoothness of function F^σ, Fig. 8 shows a construction example. Figure 8b is the fat surface construction of the input in Fig. 8a. Figure 9 gives a multiresolution construction of the hypersheet surface in the mesh direction (*h* direction). Figure 1d and 1e show two resolutions in the spline levels (*p* direction) with $\sigma = 1, 3$, respectively.

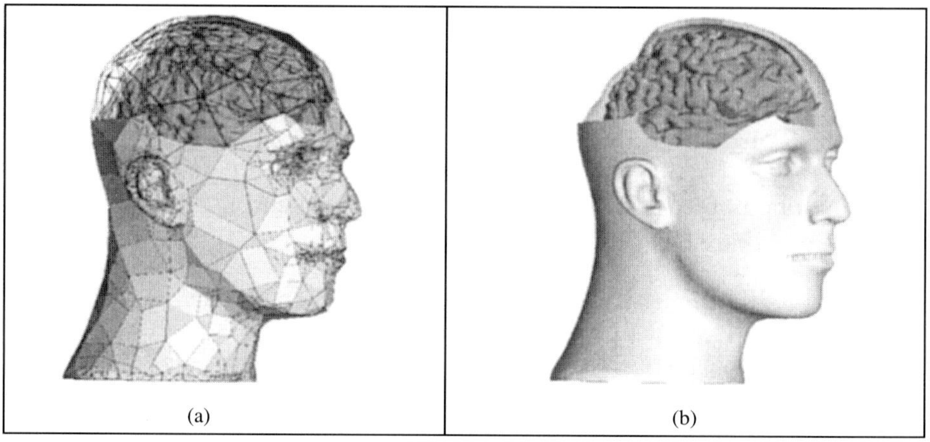

Figure 8. Smooth fat surface construction: **b** is the smoothing of **a**. Polygon **a**, that has 3296 fat triangles and 389 fat quadrilaterals, is the decimated and merged result of a mesh that has fat 25552 triangles. Note the adaptive nature: More fat triangles are used at the parts of the ears, eyes and mouth. To capture sharp features, fat triangles are not merged at the parts of the neck, eyes and mouth. The brain model consists of 40884 fat triangles

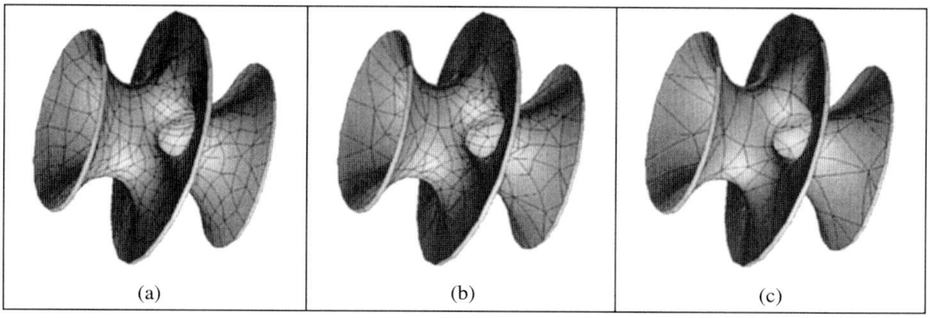

Figure 9. Different resolution constructions of smooth fat surfaces. Three mesh levels (h direction) with fixed $\sigma = 3$ are shown. From the left to the right, they have 249, 213 and 95 fat triangles and have 334, 206 and 64 fat quadrilaterals, respectively

5. Sharp Features of the Constructed Surfaces

To capture sharp features, we need to mark certain edges as sharp. To this end, we compute dihedral angle $\theta = \pi - \theta_1$ for the two incident faces, for each edge of the triangulation $T^{(0)}$ and $T^{(1)}$. If $\theta < \alpha$, then this edge is marked as a sharp edge. Here θ_1 is the angle between the two normals of the two triangles and α ia a threshold value for controlling the sharp feature. After marking the edges, the vertices also need to be marked. If there exist sharp edges incident to a vertex, then we say this vertex is sharp, otherwise, it is non-sharp. For a sharp vertex, the normal that has been assigned before needs to be re-computed. Now the triangles around a sharp vertex are divided into some groups by the sharp edges (see Fig. 10). For each group, we assign a single normal for the vertex. This normal could be computed by the weighted average of face normals. The weight is chosen to be the angle of the edges that are incident to this vertex. In the construction of surface patch for one triangle, there is only one normal is used for one vertex of the triangle. This normal is the vertex normal if the vertex is non-sharp, otherwise the normal is the group's normal.

Two examples are shown in Fig. 11. The left two are input polygons, the right shell bodies are the corresponding output. In the star-like polygon on the top-left, the left four inner and outer peak edges are selectively marked as sharp. The fat surface on the top-right exhibit the sharp feature. For the bottom-left polygon, the left four peak edges of the outer polygon are marked as sharp, no edge is marked for inner polygon. The figure on the bottom-right presents the outer-

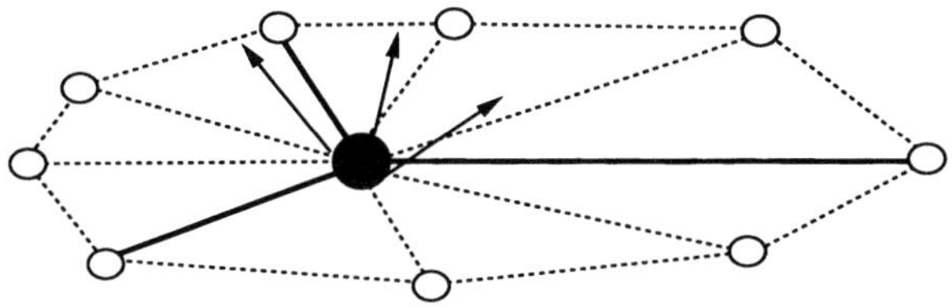

Figure 10. Grouping the triangles by the sharp edges (thick lines) and assigning one normal for each group

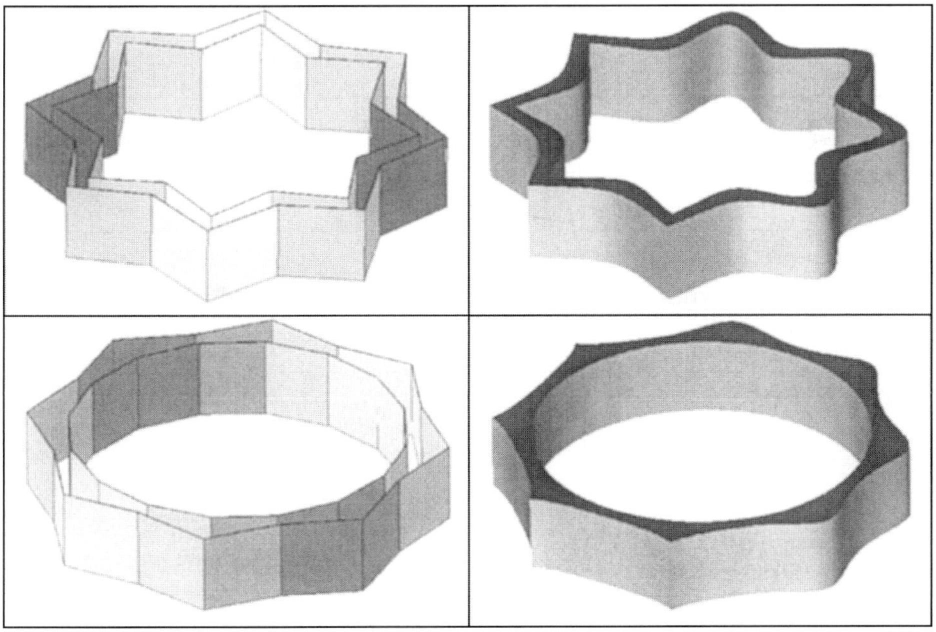

Figure 11. Left: the input polygons with some edges are marked as sharp. Right: the constructed fat surfaces with sharp features. There are four fat edges (inner and outer) on the top polygon are marked as sharp. On the bottom polygon, only four outer edges are marked as sharp

sharp, inner-smooth nature. Another example that has sharp feature is shown in Fig. 12.

6. Display of the Fat Surfaces

Often we wish to evaluate the surface $F = \alpha$ for a given $\alpha \in [-1, 1]$. Let $[V_i V_j V_k]$ be any fat triangle. Then for each $(b_1, b_2, b_3), b_i \geq 0, \sum b_i = 1$, determine $\lambda_{min}^{(\alpha)} = \lambda_{min}^{(\alpha)}(b_1, b_2, b_3)$ such that

$$F_{ijk}(b_1, b_2, b_3, \lambda_{min}^{(\alpha)}) = \alpha,$$

$$\left|\lambda_{min}^{(\alpha)} - \frac{1}{2}\right| = \min\left\{\left|\lambda - \frac{1}{2}\right| : F_{ijk}(b_1, b_2, b_3, \lambda) = \alpha\right\}.$$

The surface point is defined by $p = p(b_1, b_2, b_3, \lambda_{min}^{(\alpha)})$. The main task here is to compute $\lambda_{min}^{(\alpha)}$ for each (b_1, b_2, b_3). It follows from (4.8) that $D_i(b_1, b_2, b_3, \lambda)$ is a rational function of λ. It is of the form

$$F_0 + F_1\lambda + F_2\lambda^2 + \frac{N_0 + N_1\lambda + N_2\lambda^2 + N_3\lambda^3 + N_4\lambda^4}{D_0 + D_1\lambda + D_2\lambda^2}. \tag{6.1}$$

Hence $\phi(\lambda) := F_{ijk}(b_1, b_2, b_3, \lambda)$ is a rational function of λ. The nearest zero to $\frac{1}{2}$ of $\phi(\lambda) - \alpha$ is the required $\lambda_{min}^{(\alpha)}$.

Although $\phi(\lambda) - \alpha = 0$ is a nonlinear algebraic equation, $\phi(\lambda) - \alpha$ can be approximated by a polynomial of degree at most 2, since the rational term in (6.1) is small compared with the polynomial part. Hence, taking the root of the polynomial part as an initial value, and then using Newton iteration, we obtain the required solution.

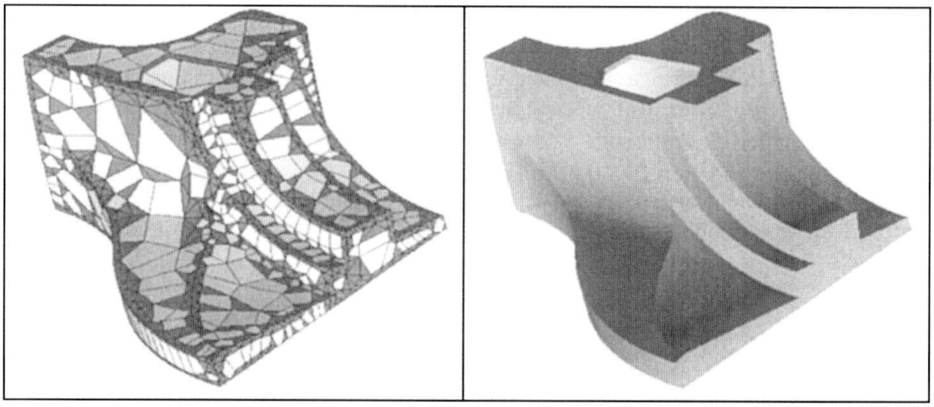

Figure 12. Left: the input polygon, that has 1914 fat triangles and 836 fat quadrilaterals, with some edges are marked as sharp. Right: the constructed fat surface with sharp features. To show the fat nature, the shell that is closed is cutaway on the top to show the interior

For the four sided polygon $[V_i V_j V_k V_l]$, the surface point $F_{ijkl}(u, v, \lambda) = \alpha$ is evaluated similarly.

7. Conclusions

Using Bézier, triangular form and tensor form trivariate spline functions, we construct a C^1 function F^σ on a collection of 3-prisms and 4-prisms, such that the contours $F^\sigma = -1$ and $F^\sigma = 1$ approximate the given input triangulation pair, which represent the inner and outer boundaries of a shell body. Apart from fitting the data clouds, the spline functions also serve to fair the shape of the constructed surface. The implementation and test examples show that the proposed method for fat surface construction is correct and fulfills our initial goals.

References

[1] Bajaj, C., Chen, J., Xu, G.: Modeling with cubic A- patches. ACM Trans. Graphics *14*, 103–133 (1995).
[2] Bajaj, C., Xu, G.: Smooth adaptive reconstruction and deformation of free-form fat surfaces. TICAM REPORT 99–08, March, 1999, Texas Institute for Computational and Applied Mathematics, The University of Texas at Austin, 1999.
[3] Bernadou, M., Boisserie, J. M.: The finite element method in thin shell theory: application to arch dam simulations. Basel: Birkhäuser, 1982.
[4] Böhm, W., Farin, G., Kahmann, J.: A survey of curve and surface methods in CAGD. Comput. Aided Geom. Des. *1*, 1–60 (1984).
[5] Dahmen, W., Thamm-Schaar, T.-M.: Cubicoids: modeling and visualization. Comput. Aided Geom. Des. *10*, 89–108 (1993).
[6] Eck, M., Hoppe, H.: Automatic reconstruction of B-spline surfaces of arbitrary topological type. In: Computer Graphics Proceedings, Annual Conference series, ACM SIGGRAPH96, pp. 325–334, 1996.
[7] Farin, G.: Curves and surfaces for computer aided geometric design: a practical guide, 2nd ed. New York: Academic Press, 1990.
[8] Hoschek, J., Lasser, D.: Fundamentals of computer aided geometric design. Natick: A. K. Peters, 1993.
[9] Piegl, L., Tiller, W.: The NURBS book. Berlin Heidelberg New York Tokyo: Springer, 1997.
[10] Sabin, M.: The use of piecewise form of numerical representation of shape. PhD thesis, Hungarian Academy of Science, Budapest, 1976.

C. L. Bajaj
Department of Computer Science
University of Texas
Austin, TX 78712 U.S.A.
e-mail: bajaj@cs.utexas.edu

G. Xu
State Key Laboratory of Scientific
and Engineering Computing ICMSEC
Chinese Academy of Sciences
Beijing, China
e-mail: xuguo@lsec.cc.ac.cn

Computing [Suppl] 14, 37–53 (2001)

Computing

Geometric Modeling of Parallel Curves on Surfaces

G. Brunnett, Chemnitz

Abstract

This paper is concerned with various aspects of the modeling of parallel curves on surfaces with special emphasis on surfaces of revolution. An algorithm for efficient tracking of the geodesics on these surfaces is presented. Existing methods for plane offset curves are adapted to generate G^1-spline approximations of parallel curves on arbitrary surfaces. An algorithm to determine singularities and cusps in the parallel curve is established.

Key Words: Parallel curves, Geodesics, surface of revolution, Indian art.

1. Introduction

Parallel curves (or offset curves) in the plane and their spline approximation have been studied intensively because of their use in path generation for NC controlled machines (see [3], [1], [4]). This paper is concerned with the more general situation of surface curves that are parallel in the sense that the tangents of the parallel curve are obtained by parallel transport along a geodesic orthogonal to the original curve.

Parallel curves are an often used stylistic feature of artistic design that appears in various contexts but especially in the design of surfaces of revolution like vases, plates etc. During his stay at Arizona State University several years ago the author encountered impressive pieces of South-West Indian Art that show extensive use of parallel curves as design elements. Inspired by these a modeling environment for the interactive design of parallel curves on surfaces of revolution was created.

This paper reports on the modeling techniques that have been realized within this software package. Despite the fact that over the last years new results about parallel curves have been established (see [8], [9]), the algorithms presented in this paper still provide efficient means for the modeling of such curves.

Section 2 provides the basics of parallel curves on surfaces and introduces the Darboux frame that is used in the discussion of cusps in the offset curve. Section 3 is concerned with the efficient computation of points on the offset curve. We show that it is more efficient to track the geodesics on a surface of revolution by applying a Runge-Kutta type of method to the system of differential equations than to make use of the fact that the geodesics can be computed by quadratures. It is also shown that the second order system of differential equations can be reduced to a first order system that has an ambiguity in the sign of one of the unknown functions. An algorithm is provided to track the geodesics based on the first order system. This method makes use of the global behaviour of geodesics on a surface of revolution.

In Section 4 it is described how to obtain a G^1-spline approximation of the offset curve by adapting established methods of the planar case to the general situation of parallel curves on surfaces.

The last section is concerned with the detection of singularities and cusps in the parallel curve which is important to obtain an accurate spline approximation of the offset curve. An algorithm to locate cusps in the offset curve on arbitrary surfaces is proposed. For parallel curves on the sphere an exact criterion for cusps and a formula that relates the geodesic curvature of the parallel to the geodesic curvature of the original curve are given.

2. Fundamentals on Parallel Curves on Surfaces

For the geometric facts cited in this section see e.g. [2] or [8].

Let $x : U \subset \mathbf{R}^2 \to \mathbf{R}^3$ denote a regular parametric surface and N the unit normal vector field of x. A curve $\hat{x} : I \subset \mathbf{R} \to \mathbf{R}^3$ is a curve on the surface x if and only if $\hat{x} = x \circ c$ where $c : I \to U$ is a plane curve in U.

The *Darbouxframe* b_1, b_2, b_3 along $\hat{x}$ is the orthogonal frame defined by

$$b_1(t) = \frac{\hat{x}'(t)}{|\hat{x}'(t)|}, \quad b_2(t) = \hat{N}(t) \times b_1(t), \quad b_3(t) = \hat{N}(t).$$

In Section 5 we will use the equations that express the derivatives b_1', b_2', b_3' in the Darboux basis b_1, b_2, b_3:

$$\begin{aligned} b_1' &= \omega\kappa_g b_2 + \omega\kappa_n b_3, \\ b_2' &= -\omega\kappa_g b_1 + \omega\tau_g b_3, \\ b_3' &= -\omega\kappa_n b_1 - \omega\tau_g b_2 \end{aligned}$$

with $\omega(t) = |\hat{x}'(t)|$.

The functions κ_g, κ_n and τ_g are called geodesic curvature, normal curvature and geodesic torsion. For the purpose of this paper it is sufficient to give geometric interpretations of these quantities.

The geodesic curvature of a surface curve $\hat{x}$ at a point $\hat{x}(t)$ is the ordinary curvature of the plane curve generated by orthogonal projection of $\hat{x}$ onto the tangent plane of x at $\hat{x}(t)$. A surface curve with identically vanishing geodesic curvature is called a *geodesic* of the surface.

The absolute value of the normal curvature of $\hat{x}$ at a point $\hat{x}(t)$ is the curvature of the intersection of x with the plane through $\hat{x}(t)$ spanned by the vectors $\hat{x}'(t)$ and $\hat{N}(t)$. While the geodesic curvature is the curvature of a surface curve from a viewpoint in the surface, normal curvature measures the curvature of the curve that is due to the curvature of the underlying surfaces. If κ denotes the ordinary curvature of the space curve $\hat{x}$ the identity $\kappa^2 = \kappa_n^2 + \kappa_g^2$ holds.

The geodesic torsion of a surface curve $\hat{x}$ at a point $\hat{x}(t)$ is the torsion of the geodesic that meets $\hat{x}$ at $\hat{x}(t)$ with common tangent direction. A *curvature line* of x, i.e. a curve with a tangent vector that points into one of the principal directions of the surface, is characterized by vanishing geodesic torsion.

Since the geodesic curvature of $\hat{x}$ can be computed using the formula

$$\kappa_g = \frac{[\hat{x}', \hat{x}'', \hat{N}]}{|\hat{x}'|^3} \tag{1}$$

a geodesic is characterized by the property that at any point of nonvanishing curvature the surface normal $\hat{N}$ lies in the osculating plane of $\hat{x}$ which is spanned by $\hat{x}'$ and $\hat{x}''$. The importance of these curves is due to the fact that a geodesic provides the path of shortest length between two points sufficiently close on the surface.

In the following we will always assume that the geodesics are arc length parametrized. In this situation $x(u(s), v(s))$ is a geodesic if and only if u, v satisfy the differential equations:

$$u'' + \Gamma^1_{11}(u')^2 + 2\Gamma^1_{12}u'v' + \Gamma^1_{22}(v')^2 = 0$$
$$v'' + \Gamma^2_{11}(u')^2 + 2\Gamma^2_{12}u'v' + \Gamma^2_{22}(v')^2 = 0$$

where the coefficients Γ^i_{jk} involve second order derivatives of x. If u, v are solutions to the system of differential equations above with initial values (u_0, v_0, u'_0, v'_0) such that

$$u'_0 x_u(u_0, v_0) + v'_0 x_v(u_0, v_0) = 1,$$

then the curve $x(u(s), v(s))$ is an arc length parametrized geodesic.

The coefficients Γ^i_{jk} are C^1, if the surface x is of differentiability C^3. Therefore it follows from the theory of ordinary differential equations that for any given point p on a C^3 surface x and for any unit vector v of the tangent space of x at p there exists a unique geodesic $g(p, v)$ that has v as the tangent vector at p. Furthermore, for $g(p, v)$ there exists an interval of maximum length l_{max} such that $g(p, v)$ can not be extended as a geodesic on an interval of length bigger than l_{max}.

For fixed value of s but for varying t the points on the geodesics $g(\hat{x}(t), b_2(t))(s)$ trace out a curve on x that has the geodesic offset s to $\hat{x}$.

Definition 1. *Suppose that the geodesics* $g(\hat{x}(t), b_2(t))$ *exist on the interval* $[0, d]$ *for each* $t \in I$, *then the curve* $\hat{x}_d : I \to \mathbf{R}^3$ *defined by* $\hat{x}_d(t) := g(\hat{x}(t), \hat{b}_2(t))(d)$ *is called the* offset curve *or the* parallel curve *of geodesic distance* d *to* $\hat{x}$ *on* x.

The name parallel curve for $\hat{x}_d$ refers to the fact that for any t the tangent vector $\hat{x}_d'(t)/|\hat{x}_d'(t)|$ is obtained by parallel transport $\hat{x}'(t)/|\hat{x}'(t)|$ along $g(\hat{x}(t), b_2(t))$. Therefore $\hat{x}_d$ is an orthogonal trajectory to the family $g(\hat{x}(t), b_2(t))$ of geodesics.

Example. In the case that x is a plane the vector b_2 is the normal vector n of the curve $\hat{x}$ and Definition 1 reduces to the well known formula

$$\hat{x}_d(t) := g(\hat{x}(t), n(t))(d) = \hat{x}(t) + dn(t).$$

3. Tracking the Geodesics on a Surface of Revolution

According to the previous section a geodesic $x(u(s), v(s))$ satisfies a system of differential equations of second order. Since the cases where this system of differential equations can be explicitely integrated are rare, a numerical solution of the system is in general the only way to compute points on a geodesic. Examples of geodesics special surfaces can be found in [8], II, pp. 222–234.

For the class of Liouville surfaces (see also [8]), which include surfaces of revolution, the geodesics can be integrated by using quadratures only. In the following we discuss the efficient tracking of a geodesic on a surface of revolution.

On a surface x of the form

$$x(u, v) = (f(v)\cos(u), f(v)\sin(u), g(v))$$

the differential equations for a geodesic $x(u(s), v(s))$ are given by

$$u'' + 2\frac{f'}{f}u'v' = 0, \tag{2}$$

$$v'' - \frac{ff'}{(f')^2 + (g')^2}(u')^2 + \frac{f'f'' + g'g''}{(f')^2 + (g')^2}(v')^2 = 0. \tag{3}$$

The meridians of a surface of revolution are always geodesics while the parallels are geodesics only for $f'(v) = 0$. A part of a geodesic that is neither a meridian nor a parallel has a representation of the form

$$u(v) = c\int_0^v \frac{1}{f}\sqrt{\frac{(f')^2 + (g')^2}{f^2 - c^2}}dv + u_0 \tag{4}$$

with $c \in \mathbf{R}$.

For the efficient computing of points on a parallel curve formula (4) is not very beneficial. The main reason for this is that the value v for which the length of the geodesic equals a given value d is unknown and has to be determined by locating the zero of a rather expensive function:

$$l(v) - d = \int_0^v |f| \sqrt{\frac{f^2 - c^2}{(f')^2 + (g')^2}} dv - d. \tag{5}$$

The Timing-function of the system MATHEMATICA was used to compare the CPU time for one evaluation of the above integral with the time spend for one complete step in the Runge-Kutta method of fourth order. To compute (5) the adaptive integration routine 'Nintegrate' of MATHEMATICA was used. We found that about eight successive steps of the Runge-Kutta method can be performed in the same time that is needed for one evaluation of (5). Since a root-finding procedure has to be applied to the function (5), including the generation of startpoints, the computation of one point of the offset curve based on (4) involves several evaluations of (5). The situation is worsened by the fact that the integrand of (5) obviously has poles for $f(v)^2 = c^2$. Therefore it is necessary to compute all zeros of the function $f^2 - c^2$ and eventually to split the integral.

For these reasons it is strongly recommended to rely on a numerical integration of the system of differential equations rather than to use (4) for the computation of points on the offset curve. The superiority of Eqs. (2), (3) is due to the fact that their solutions $(u(s), v(s))$ for initial values (u_0, v_0, u_0', v_0') with $|u_0' x_u(u_0, v_0) + v_0' x_v(u_0, v_0)| = 1$ yield arc length parametrized curves $x(u(s), v(s))$. Therefore the integration of (2), (3) has to be performed to the point $s = d$ and no rootfinding procedure is necessary to find the desired point on the geodesic.

However, in the numerical solution of the second order system (2), (3) in each step the vector $(u(s), v(s), (u'(s), v'(s))$ is approximated by a vector $(\tilde{u}, \tilde{v}, \tilde{u}', \tilde{v}')$ according to the Runge-Kutta scheme. After several steps of the method the vector $\tilde{u}' x_u(\tilde{u}, \tilde{v}) + \tilde{v}' x_u(\tilde{u}, \tilde{v})$ that approximates the unit tangent vector of the geodesic will have a length l that differs from 1. As the property of arc length parametrization of the geodesic is crucial to determine the correct end point on the curve this may produce a serious error in the location of the points of the parallel curve. Figure 1 illustrates the situation: the light blue curve is a highly accurate approximation of the true offset curve while the white curve was computed based on the Runge-Kutta scheme applied to (2), (3).

A drastic improvement was obtained by scaling the vector $(\tilde{u}', \tilde{v}')$ by $1/l$ in order to normalize the tangent vector after each step of the numerical integration. The curve based on this method is displayed in dark blue in Fig. 1. Note that the error depends on the curvature of the surface that is highly curved at the top but mildly curved at the bottom. For all curves ten steps of the Runge-Kutta method have been performed to compute one point of the offset curve.

The algorithm that was used to compute the light blue curve in Fig. 1 is based on the fact that the system (2), (3) can be transformed into a system of first order. Obviously (2) can be integrated to

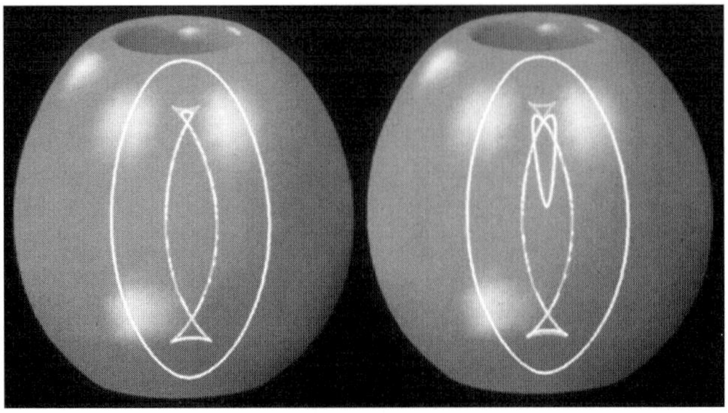

Figure 1. Comparison of three algorithms to approximate the offset curve

$$f(v)^2 u' = c \tag{6}$$

where c is a constant. Equation (6) is commonly used to prove Clairaut's relation for the angle between a geodesic and a parallel circle on a surface of revolution (see [8]).

A second equation for u' and v' can be derived if one takes into account that the geodesic is arc length parametrized. The equation

$$|\hat{x}'(s)| = |u' x_u(u, v) + v' x_u(u, v)| = 1$$

yields

$$(u')^2 f(v)^2 + (v')^2 ((f'(v))^2 + (g'(v))^2)$$

or together with (6)

$$(v')^2 = \frac{1 - (c/f(v))^2}{(f'(v))^2 + (g'(v))^2}. \tag{7}$$

Instead of using (2), (3) we may therefore use the system formed by (6) and (7) to compute the geodesics. As (7) does not yield the sign of v' we have to complete the equations by a strategy that provides the missing information.

First, we consider initial conditions (u_0, v_0, u_0', v_0') for the geodesic with $v_0' \neq 0$. In this case we only have to figure out under which circumstances the sign of v_0' has to be changed.

(7) implies that for all points (u, v) of a (real) geodesic with constant c according to (6) the relation $f(v)^2 \geq c^2$ is satisfied. Furthermore, v' vanishes along the geodesic if and only if $f(v)^2 = c^2$, i.e. if the geodesic intersects a parallel circle of radius $|c|$.

If the coordinate line $v = v_c$ is the parallel of radius $|c|$ closest to the startpoint (u_0, v_0) then we have to distinguish two different scenarios.

If $f'(v_c) = 0$, the parallel $v = v_c$ is itself a geodesic. In this case the considered geodesic will come arbitrary close to $v = v_c$ (because v is monotone and v' gets small only if $f(v)^2 - c^2$ gets small) but can never intersect the parallel because of the uniqueness of the geodesics. Thus it will asymptotically approach the parallel circle of radius $|c|$.

If $f'(v_c) \neq 0$ the function $f^2 - c^2$ has a zero with sign chance for $v = v_c$ which means that f^2 is smaller than c^2 on the other side of the parallel $v = v_c$. Therefore the geodesic will turn backwards at $v = v_c$ into the region where $f(v)^2 \geq c^2$. Figure 2 illustrates this situation. The curve drawn in black is a single geodesic that has been tracked along its way on the surface.

These facts about the behaviour of the geodesics enter the algorithm for computing the geodesics based on the system (6), (7) as follows. While computing points on a geodesic we keep track of the value $\delta = f^2 - c^2$ after each step of the integration method. If δ gets smaller than a prescribed tolerance at some value $\bar{v}$ we compute the root v_r of the function $f^2(v) - c^2$ using $\bar{v}$ as the initial value. Then, we evaluate f' at v_r. If $f'(v_r) = 0$ we simply continue with the numerical integration of the first order differential equations. If $f'(v_r) \neq 0$, we proceed differently because we know that the geodesic will turn after its intersection with the parallel circle of radius $|c|$.

Let $(u_i, v_i) i = 1, \ldots, n-1$ denote the computed sequence of points on the geodesic and let δ be smaller than the tolerance for the first time for $i = n - 1$. Our strategy is to approximate the intersection point of the geodesic and the parallel circle of radius $|c|$ by the intersection point of the line $v = |c|$ in the parameter space and the tangent of the curve $(u(s), v(s))$ at (u_{n-1}, v_{n-1}). Therefore we compute the factor λ such that $v_{n-1} + \lambda v'_{n-1} = v_r$ and set

$$(u_n, v_n) = (u_{n-1}, v_{n-1}) + \lambda(u'_{n-1}, v'_{n-1}).$$

Figure 2. Intersection of a geodesic with a parallel of radius $|c|$

This approach will provide a good approximation to the intersection point if the tolerance is chosen sufficiently small.

For symmetry reasons the part of the geodesic after the intersection point is simply a reflection of the portion of the curve before the intersection point. Therefore we set

$$(u_{n-1}, v_{n-1}) = (2u_n - u_{n-1}, v_{n-1})$$

and then continue the tracking of the geodesic using a numerical integration of the system of differential equations formed by (6) and (7) with a different sign of v'.

Note that if, may happen that for the last computed point on the geodesic δ is bigger than the tolerance but the next Runge-Kutta step already involves points with $\delta < 0$. We take care of this situation by stepping back to the last computed point and reducing the stepsize in the numerical integration scheme.

In the case that the initial conditions of the geodesic are such that $v_0' = 0$ a numerical integration of the system (6), (7) would produce a sequence of points that all lie on the parallel circle of radius $r = |c|$. This parallel is only a geodesic if $f'(v_0) = 0$ and therefore (6), (7) can be used only in this case. If $f'(v_0) \neq 0$ we use (2), (3) to compute the first point on the geodesic that deviates from the parallel $r = |c|$ and then continue to track the geodesic with the system (6), (7).

4. Spline Approximation of Parallel Curves on Surfaces

Spline approximations of parallel curves on surfaces can be obtained by adapting well established methods for plane offset curves (see [1], [7]) to the parameter domain of the surface.

Let $\hat{x}(t) = x(u(t), v(t))$ be the parametrization of a surface curve to be offsetted and denote the parallel curve with $\hat{x}_d(t) = g(\hat{x}(t), b_2(t))(d)$. The numerical integration of the geodesic for a fixed t yields a vector (u, v, u', v') such that $x(u, v)$ approximates $\hat{x}_d(t)$ and $V = u'x(u, v)$ approximates the tangent vector of the geodesic $g(\hat{x}(t), b_2(t))$ at $x(u, v)$.

According to Section 2 the tangent vector T_d of $\hat{x}_d$ at t is given by $T_d =$ sign $(d)V/|V| \times N$ where N is the unit normal vector of x. As x is a regular surface the matrix (x_u, x_v) has a quadratic submatrix of rank 2 that is denoted by

$$\begin{pmatrix} x_u^{(i)} & x_v^{(i)} \\ x_u^{(j)} & x_v^{(j)} \end{pmatrix}$$

with $i, j \in 1, 2, 3$ and $i \neq j$. Solving the linear system

$$\begin{aligned} ax_u^{(i)} + bx_v^{(i)} &= T_d^{(i)}, \\ ax_u^{(j)} + bx_v^{(j)} &= T_d^{(j)} \end{aligned}$$

we obtain the direction (a, b) in the parameter domain that corresponds to the direction T_d on the surface.

Therefore a G^1-spline approximation of the offset curve x_d can be constructed using spline segments of the form $x \circ s_i(t)$ with

$$s_i(t) = \begin{pmatrix} u_i \\ v_i \end{pmatrix} F_0(t) + \begin{pmatrix} u_{i+1} \\ v_{i+1} \end{pmatrix} F_1(t) + \alpha \begin{pmatrix} u'_i \\ v'_i \end{pmatrix} G_0(t) + \beta \begin{pmatrix} u'_{i+1} \\ v'_{i+1} \end{pmatrix} G_1(t)$$

where $\begin{pmatrix} u_i \\ v_i \end{pmatrix}, \begin{pmatrix} u_{i+1} \\ v_{i+1} \end{pmatrix}$ are the parameters of two points on the parallel curve and $\begin{pmatrix} u'_i \\ v'_i \end{pmatrix}, \begin{pmatrix} u'_{i+1} \\ v'_{i+1} \end{pmatrix}$ are the directions in the parameter domain that correspond to the tangent vectors of the parallel curve at these points. The functions F_k, G_k denote the cubic Hermite blending functions.

For the plane case $x = \text{id}$ various methods have been proposed to determine the free parameters α and β. Klass used these parameters to interpolate the curvatures of the offset curve at the end points of the segment (see [7]) while Arnold imposed the condition $\hat{x}_d(0.5) = s(0.5)$ on the cubic spline segment (see [1]). Hoschek used a least squares fit to minimize the deviation of the spline from a whole sequence of points on the offset curve (see [4]).

Klass' method can only in special cases be adapted to the case of a parallel curve on a surface. One of the reasons is that this method requires an explicit formula for the curvature of the offset curve. (Such a formula can be established if the surface is a sphere; see Section 5). Note that this method involves the solution of a nonlinear system of two equations.

Arnold's method is linear but tends to create unbalanced segments with abrupt changes close to the forced interpolation point $\hat{x}_d(0.5)$. Furthermore, in situations where the data is nearly linear it causes extreme overshooting of the spline segment. This effect does not disappear after subdivision of the segment. To overcome the problem of overshooting by a refinement strategy it is necessary to subdivide to a level where it is appropriate to use line segments to fit the data.

As the computation of points on the parallel curve is the most expensive step of the algorithm the least squares approach was implemented using only two points in the interior of the spline segment. The curves obtained by this method look more balanced than those based on the interpolation strategy. The problem of overshooting in nearly linear situations does not occur. However, to obtain a nice curve fit in a highly curved segment it is necessary to apply the parameter optimization proposed by Hoschek in [5]. Figures 3 and 4 show the different curves obtained by the least squares method without and with parameter optimization.

Figures 5 and 6 show spline approximations to offset curves on surfaces. The spline in Fig. 6 has several cusps which have been determined by the method

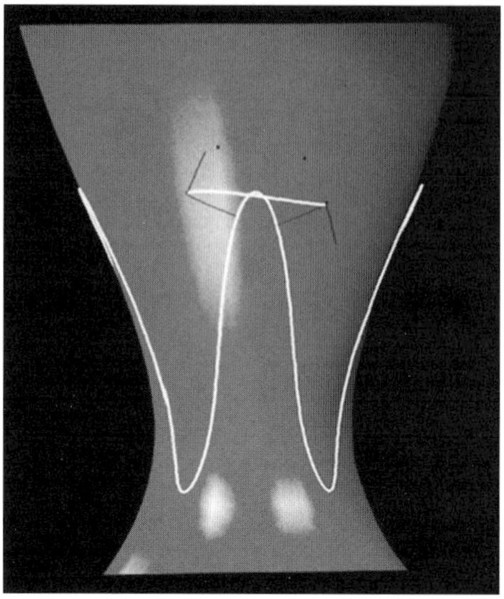

Figure 3. Segment of offset-curve without parameter optimization

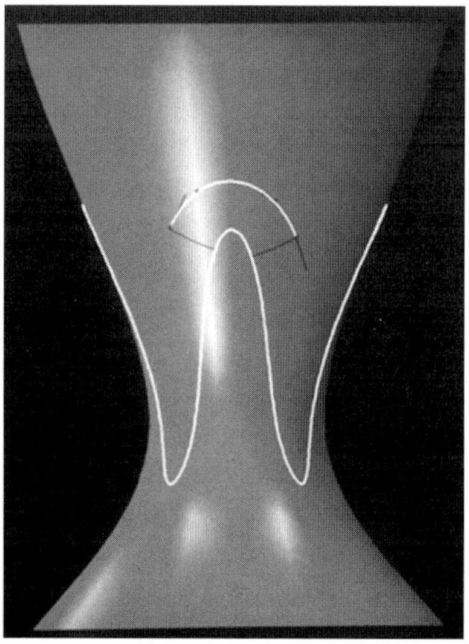

Figure 4. Segment of offset-curve with parameter optimization

described in the next section. In both pictures the original curve and its offset curve are displayed in with while the spline approximation of the parallel curve is drawn in light blue. The endpoints of the geodesics displayed in black are breakpoints of the spline.

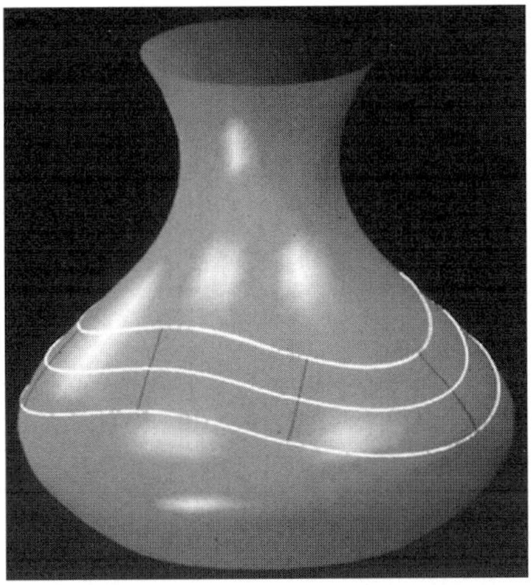

Figure 5. Spline approximation of a parallel curve

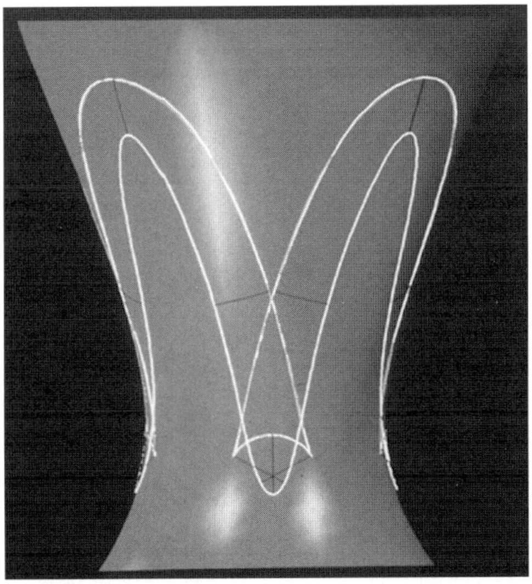

Figure 6. Spline approximation of a parallel curve

5. Detection of Cusps and Singularities

Let $\hat{x}$ denote a differentiable curve. A curve point $\hat{x}(t_s)$ is called singular if $\hat{x}'(t_s) = 0$. We assume further that all singularities of x are isolated points. In this situation a point $x(t_c)$ is called a cusp if for the tangent vector $T = x'/|x'|$

$$\lim_{t \to t_c^-} T(t) \neq \lim_{t \to t_c^+} T(t).$$

According to this definition a singularity in the curve may or may not be a cusp but since $\hat{x}$ is differentiable a cusp is always a singularity. An offset curve

$$\hat{x}_d(t) = g(\hat{x}(t), b_2(t))(d)$$

parallel to a C^2 curve $\hat{x}$ on a C^3 surface x is differentiable with respect to t. This follows immediately from standard theorems for geodesics, e.g. Theorem 1a, Section 4.7 of [2]. Therefore a point t_c in the offset curve that is a cusp has to be a singularity.

The appearance of cusps in parallel curves is a phenomenon that is well-known in the planar case and algorithms for locatin singularities and cusps have been proposed (see [1], [6]). Our first objective is to extend the singularity criterion used by Arnold in [1] to a criterion for cusps.

Theorem 2. *The offset curve $c_d(t) = c(t) + dn(t)$ of a plane curve c with curvature function κ has a cusp at t if and only if the function $1 - d\kappa$ has a zero with sign change at t.*

Proof: Differentiating c_d one obtains the formula

$$c'_d(t) = (1 - d\kappa(t))c'(t).$$

For the tangent vector T_d of c_d we get

$$\begin{aligned} T_d(t) &= \frac{c'_d(t)}{|c'_d(t)|} \\ &= \frac{1 - d\kappa(t)}{|1 - d\kappa(t)|} T(t) \\ &= sign(1 - d\kappa(t))T(t) \end{aligned}$$

if T denotes the unit tangent vector of x. Therefore a cusp occurs if and only if $1 - d\kappa$ changes sign at t. □

The detection of cusps is important for the correct modeling of the offset curve. But it depends on the applications how accurately the critical point has to be determined. Very often cusps occur in a part of the parallel curve that lies in a region of collision with the original curve. Figure 6 shows the most frequent situation that two cusps appear in a loop that will be removed from the offset curve in a post process. In this case the detection of the cusps serves only the purpose of modeling the loop correctly because a poorly modeled loop may lead to an avoidable error in the computation of the intersection point in the curve. For this application it is sufficient to find a point close to the cusp. This is our next objective.

As a rough approximation of the offset curve $\hat{x}_d(t) = g(\hat{x}(t), b_2(t))(d)$ to the curve $\hat{x}$ on the surface x we consider the curve y_d generated by a constant offset d in the direction b_2:

$$y_d(t) = \hat{x}(t) + db_2(t). \tag{8}$$

y_d is a parallel curve to $\hat{x}$ in the ruled surface R formed by the family of straight lines in the direction b_2 along the curve $\hat{x}$.

This ruled surface is in fact a torse because the determinant $[\hat{x}', b_2', b_2]$ vanishes and may therefore be developed into the plane. Since the geodesic curvature of $\hat{x}$ is the curvature of the developed curve, the statement of Theorem 2 will hold for parallel curves on R, if we substitute curvature by geodesic curvature.

Theorem 3. *The offset curve $y_d(t) = \hat{x}(t) + db_2(t)$ of a curve $\hat{x}$ on the ruled surface R has a cusp at t if and only if the function $1 - d\kappa_g$ has a zero with sign change at t.*

Proof: R shares with x the same normal vector $\hat{N}$ along $\hat{x}$ and therefore the Darboux frame of $\hat{x}$ with respect to x and R are identical. Differentiating (8) and expressing $b_2'(t)$ in the Darboux frame of $\hat{x}$ yields

$$y_d'(t) = (1 - d\kappa_g)\hat{x}'(t) + r_g|x'(t)|\hat{N}(t).$$

The generators of R and their orthogonal trajectories are the lines of curvature of the surface (see e.g. [10], III, p. 27). Therefore r_g vanishes identically. The rest of the proof is in complete analogy to the planar case. □

Based on Theorem 3 we used the criterion

$$\frac{[\hat{x}', \hat{x}'', \hat{N}]}{|\hat{x}'|^3} = 1/d \tag{9}$$

to compute start point for the detection of singularities and cusps on nonruled surfaces. As the ruled surface R is only a rough approximation of the actual surface x it could only be expected that the point determined by the criterion (9) lies in some neighborhood of the singularity. However, the closeness of the compute points to the singularities observed in the tests of the method was surprisingly high even for strongly curved surfaces.

Figure 7 illustrates this statement by displaying the geodesics starting at the points on the original curve obtained by (9). The offset curves displayed are computed with distances 0.6, 1.2 and 1.8. We observe that the cusps and the endpoints of the drawn geodesics can be visually distinguished only for high distances. Figure 8 illustrates the same situation on a non-convex surface.

Since an exact criterion for a cusp in an offset curve on an arbitrary surface is not available it is difficult to establish a formal proof for this phenomenon. However, the following two points will provide strong formal arguments for its occurrence.

- Equally spaced points in the parameter interval of the parallel curve are differently spaced along the parallel curve according to its parametrization. In the neighborhood of a singularity the points lie very close together. Therefore if the parameter value t computed according to (9) lies in the vicinity of the parameter value of the singularity, the point $\hat{x}_d(t)$ will lie very close to the singularity itself.

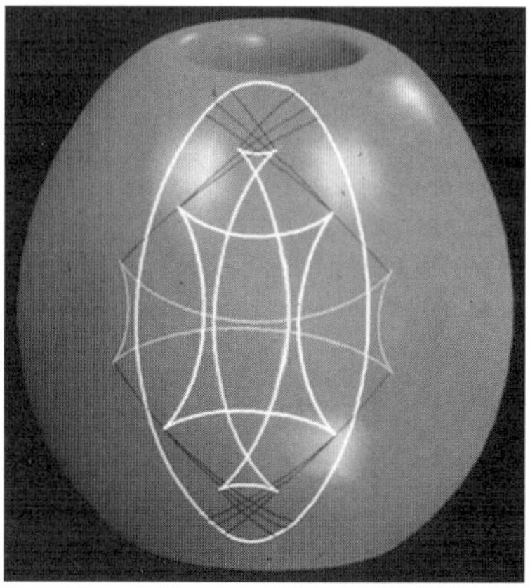

Figure 7. Detecting cusps in the offset curve

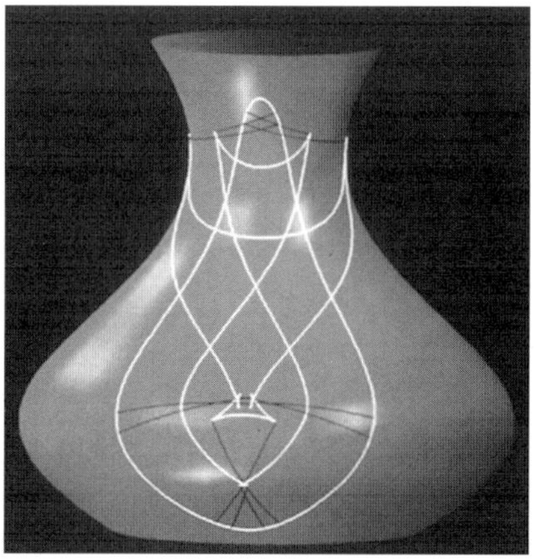

Figure 8. Detecting cusps in the offset curve

- The visual effect of parallel curves is especially striking if the curves lie close together. This fact bounds the distance d that controls the accuracy of the approximation.

For parallel curves on the sphere it is possible to derive an exact cusp criterion.

Theorem 4. *Let x be a parametrization of a part of the sphere of radius r with normal vector $N = (1/r)x$ and $\hat{x}$ a curve on x with geodesic curvature κ_g. The parallel curve of distance d to $\hat{x}$ is given by*

$$\hat{x}_d(t) = \cos(d/r)\hat{x}(t) + \sin(d/r)\hat{x}(t) \times \frac{\hat{x}'(t)}{|\hat{x}'(t)|}.$$

$\hat{x}_d$ has a cusp at t if and only if the function

$$\kappa_g - (1/r)\cot(d/r)$$

has a zero with sign change at t. *The geodesic curvature* κ_g *of* $\hat{x}_d$ *is related to the geodesic curvature of* $\hat{x}$ *by*

$$\bar{\kappa}_g(t) = \frac{\kappa_g(t)\cos(d/r) + (1/r)\sin(d/r)}{|\cos(d/r) - r\kappa_g(t)\sin(d/r)|}.$$

Proof: A geodesic on a sphere is an arc length parametrized great circle. Therefore the offset curve $\hat{x}_d(t) = g(\hat{x}(t), b_2(t))(d)$ is given by

$$\hat{x}_d(t) = \cos(d/r)\hat{x}(t) + r\sin(d/r)b_2(t)$$

where $b_2(t) = \hat{N}(t) \times \hat{b}_1(t)$ and $\hat{N}(t) = (1/r)\hat{x}(t)$.

Thus

$$\begin{aligned}\hat{x}_d'(t) &= \cos(d/r)\hat{x}'(t) + r\sin(d/r)b_2'(t)\\ &= \cos(d/r)\hat{x}'(t) + r\sin(d/r)(-\omega(t)\kappa_g(t)b_1(t) + \omega(t)r_g(t)b_3(t)).\end{aligned}$$

Since all curves on a sphere are lines of curvature, r_g vanishes identically and we obtain the equation

$$\hat{x}_d'(t) = (\cos(d/r) - r\sin(d/r)\kappa_g(t))\hat{x}'(t). \tag{10}$$

$\hat{x}_d'(t)$ can only vanish if $\sin(d/r) \neq 0$ and we may therefore divide by $\sin(d/r)$. In analogy to the proof of Theorem 2 a cusp occurs if and only if the function $\kappa_g(t) - (1/r)\cot(d/r)$ has a zero with sign change at t.

Since the geodesic curvature is parameter invariant we may assume that $\hat{x}$ is arc length parametrized. Then, differentiating (10) and expressing all vectors in the Darboux frame yields

$$\begin{aligned}\hat{x}_d'' = &-r\sin(d/r)b_1\\ &+ (\cos(d/r) - r\sin(d/r)\kappa_g)\kappa_g b_2\\ &+ (\cos(d/r) - r\sin(d/r)\kappa_g)\kappa_n\hat{N}.\end{aligned}$$

Note that due to the parametrization $\hat{x} = (1/r)\hat{N}$ of the sphere any surface curve has normal curvature $\kappa_n = -1/r$. Putting the expressions for $\hat{x}_d'$, $\hat{x}_d''$ and

$$\hat{N}_d = \cos(d/r)\hat{N} + \sin(d/r)b_2$$

into formula (1) one obtains the claimed relation for the geodesic curvature. □

We will now use the example of the sphere to demonstrate that for typical values of d the criterion $\kappa_g(t) = 1/d$ yields a point that is a very good approximation for the singularity in the offset curve on a surface.

First, we need to understand the range for the distance d in which parallelity of curves will have a visual appealing effect. In order to be able to see two curves simultaneously on a sphere of radius r their distance has to less than πr. For a visually striking use of parallel curves their distance will be typically smaller than $1/10$ of that value.

Consider the first terms in the Taylor expansion of the cot function

$$\begin{aligned}\frac{1}{r}\cot\left(\frac{d}{r}\right) &= \left(\frac{r}{d} - \left(\frac{1}{3}\frac{d}{r} + \frac{1}{45}\frac{d^3}{r^3} + \cdots\right)\right)\frac{1}{r}\\ &= \frac{1}{d} - \left(\frac{1}{3}\frac{d}{r^2} + \frac{1}{45}\frac{d^3}{r^4} + \cdots\right).\end{aligned}$$

If we set $d = \pi r/f$ with some factor f we obtain for the ratio of the first two terms in the expansion the expression $3f^2/\pi^2$.

If we assume a value of $f = 10$ that corresponds to a high value of $d = \pi r/10$, we calculate that the first term in the expansion is more than the first term in the expansion is more than thirty times bigger than the second term. This illustrates the usefullness of criterion (9) to compute an initial approximation for the cusp in a parallel curve on a general surface.

Figure 9 shows three offset curves of distances 0.3, 0.6 and 0.9 on a sphere of radius 1. To illustrate the difference between criterion (9) and the exact criterion

$$\frac{[\hat{x}', \hat{x}'', \hat{N}]}{|\hat{x}'|^3} = (1/r)\cot(d/r) \tag{11}$$

the geodesics starting at points on the original curve which were computed with (9) resp. (11) are displayed. We observe that for the offset curve of radius 0.3 the

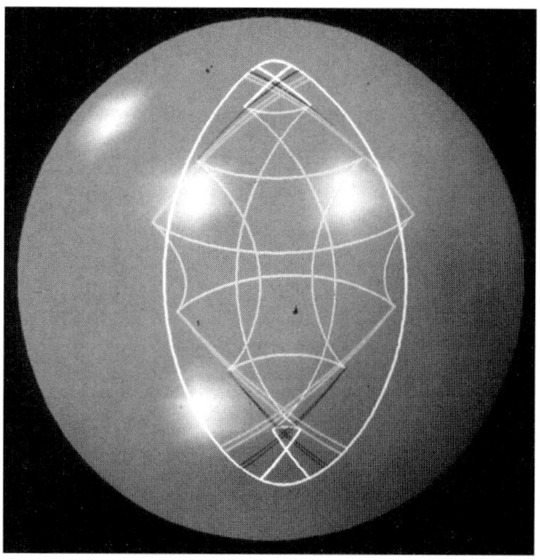

Figure 9. Cusps detection on the sphere

different geodesics almost coincide. For the other offset curves the geodesics according to (9) and (11) can be distinguished at the start points but they seem to converge as they approach the parallel curve.

If the application requires to locate a cusp precisely, an iterative method has to be used to detect it. In the first step we use the criterion (9) to obtain a point close to the singularity. In the second step we perform a steepest descent method to find the minimum of the function $f(t) = (x_d(t+h) - x_d(t))^2$ with a fixed small displacement $h \in \mathbf{R}$ to locate the singularity. The choice of the function f reflects the fact that the curve points with equally spaced parameter values come closer and closer together if the singularity is approached.

References

[1] Arnold, R.: Quadratische und kubische Offset-Bezierkurven. Dissertation, Universität Dortmund, 1986.
[2] do Carmo, M. P.: Differentialgeometrie von Kurven und Flächen. Leipzig: Vieweg, 1983.
[3] Faux, I. D., Pratt, M. J.: Computational geometry for design and manufacture. Ellis Horwood Ltd., 1979.
[4] Hoschek, J.: Spline approximation of offset curves. CAGD *5*, 33–40 (1988).
[5] Hoschek, J.: Intrinsic parametrization for approximation. CAGD *5*, 27–31 (1988).
[6] Hoschek, J.: Offset curves in the plane. CAD *17*, 77–82 (1985).
[7] Klass, R.: A offset spline approximation for planar cubic splines. CAD *15*, 297–299 (1983).
[8] Kunze, R., Wolter, F.-E., Rausch, T.: Geodesic Voroni diagrams on parametric surfaces, CGI'97, IEEE, Comp. Soc. Press Conf. Proc., pp. 230–237, 1997.
[9] Rausch, T., Wolter, F.-E., Sniehotta, O.: Computation of medical curves on surfaces, Conf. Math. of surfaces VII, IMA Conf. Series, pp. 43–68, 1997.
[10] Strubecker, K.: Differentialgeometrie I-III. Sammlung Göschen, Berlin: de Gruyter 1969.

Guido Brunnett
Computer Science Department
Technical University Chemnitz
D-09107 Chemnitz
Germany
e-mail: brunnett@informatik.tu-chemnitz.de

Computing [Suppl] 14, 55–72 (2001)

Computing

Computing Volume Properties Using Low-Discrepancy Sequences

T. J. G. Davies, and **R. R. Martin,** Cardiff, and **A. Bowyer,** Bath

Abstract

This paper considers the use of low-discrepancy sequences for computing volume integrals in solid modelling. An introduction to low-discrepancy point sequences is presented which explains how they can be used to replace random points in Monte Carlo methods. The relative advantages of using low-discrepancy methods compared to random point sequences are discussed theoretically, and then practical results are given for a series of test objects which clearly demonstrate the superiority of the low-discrepancy method when used in a simple approach. Finally, the performance of such methods is assessed when used in conjunction with spatial subdivision in the svLis geometric modeller.

Key Words: Solid modelling, volume computation, mass properties, low-discrepancy sequences.

1. Low-Discrepancy Sequences

Monte Carlo methods of integration are used widely for calculating volume integrals in solid modelling. The Monte Carlo method uses randomly generated points inside a box enclosing an object of interest to calculate volume integrals. For example, the volume of the object can be estimated as the ratio of the number of points that are contained within the object to the total number of points generated, multiplied by the volume of the box. Naturally, such a method is subject to errors because of the random nature of the sampling, and in particular we cannot guarantee that all parts of space will be sampled equally well. Quasi-Monte Carlo methods [6] use *pseudo-random* sequences of numbers, called *low-discrepancy* sequences, for computing multi-dimensional integrals, where here pseudo-random indicates that the sampling is to be done in a rather more structured manner.

The key idea is the one of *discrepancy*, which is a measure of how uniformly the points sample the space [5]. (A simple introduction to low-discrepancy methods, in the context of applications to financial problems, can be found in [3].) Two sets of 200 points in two dimensions are shown in Fig. 1. Those on the left were generated using a random number generator, while those on the right were generated using a low-discrepancy sequence. Clearly, there are some large 'holes' in the random sampling, while the holes in the low-discrepancy sampling are less pronounced. Note also, however, that the low-discrepancy samples do not form a regular grid. Such a grid can give large errors when used for volume integral computation, in cases where the object is just a little larger or a little smaller than

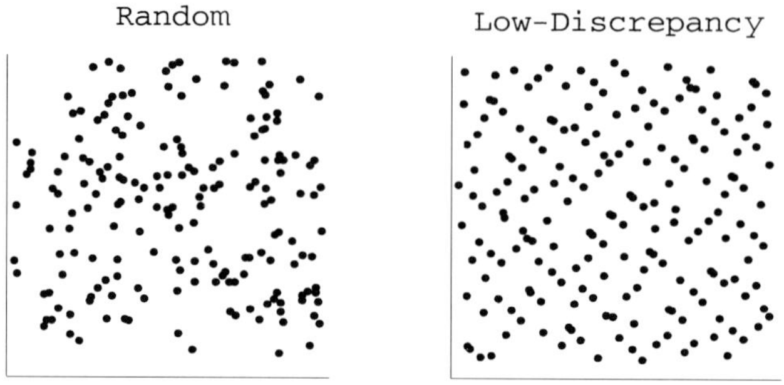

Figure 1. Comparison of points generated randomly and using low-discrepancy sequences

the grid spacing, for example. This problem does not arise for the low-discrepancy point sequences.

To understand discrepancy, let us first consider one dimension. Take the interval $[0, 1]$ and let E be any subset of this interval, defined by the characteristic function

$$f_E(x) = \begin{cases} 0 & \text{if } x \notin E \\ 1 & \text{if } x \in E. \end{cases} \tag{1}$$

Now define

$$A(E, N) = \sum_{n=1}^{N} f_E(x_n), \tag{2}$$

where $x_1, x_2, \ldots, x_N$ are N numbers in $[0, 1]$. Thus $A(E, N)$ is the number of the x_N which are in E. The *discrepancy* D_N of the N numbers $x_1, x_2, \ldots, x_N$ is

$$D_N = \sup_J \left| \frac{A(J, N)}{N} - |J| \right|, \tag{3}$$

where J now runs through all subintervals of $[0, 1]$, and $|J|$ is the length of J. Thus D_N is the biggest possible error when estimating the length of any interval J by sampling using the given set of x_N and using $A(J, N)/N$ as the estimate of its length.

More generally, if f is any function with bounded variation $V(f)$ on I it can be shown that

$$\left| \frac{1}{N} \sum_{n=1}^{N} f(x_n) - \int_0^1 f(t)dt \right| \leq V(f) D_N^*, \tag{4}$$

where D_N^* has a slightly different definition of discrepancy based only on those intervals whose left hand ends start at 0.

Similar definitions and results apply in m dimensions where the intervals are replaced by rectangular parallelepipeds. It can be shown that the two different definitions of discrepancy are of the same order for fixed m:

$$D_N^* \leq D_N \leq 2^m D_N^*. \tag{5}$$

Making use of this concept of discrepancy relies on the fact that there are known algorithms (see later) for generating sequences of points in m dimensions which have low discrepancy. In particular, the discrepancies of such sequences are smaller than the expected discrepancies for a random set of points.

In light of these remarks, we would expect the use of such sequences to have an advantage in calculating volume integrals in solid modelling, even where the volumes to be integrated over are not axis-aligned polyhedra, but are perhaps mechanical components with more general planar and curved faces. The experimental tests which we present in the rest of this paper examine the extent to which this expectation is justified. Initial results using a simple algorithm show a significant advantage for the low-discrepancy methods. Further results then illustrate the performance gains which are achieved when the method is used in a real CSG solid modeller. In practice, these use recursive subdivision methods to speedily classify large regions of space as inside or outside the object, and only carry out detailed volume calculations in smaller boxes near the boundary of the object.

The main purpose of this paper is to draw the attention of the geometric modelling community to the potential advantages of using low-discrepancy sequences for volume integration.

2. Theoretical Advantage

Following an observation made by Woodwark [9] we may note the following, in the case of randomly generated points. If N trials are made of a random event whose probability of success is p, then the expected number of successes is Np, and the standard deviation in that number is $\sqrt{Np(1-p)}$. Thus, when using points generated randomly in a Monte Carlo method to estimate volumes in this way, we would expect a relative error in the volume of a size comparable to

$$\frac{\sqrt{Np(1-p)}}{Np} = \sqrt{\frac{(1-p)}{Np}}, \tag{6}$$

which is

$$O(N^{-\frac{1}{2}}) \tag{7}$$

in the number of sample points.

However, when using low-discrepancy sequences, it is possible in m dimensions to generate sequences of points whose discrepancy is $O(\log^m N)$, so giving an expected relative error in volume (see Eq. 4) of

$$O(N^{-1} \log^m N). \tag{8}$$

Clearly, asymptotically, this means that the expected error for low-discrepancy sequences is lower than that for random points.

In practice, there are two additional considerations. Firstly, for small N, what are the relative slopes of these functions? As can be seen in Fig. 2 for the case of three dimensions (the main case of interest for geometric modelling), while $O(N^{-\frac{1}{2}})$ may decrease slightly quicker with N for N between 100 and 1000 points, clearly by the time N is above 10000 points, $O(N^{-1} \log^3 N)$ is decreasing more rapidly (Fig. 2 uses logs to base 10).

Secondly, there is the question of the constants of proportionality in these different functions. (This corresponds to a relative vertical shift in the two curves in Fig. 2 – whereupon we get the question of at what value of N the $O(N^{-1} \log^3 N)$ graph overtakes the $O(N^{-\frac{1}{2}})$ graph.) This depends on the particular low-discrepancy sequence used, and for example it is well known that Sobol's point generation method [8] has a worse constant of proportionality than Niederreiter's [7]. We offer no further theoretical analysis on this point, but as the results show later, the constants of proportionality are such that low-discrepancy sequences have an advantage even for quite small N.

3. Initial Point Sequences and Test Data

We performed experiments to compute the volumes of objects using random points and two different low-discrepancy point sequences in Monte Carlo and quasi-Monte Carlo methods. For random points, the built in UNIX random number generator was used. Although in principle pseudo-random number generators of this type can exhibit undesirable lattice structures in higher dimensions

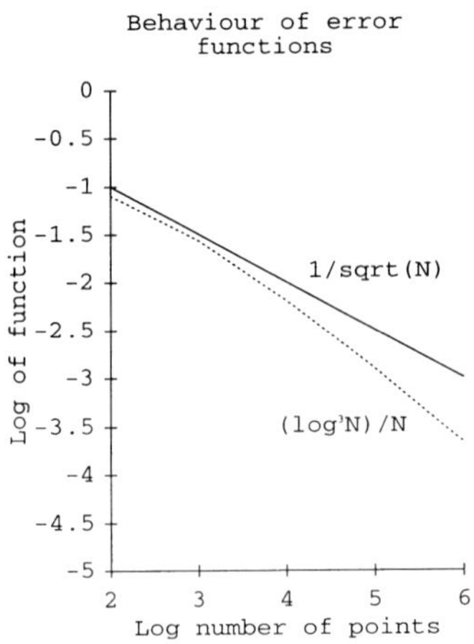

Figure 2. Comparison of error for Monte Carlo and low-discrepancy methods

[6], we did not observe such effects here. The two low-discrepancy sequences used were Sobol's (for theory see [8]) and Niederreiter's (for theory see [7]). In both cases, implementations from *Collected Algorithms of the ACM* were used: for Sobol's method, see [2], and for Niederreiter's method, [4].

Various forms of Niederreiter's method exist. We used the base 2 method, which can be implemented more efficiently.

A small collection of test objects was compiled, comprising three simple shapes, and three more complex mechanical components. Objects 4 and 5 were supplied by J. Corney of Heriot-Watt University; the objects are available on the Web in the NIST Repository: `http://repos.mcs.drexel.edu`. Object 6 was supplied by A. Safa of Intergraph Italia. These objects are described below, as are the bounding boxes used for the volume calculations (note that these are not always as tight as possible).

- Object 1: Sphere, radius 1.0. Bounding box used: $2 \times 2 \times 2$.
- Object 2: L-shaped block, width 2, height 2, length 3 with a block of width 1, height 1 and length 3 removed from the top right corner. Bounding box used: $4.5 \times 6.5 \times 4.5$.
- Object 3: Block with cylindrical hole, width 2, height 2 length 3, with vertical cylindrical hole of diameter 1.0 through the centre. Bounding box used: $4.5 \times 6.5 \times 4.5$.
- Object 4: HW1: A mechanical object – see Fig. 3. Bounding box used: $318 \times 148 \times 30$.
- Object 5: HW2: Another mechanical object – see Fig. 4. Bounding box used: $123.709 \times 117.919 \times 475$.
- Object 6: A valve – see Fig. 5. Bounding box used: $0.237 \times 0.165 \times 0.1675$.

4. Initial Experiments

The volumes of the objects were computed in each case in three distinct ways: using random points, and using low-discrepancy point sequences generated by

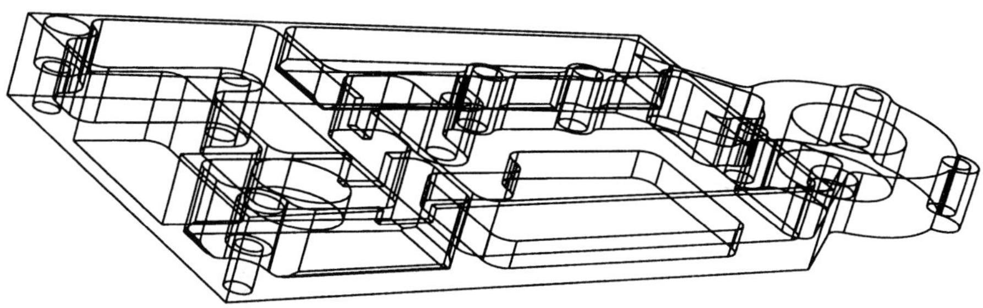

Figure 3. HW1 object

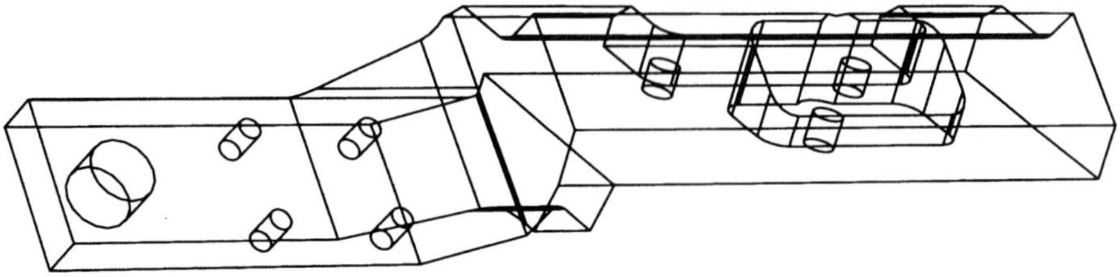

Figure 4. HW2 object

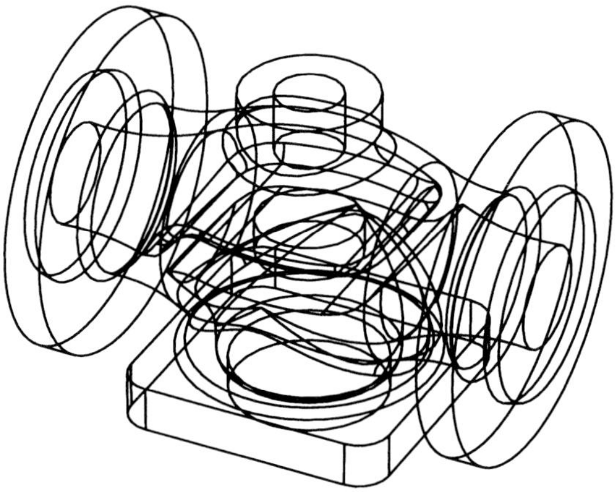

Figure 5. Valve object

Sobol's method and then by Niederreiter's method. Each volume was calculated by generating points lying inside a rectangular box enclosing the object, using point-membership classification to decide if each point was in the object, and then using the formula:

$$V_{obj} = V_{box}\left(\frac{N_{in}}{N}\right). \tag{9}$$

V_{obj} is the estimated volume of the test object, V_{box} is the volume of the box, N_{in} is the number of points found in the object and N is the total number of points generated.

The experiment carried out on each object, for each method, was to compute the volume of the object for an increasing number of points, and in each case to observe the fractional error in the computed volume relative to the true value. For Objects 1–3 the error was calculated at 10^2, 10^3, 10^4, 10^5 and 10^6 points. For Objects 4–6 the error was calculated at 10^2, 10^3, 10^4 and 10^5 points.

For Object 2, the L-shaped block, errors were also calculated every 100 points up to 10^5 points to investigate the behaviour of the low-discrepancy sequences in more detail.

Values used for the true volume of the object were computed theoretically for Objects 1–3, and found accurately using a commercial solid modeller for Objects 4–6.

5. Initial Results

5.1. Timing Observations

Using UNIX timing functions, it was found that when using all three methods, the point-classification step was much slower than generating the points, and practice there was no observable time disadvantage in using any of the three methods to generate an equal number of sample points.

5.2. Sobol's Method

Experiments with the Sobol point generator proved disappointing, and gave results which were not much better than those achieved using random points. We thus do not present these results here. On the other hand, the Niederreiter point generator achieved impressive improvements over random point generation, and we give these results in detail below. As mentioned earlier, it is already known that Sobol's sequences do not have such good properties as Niederreiter's which was borne out by our own experimental observations.

5.3. Errors from Random Points

Note that each run of a Monte Carlo method with differing random points will give differing results, with differing errors. Using any one run of the Monte Carlo method as an indication of the errors obtained may thus be misleading. Instead, we have a theoretical estimate (see Eq. 6) of how big that error is. We thus did a preliminary investigation to see if using the UNIX random number generator did produce relative errors in computed volumes of this size. For various numbers of points, we computed the volume ten times, and found the standard deviation in the volume computed. These are compared in Table 1 to the standard deviations predicted by Eq. (6). As can be seen, the errors in practice match well to those predicted theoretically.

Table 1. Errors in Monte Carlo method versus number of points

No. of points	Experimental	Theoretical
10	1.395	1.263
100	0.427	0.4
1000	0.082	0.1263
10000	0.052	0.04
100000	0.0125	0.0126

Thus, in the following section, we use the standard deviations predicted by theory for random points as 'typical' errors for comparison with errors from the low-discrepancy methods, to avoid statistical fluctuations in the Monte Carlo method affecting the comparison.

In contrast, note that only one result is possible for a given number of sample points using a given low-discrepancy sequence, as it is a well defined sequence of points.

5.4. *Results for Each Test Object*

Figures 6–11 shows the results obtained in our tests. In each graph, the relative error in computed volume is plotted versus the number of point samples used to calculate the volume (the graphs are plotted on a logarithmic scale using logs to base 10). In each case, theoretical errors for a Monte Carlo method based on random points are compared to the actual relative errors obtained using Niederreiter's method.

From the results obtained for each object, it is clear that in general the Niederreiter method gives a distinct advantage over the Monte Carlo method, in that many fewer points are needed to achieve a given accuracy, even for quite small numbers of points (more than a few hundred). We can also see that in most cases, at least, the graph for Niederreiter's points has a steeper slope than that for the Monte Carlo method on average: in each case we have found the best-fit straight line through this graph, and presented its slope in Table 2. The corresponding

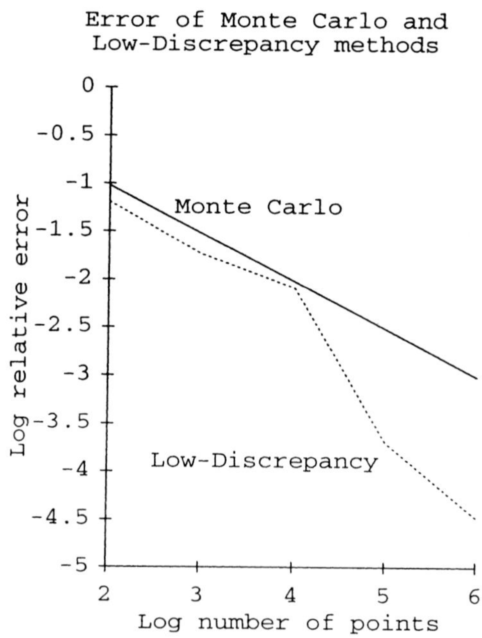

Figure 6. Accuracy versus number of points for the sphere

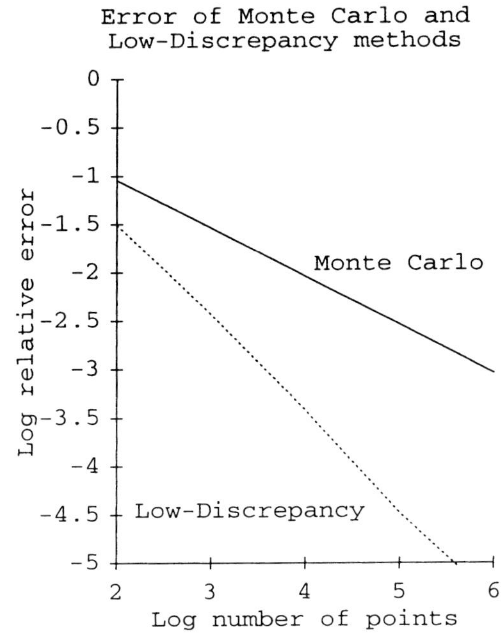

Figure 7. Accuracy versus number of points for the L-shaped block

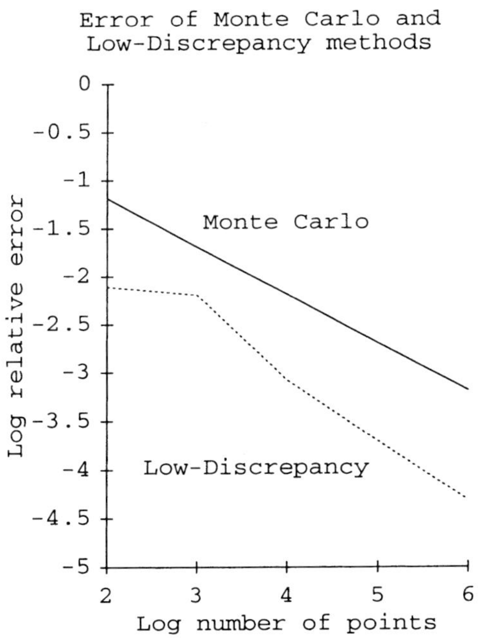

Figure 8. Accuracy versus number of points for the block with a cylindrical hole

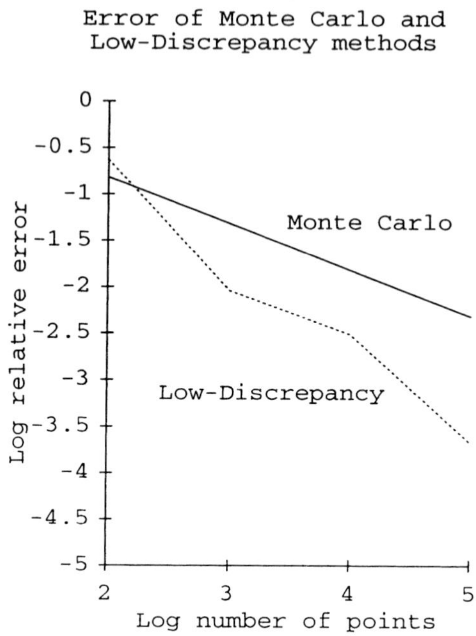

Figure 9. Accuracy versus number of points for the HW1 Object

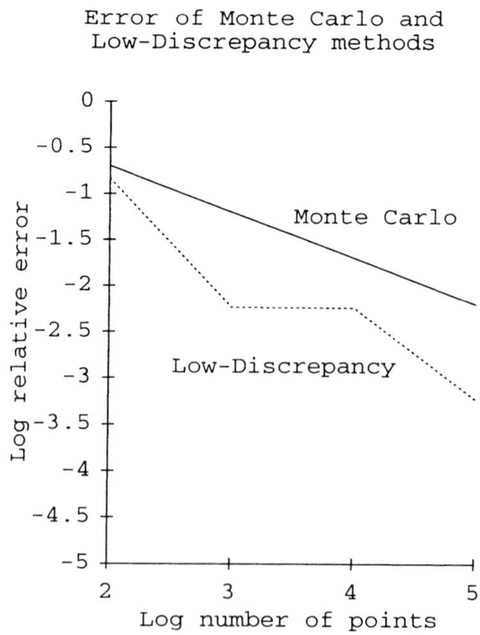

Figure 10. Accuracy versus number of points for the HW2 Object

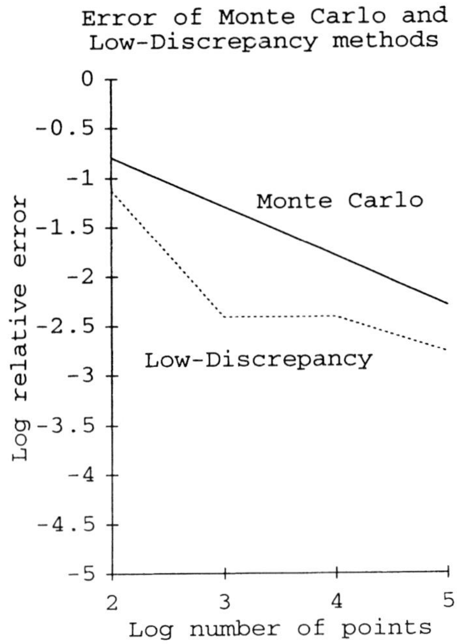

Figure 11. Accuracy versus number of points for the Valve object

slope of the Monte Carlo graph is 0.5 in each case. This means that, as more points are chosen, the relative advantage of the low-discrepancy method increases relative to the Monte Carlo method. (For the Valve object, the gradient is in fact less than that of the Monte Carlo graph. Nevertheless, the low-discrepancy method is more accurate for this object for any given number of points in the experimental range than the Monte Carlo method. These graphs have been drawn from a small number of samples, which probably explains the low slope found in this particular case.)

5.5. Detailed Results

Figure 12 is the more detailed graph for the L-shaped block showing errors every 100 points. A best-fit line was drawn; the gradient of this line is 0.79. While there are considerable fluctuations in the errors as the number of sample points varies, nowhere is the actual error more than 10 times greater than the trend, shown by

Table 2. Gradients of low-discrepancy graphs

Object	Low-discrepancy slope
Sphere	0.85
L-shaped block	0.98
Block with hole	0.59
HW1	0.96
HW2	0.72
Valve	0.49

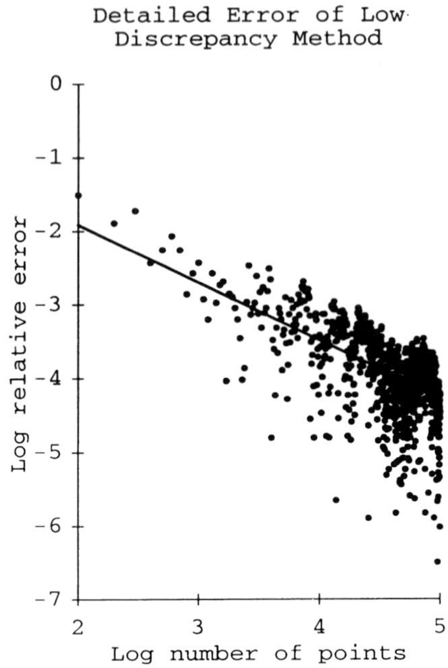

Figure 12. Detailed graph of errors for the low-discrepancy method for the L-shaped block

the best-fit line, and often, the method does much better than the trend. Because the error does not vary smoothly with the number of sample points, it would in general be difficult to give *guarantees* of the error obtained in computing volume integrals using low-discrepancy sequences (note that Eq. (4) is only directly relevant for rectangular parallelepipeds), although clearly reasonable estimates of likely errors can be given.

5.6. Relative Efficiency

Comparing the Monte Carlo method with Niederreiter's method, the actual results achieved are very impressive. In the case of Object 4 (the HW1 object), more than 20000 points are needed in the Monte Carlo method to achieve an accuracy of 1%, while fewer than 1000 points from the Niederreiter sequence are needed, for example. Table 3 shows for each object the relative efficiency of each method for each test object. In each case, the number of test points approximately needed

Table 3. Sample points approximately needed for 1% accuracy of volume for each object

Object	Monte Carlo	Niederreiter	Ratio
Sphere	8110	5136	1.6
L-shaped block	8000	355	22.6
Block with hole	4782	72	66.4
HW1	24547	893	27.5
HW2	39355	662	59.4
Valve	25148	437	57.5

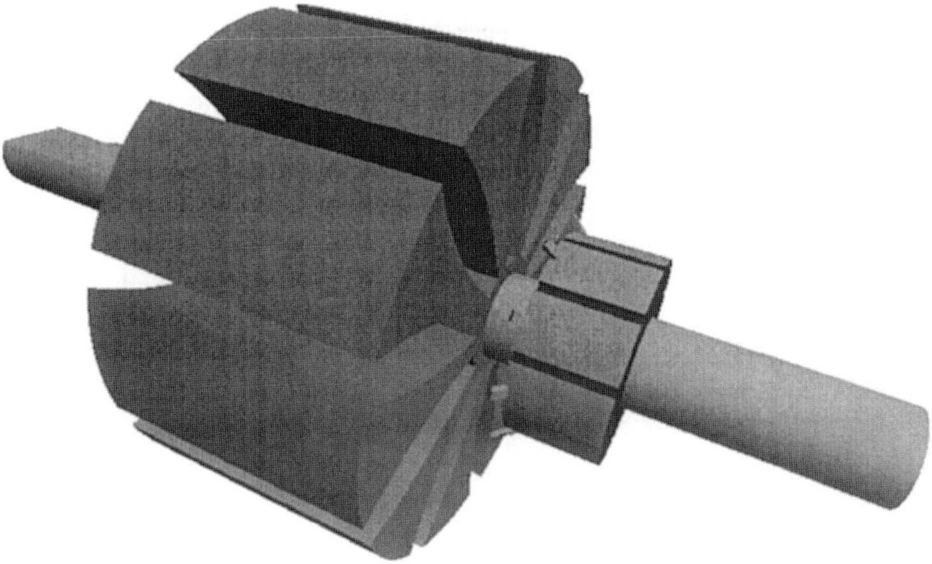

Figure 13. The svLis model of the electric motor armature

to achieve 1% accuracy is shown, and the relative advantage of the low-discrepancy method computed. As can be seen from Fig. 6, the results for the sphere are somewhat unlucky due to an upturn in the low-discrepancy curve just around 1% error, and the other results are probably more representative.

6. Real Modeller Experiments

A further series of experiments was also performed using the svLis geometric modeller [1], to ascertain whether the clear advantages of low discrepancy methods demonstrated in the simpler initial tests would also be obtained when used in a more realistic setting. We used two test objects in svLis: a hemisphere of radius 5 with a consequent true volume of 261.7994 cubic units, and the electric motor armature shown in Fig. 13. Despite the armature's complexity, it was possible (after a considerable amount of work with pencil and paper...) to calculate its volume analytically; it was 47734 cubic millimetres.

In a practical CSG geometric modeller, Monte Carlo methods are combined with a recursive subdivision scheme [10]. If (the part of) the object inside a rectangular box is deemed to be too complicated, the box is subdivided into two sub-boxes[1], and these are considered in turn. Some boxes may be rapidly classified as entirely inside or outside the object, while others may contain fewer bounding surfaces of

[1]This division can either be a cut that halves the longest side of the box, or an attempt may be made to estimate the shape of the box's contents and the division made at a place that minimizes the complexity of the two-sub boxes created. SvLis supports both of these, and we used the simpler halving scheme for this work.

the object than the original box, also simplifying the problem. Boxes entirely inside the object have their exact volume added to the total directly, and are then subsequently ignored. Recursion stops when some direct method of computing the volume in the smaller boxes is able to produce an answer with sufficient speed and accuracy. Detailed volume calculations are thus generally only necessary in small boxes that contain the boundary of the object.

We performed two types of experiment. In the first, the amount of subdivision was fixed, and the number of sample points used to compute the volume was varied. In the second, the number of sample points used to compute the volume was kept fixed, but the depth of subdivision was varied.

6.1. Increasing Numbers of Sample Points

In the first set of experiments, we divided the model to a certain depth of division tree, then allocated increasing numbers of points to the resulting leaf boxes containing the surfaces of the objects. This was comparable with the simpler experiments described above, except that the subdivision concentrated the allocated points at the surfaces of the objects, their interiors having already been exactly classified and summed. Figure 14 shows the results. For each object all leaf boxes were congruent to each other. There were 7355 leaf boxes for the hemisphere, with a total volume of 63.234 cubic units, and there were 35186 leaf boxes for the armature, with a total volume of 36503 cubic millimetres. The primary reason for these different characteristics is that the armature is highly non-convex, of course. But in addition, svLɪs classifies boxes using interval arithmetic, which is conservative: it is guaranteed to find all surface-containing boxes, but it may also classify some boxes *near* the surface as containing surface. The classification *is* exact for simple shapes like planes cylinders and spheres, but the radial pillars for the windings on the armature are exponential curves, and the classification is conservative for those.

The volume of known solid boxes was 230.411 cubic units for the hemisphere and 30002 cubic millimetres for the armature.

In each case the Niederreiter low-discrepancy sequence performed significantly better than the uniform random number generator. Regression on the data (the straight lines on the graphs) gives the following results for the hemisphere:

$$\log(V_{error}/V) = -0.5246 \log(P_{box}) - 3.2563 \quad \text{(uniform)}$$
$$\log(V_{error}/V) = -0.6383 \log(P_{box}) - 3.4305 \quad \text{(Niederreiter)}$$

where V is the true volume, V_{error} is the absolute value of the error, and P_{box} was the number of points allocated to each leaf box.

The following were the results of regression for the armature:

$$\log(V_{error}/V) = -0.5084 \log(P_{box}) - 2.8986 \quad \text{(uniform)}$$
$$\log(V_{error}/V) = -0.6196 \log(P_{box}) - 2.9869 \quad \text{(Niederreiter).}$$

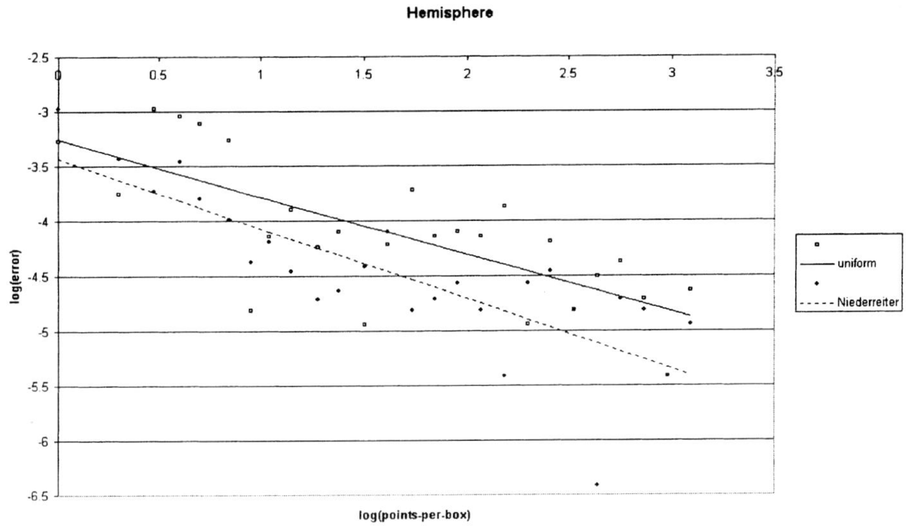

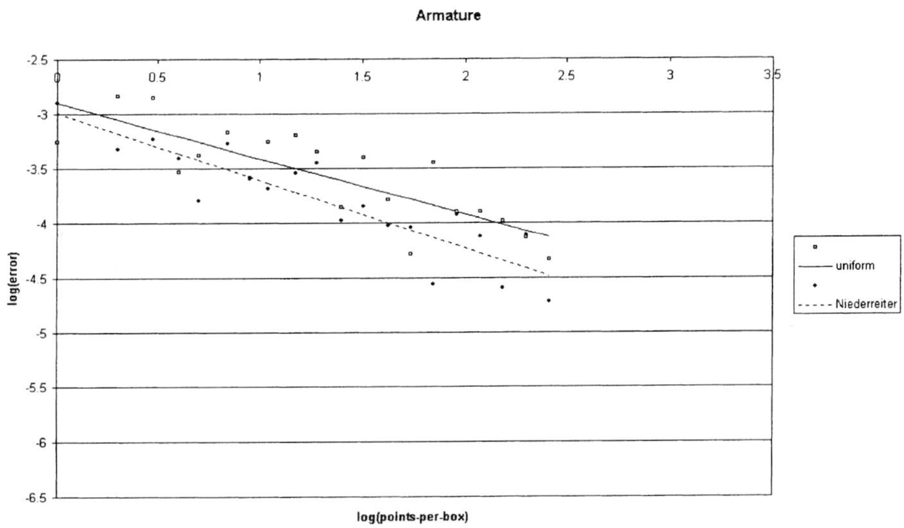

Figure 14. Error in volume estimation of the hemisphere (top) and the armature (bottom) versus the total number of points used. The vertical axis shows $\log_{10}(V_{error}/V)$; the horizontal axis shows $\log_{10} P_{box}$. Solid lines are using the uniform random number generator, and the dotted lines are for Niederreiter low-discrepancy sequences

6.2. Varying Amounts of Subdivision

In the second set of experiments the total number of sample points used overall was kept constant (at 2×10^5), and then we varied the degree to which the bounding volume was subdivided. Clearly, the modeller needs to make some decision in the recursive subdivision process as to when it is more efficient to perform a further round of subdivision, and when it is better to compute the volume of the part of the object remaining in a region directly, using a low discrepancy method in this case. (Analytical methods could be used instead when the geometry in a box is very simple.) The results of this experiment are given in Fig. 15.

As the subdivided boxes became smaller, less points were used in each box, but as large regions of the volume were already exactly classified as in or out, the sample points were allocated more to places near the boundary of the object. Note that as

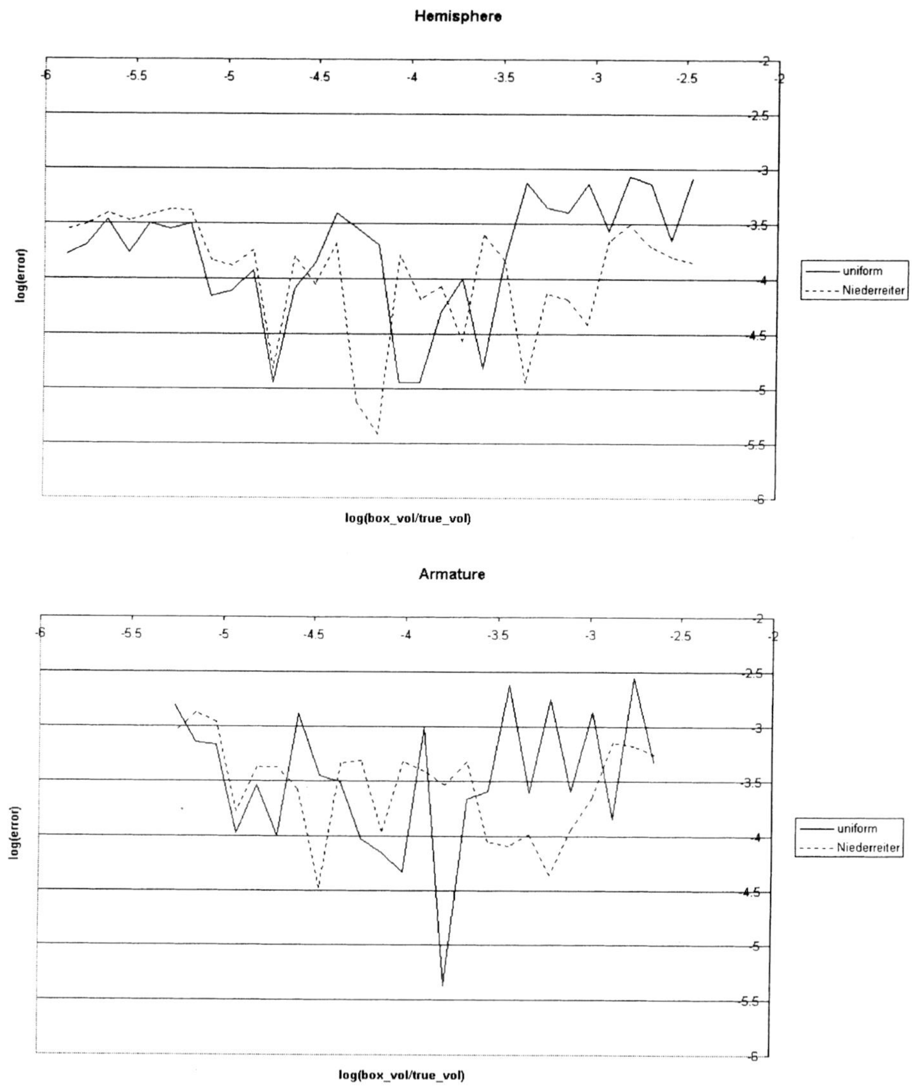

Figure 15. Error in volume estimation of the hemisphere (top) and the armature (bottom) for a constant number of points and varying depth of division. The vertical axes gives error as before; the horizontal axes gives $\log_{10}$ of the volume of the leaf boxes divided by the true volume of the object

we go towards a limit with fewer and fewer sample points in each smaller box, we would intuitively expect the advantage of the low discrepancy method over a random point distribution to vanish, as the regularity is more important when many sample points are placed in a volume. At the left-hand end of the graphs in Fig. 15 there are only two points in each box, and the low-discrepancy method has no advantage. But at the right hand ends of the graphs, which represent much less work for the modeller in doing the box division, the low-discrepancy point errors are less than the uniformly-random point errors. The number of points-per-box at the right-hand ends is about 400 for the hemisphere and 130 for the armature.

For both objects (and for both uniform and low-discrepancy techniques) the lowest errors occur at a depth of division that creates leaf boxes of about 10^{-4} of the volume of the object. However, the low-discrepancy sequence method maintains its accuracy better to the right-hand end of the graphs where the division is coarser.

7. Conclusions

It is clear from the initial results and graphs that using Niederreiter low-discrepancy point sequences in a quasi-Monte Carlo method is much better than using random points for computing volumes for all the initial test objects, even for a small number of points. Furthermore such low-discrepancy point sequences can be generated at negligible extra cost compared to random point sequences of the same number of points, when taking the overall computational time into account.

The tests using the svLis CSG modeller that combined the techniques with a recursive box division of the object space to pre-classify exactly parts of the objects whose volume was being estimated again showed significant advantages for the low-discrepancy techniques. In all cases the execution times for the experiments using the uniform random number generator were almost identical to those for the low-discrepancy volume estimator, so there is no additional computational cost in using the latter (apart from the fact that the compiled code is a few kilobytes larger – not a significant consideration in a geometric modeller that has an executable image of 1.5 megabytes).

We fully expect low-discrepancy sequences to be adopted in the future for computing volume integrals in solid modelling.

Acknowledgements

We would like to thank the Nuffield Foundation for funding T. Davies in this work with a bursary under program NUF-URB97. We would also like to thank J. Corney of Heriot Watt University for supplying Objects 4 and 5 for this research, and A. Safa of Intergraph Italia for supplying Object 6. Finally, we would also like to thank the organizers of this meeting for the opportunity to present this work.

References

[1] Bowyer, A.: Svlis set-theoretic kernel modeller: introduction and user manual information Geometers, 1995. See also `http://www.bath.ac.uk/~ensab/G_mod/Svlis/`.

[2] Bratley, P., Fox, B. L.: ALGORITHM 659. Implementing Sobol's quasi-random sequence generator. ACM Trans. Math. Softw, *14*, 88–100 (1988).
[3] Cipra, B.: In math we trust. In: What's happening in the mathematical sciences 1995–1996, pp. 100–111. American Mathematical Society 1996.
[4] Fox, B. L., Niederreiter, H.: ALGORITHM 738. Programs to generate Niederreiter's low-discrepancy sequences. ACM Trans. Math. Software *20*, 494–495 (1994).
[5] Matousek, J.: Geometric discrepancy. Berlin Heidelberg New York Tokyo: Springer, 1999.
[6] Niederreiter, H.: Quasi-Monte Carlo methods and pseudo-random numbers. Bull. Am. Math. Soc. *84*, 957–1041 (1978).
[7] Niederreiter, H.: Low-discrepancy and low-dispersion sequences. J. Number Theory *30*, 51–70 (1988).
[8] Sobol, I. M.: On the distribution of points in a cube and the approximate evaluation of integrals. USSR Comput. Math. Phys. *7*, 51–70 (1988).
[9] Woodwark, J. R.: Exercise. In: Starting work on solid models. Oxford: Geometric Modelling Society Course, 1992.
[10] Woodwark, J. R., Quinlan, K. M.: Reducing the effect of complexity on volume model evaluation. Comput. Aided Des. *14*, 89–95 (1982).

T. J. G. Davies
R. R. Martin
Department of Computer Science
Cardiff University
Cardiff CF103XG
Wales, U.K.

A. Bowyer
Department of Mechanical Engineering
University of Bath
Bath BAZ 7AY
U.K.
e-mail: a.bowyer@bath.ac.uk

Computing [Suppl] 14, 73–88 (2001)

Computing

Bisectors and α-Sectors of Rational Varieties

G. Elber, **G. Barequet**, Haifa, and **M. S. Kim**, Seoul

Abstract

The bisector of two rational varieties in $\mathbb{R}^d$ is, in general, non-rational. However, there are some cases in which such bisectors are rational; we review some of them, mostly in $\mathbb{R}^2$ and $\mathbb{R}^3$. We also describe the α-sector, a generalization of the bisector, and consider a few interesting cases where α-sectors become quadratic curves or surfaces. Exact α-sectors are non-rational even in special cases and in configurations where the bisectors are rational. This suggests the pseudo α-sector which approximates the α-sector with a rational variety. Both the exact and the pseudo α-sectors identify with the bisector when $\alpha = 1/2$.

AMS Subject Classifications: 14G40, 14H45, 14H50, 14J25, 14QOS.

Key Words: Bisector, α-sector, rational variety.

1. Introduction

Given m different objects $O_1, \ldots, O_m$, the Voronoi region of an object O_i $(1 \leq i \leq m)$ is defined as the set of points that are closer to the object O_i than to any other object O_j $(j \neq i)$. The boundary of each Voronoi region is composed of portions of bisectors, i.e., the set of points that are equidistant from two different objects O_i and O_j $(i \neq j)$. The medial axis of an object is defined as the set of interior points for which the minimum distance to the boundary corresponds to two or more different boundary points; that is, the medial axis is the self-bisector of the boundary of an object.

The concepts of Voronoi diagram and medial axis greatly simplify the design of algorithms for various geometric computations, such as shape decomposition [1], finite-element mesh generation [19, 20], motion planning with collision avoidance [13], and NC tool-path generation [14]. When the objects involved in these applications have freeform shapes, the bisector construction for rational varieties is indispensable. Unfortunately, the bisector of two rational varieties is, in general, non-rational. Moreover, even the bisector of two simple geometric primitives (such as spheres, cylinders, cones, and tori) is not always simple.

In the first part of this paper we review some important special cases where the bisectors are known to be rational. Farouki and Johnstone [10] showed that the bisector of a point and a rational curve in the same plane is a rational curve. Elber and Kim [4] showed that in $\mathbb{R}^3$ the bisector of two rational space curves is a rational surface, whereas the bisector of a point and a rational space curve is a rational ruled surface (which is also developable [16]). Moreover, the bisector of a point and a rational surface is also a rational surface [6]. Although the bisector of two rational surfaces, in general, is non-rational, there are some special cases in which the bisector is a rational surface. Dutta and Hoffmann [2] considered the bisector of simple CSG primitives (planes, spheres, cylinders, cones, and tori). Note that these CSG primitives are surfaces of revolution. When two CSG primitives have the same axis of rotation, their bisector is a quadratic surface of revolution, which is rational. Elber and Kim [6] showed that the bisector of a sphere and a rational surface with a rational offset is a rational surface; moreover, the bisector of two circular cones sharing the same apex is also a rational conic surface with the same apex. In a recent work, Peternell [16] investigated algebraic and geometric properties of curve-curve, curve-surface, and surface-surface bisector surfaces. Based on these properties, Peternell [16] proposed elementary bisector constructions for various special pairs of rational curves and surfaces, using dual geometry and representing bisectors as envelopes of symmetry lines or planes.

This paper outlines the computational procedures that construct the rational bisector curves and surfaces discussed above (except some material discussed by Peternell [16]). The basic construction steps are important since a similar technique will be employed in extending the bisector to a more general concept, the so-called *α-sector*. Instead of taking an equal distance from two input varieties, the α-sector allows different relative distances from the two varieties. Even in the simple case of a point and a line, the α-sector may assume the form of any type of conic, depending on the value of α $(0 < \alpha < 1)$. Exact α-sectors are non-rational even in the special cases where the bisectors are rational. We also present the pseudo α-sectors which approximate exact α-sectors with rational varieties. Both the exact and pseudo α-sectors reduce to bisectors when $\alpha = 1/2$.

The rest of this paper is organized as follows. In Section 2, we consider special cases where the bisectors of two varieties are rational curves and surfaces (in $\mathbb{R}^2$ and $\mathbb{R}^3$, respectively). In Section 3, we consider bisectors in higher dimensions. In Section 4, we extend the bisector ('1/2-sector') to the more general concept of α-sector. We conclude this paper with some final remarks in Section 5.

2. Rational Bisectors

There are some special cases in $\mathbb{R}^2$ and $\mathbb{R}^3$ where the bisector has a simple closed form or a rational representation. In this section we survey some important results already known.

2.1. Point-Curve Bisectors in $\mathbb{R}^2$

Farouki and Johnstone [10] showed that the bisector of a point and a rational curve in the plane is a rational curve. Consider a fixed point $Q \in \mathbb{R}^2$ and a regular

C^1 rational curve $C(t) \in \mathbb{R}^2$. Let $\mathscr{B}(t)$ denote the bisector point of Q and $C(t)$. Then we have

$$\left\langle \mathscr{B}(t) - C(t), \frac{dC(t)}{dt} \right\rangle = 0, \tag{1}$$

$$\|\mathscr{B}(t) - Q\| = \|\mathscr{B}(t) - C(t)\|, \tag{2}$$

where $\|\cdot\|$ denotes the length of a vector (in the L_2 norm).

Equation (1) means that the bisector point $\mathscr{B}(t)$ belongs to the normal line of the curve $C(t)$, while Eq. (2) implies that $\mathscr{B}(t)$ is at an equal distance from Q and $C(t)$. We can square both sides of Eq. (2) and cancel out $\|\mathscr{B}(t)\|^2$, to obtain the equation

$$\langle \mathscr{B}(t), C(t) - Q \rangle = \frac{\|C(t)\|^2 - \|Q\|^2}{2}. \tag{3}$$

Equations (1) and (3) are *linear* in $\mathscr{B}(t)$. Using Cramer's rule, we can solve these equations for $\mathscr{B}(t) = (b_x(t), b_y(t))$ and compute a rational representation of $\mathscr{B}(t)$. Note that the resulting bisector curve $\mathscr{B}(t)$ has its supporting foot points at Q and $C(t)$. In other words, the bisector curve $\mathscr{B}(t)$ has the same parameterization as the original curve $C(t)$.

2.2. *Point-Curve, Curve-Curve, and Point-Surface Bisectors in* $\mathbb{R}^3$

Elber and Kim [4] showed that the bisector of two rational space curves is a rational surface; moreover, the bisector of a point and a rational space curve in $\mathbb{R}^3$ is a rational ruled surface. Consider a fixed point $Q \in \mathbb{R}^3$ and a regular C^1 rational space curve $C(t) \in \mathbb{R}^3$. Let $\mathscr{B}(t)$ be the bisector point of Q and $C(t)$. Then we have

$$\left\langle \mathscr{B}(t) - C(t), \frac{dC(t)}{dt} \right\rangle = 0, \tag{4}$$

$$\|\mathscr{B}(t) - Q\| = \|\mathscr{B}(t) - C(t)\|. \tag{5}$$

Since $\mathscr{B}(t)$ is a three-dimensional point, there is one degree of freedom in these equations.

Consider a fixed location $C(t_0)$ on the space curve $C(t)$. Clearly $\mathscr{B}(t_0) \in \mathscr{P}_n(t_0)$, where $\mathscr{P}_n(t_0)$ is the normal plane of the curve at the fixed point $C(t_0)$. Furthermore, $\mathscr{B}(t_0)$ is at an equal distance from Q and $C(t_0)$. Hence, $\mathscr{B}(t_0)$ must belong to the plane $\mathscr{P}_d(t_0)$ which bisects Q and the point $C(t_0)$. Any point on the line $\mathscr{L}_{nd}(t_0) = \mathscr{P}_n(t_0) \cap \mathscr{P}_d(t_0)$ satisfies both Eqs. (4) and (5). Thus, the bisector surface $S(u, t)$ of the point Q and the curve $C(t)$ must be a ruled surface, where each ruling line $\mathscr{L}_{nd}(t)$ is parameterized by a linear parameter u. Figure 1a shows an

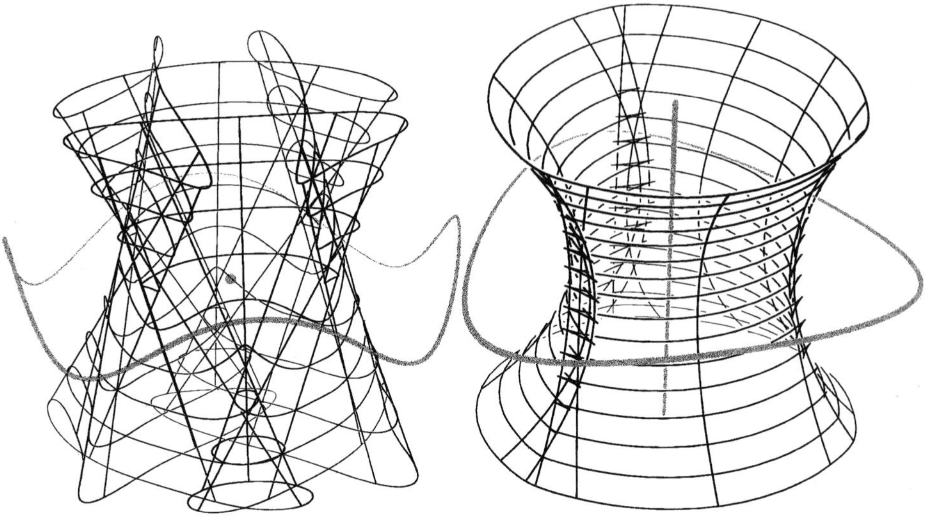

Figure 1. **a** The bisector surface of a point and a space curve in $\mathbb{R}^3$. **b** The bisector surface of a line and a round triangular periodic cubic curve in $\mathbb{R}^3$. The original curves are shown in gray

example of such a rational ruled bisector surface generated in this case from a point and a periodic rational space curve in $\mathbb{R}^3$. Based on the concept of dual geometry, Peternell [16] showed that the ruled surface $S(u, t)$ is in fact a developable surface.

The bisector surface (in $\mathbb{R}^3$) of two regular C^1 rational space curves $C_1(u)$ and $C_2(v)$ is also rational. Let $\mathscr{B}(u, v)$ be the bisector point of $C_1(u)$ and $C_2(v)$. Then, the bisector must satisfy the following three equations:

$$\left\langle \mathscr{B}(u, v) - C_1(u), \frac{dC_1(u)}{du} \right\rangle = 0, \tag{6}$$

$$\left\langle \mathscr{B}(u, v) - C_2(v), \frac{dC_2(v)}{dv} \right\rangle = 0, \tag{7}$$

$$\|\mathscr{B}(u, v) - C_1(u)\| = \|\mathscr{B}(u, v) - C_2(v)\|. \tag{8}$$

Equations (6) and (7) mean that the bisector point $\mathscr{B}(u, v)$ is simultaneously contained in the two normal planes of $C_1(u)$ and $C_2(v)$, while Eq. (8) implies that $\mathscr{B}(u, v)$ is at an equal distance from $C_1(u)$ and $C_2(v)$.

The constraints in Eqs. (6)–(8) are all *linear* in $\mathscr{B}(u, v)$. (Note that the quadratic terms in Eq. (8) cancel out.) Using Cramer's rule, we can solve these equations for $\mathscr{B}(u, v) = (b_x(u, v), b_y(u, v), b_z(u, v))$ and compute a rational surface representation of $\mathscr{B}(u, v)$. The resulting bisector surface follows the parameterization of the two original curves. In other words, for each point on the first curve, $C_1(u_0)$, and each point on the second curve, $C_2(v_0)$, $\mathscr{B}(u_0, v_0)$ is the bisector point. Figure 1b shows a rational bisector surface of a line and a rounded triangular periodic cubic curve in $\mathbb{R}^3$.

The bisector of a point and a rational surface in $\mathbb{R}^3$ is also rational [6]. Consider a fixed point $Q \in \mathbb{R}^3$ and a regular C^1 rational surface $S(u, v) \in \mathbb{R}^3$. Let $\mathscr{B}(u, v)$ be the bisector point of Q and $S(u, v)$. Then we have,

$$\left\langle \mathscr{B}(u,v) - S(u,v), \frac{\partial S(u,v)}{\partial u} \right\rangle = 0, \tag{9}$$

$$\left\langle \mathscr{B}(u,v) - S(u,v), \frac{\partial S(u,v)}{\partial v} \right\rangle = 0, \tag{10}$$

$$\|\mathscr{B}(u,v) - Q\| = \|\mathscr{B}(u,v) - S(u,v)\|. \tag{11}$$

The constraints in Eqs. (9)–(11) are also all *linear* in $\mathscr{B}(u, v)$. Using Cramer's rule again, we can solve these equations for $\mathscr{B}(u,v) = (b_x(u,v), b_y(u,v), b_z(u,v))$ and compute a rational surface representation of $\mathscr{B}(u, v)$. The resulting bisector surface follows the parameterization of the original surface. Figure 2a shows the rational bisector surface of a torus and a point located at the center of the torus.

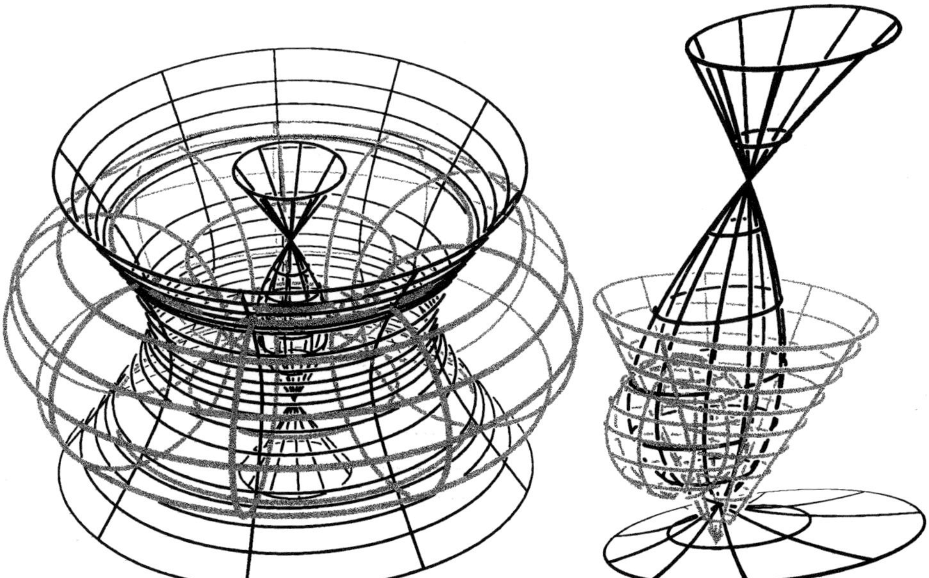

(a) (b)

Figure 2. a The bisector of a torus and a point at the center of the torus, in $\mathbb{R}^3$. **b** The bisector of a cone and a sphere in $\mathbb{R}^3$. Original surfaces are shown in gray. Both bisector surfaces are infinite

2.3. *Special Cases of Surface-Surface Bisectors in* $\mathbb{R}^3$

In general, the bisector of two rational surfaces is non-rational in $\mathbb{R}^3$, as we have already noted. However, there are some special cases where the bisector surface is rational. For example, when one of the initial surfaces is a sphere, the problem reduces to finding the bisector of a point and an offset surface. Thus, the bisector is rational when the offset surface is rational. This special case is discussed in Section 2.3.1. Moreover, when the two surfaces are given as surfaces of revolution sharing a common axis of rotation, the problem reduces to finding the planar bisector of the generating curves of the two surfaces. The bisector surface is rational if and only if the bisector of the two generating curves is rational. This special case is discussed in Section 2.3.2. The bisector of two conic surfaces sharing the same apex is closely related to the bisector of two spherical curves; Section 2.3.3 considers the bisectors of points and curves on the unit sphere. A plane is a special case of a cone with $\frac{\pi}{2}$ as its spanning angle. Moreover, the set of all planes is closed under the offset operation. Section 2.3.4 combines the results of Sections 2.3.2 and 2.3.3 to compute the line-plane and cone-plane bisectors.

2.3.1. Sphere-Surface Bisectors in $\mathbb{R}^3$

In Section 2.2 we showed that the bisector of a point and a rational surface in $\mathbb{R}^3$ is a rational surface; this immediately implies that the bisector of a sphere and a surface with a rational offset is also a rational surface. Simultaneously offsetting both varieties by the same distance does not change the bisector of the two varieties. Figure 2b shows the bisector surface of a sphere and a cone computed by offsetting the cone by the radius of the sphere.

Pottmann [17] classified the class of all rational curves and surfaces that admit rational offsets. An important subclass of all polynomial curves having rational offsets includes the Pythagorean Hodograph (PH) curves [9]. Simple surfaces (that is, planes, spheres, cylinders, cones, and tori), Dupin cyclides, rational canal surfaces, and non-developable rational ruled surfaces, all belong to this special class of rational surfaces with rational offsets [3, 15, 18]. Thus, our results can be used to construct a wide range of bisectors in $\mathbb{R}^2$, where one curve is a circle and the other is a rational curve having rational offsets, and in $\mathbb{R}^3$, where one surface is a sphere and the other is a rational surface having rational offsets.

Even the simple rational bisector of two spheres, or the bisector of a point and a sphere, has many important applications in practice. The bisector of two spheres of different radii can be used for finding an optimal path of a moving object (e.g., an airplane) which attempts to avoid radar detection. Different radar devices have different intensities, and thus their regions of influence may be modeled by spheres of different radii. The optimal path lies on the bisector surface of the spheres.

2.3.2. Special Cases of Simple Surfaces with Rational Bisectors in $\mathbb{R}^3$

Dutta and Hoffmann [2] considered the bisectors of simple surfaces (CSG primitives), such as natural quadrics and tori, in particular configurations. Note that these CSG primitives are surfaces of revolution which can be generated by rotating lines or circles about an axis of rotation. When two primitives share the same axis of rotation, their bisector construction essentially reduces to that of the generating curves of two primitives. The bisectors of lines and circles are conics, which are rational. Thus, the bisector of two primitives sharing the same axis of rotation is a rational quadratic surface of revolution.

We can extend this result to a slightly more general case. Consider a rational surface of revolution generated by a planar curve with a rational offset. When the axis of rotation is identical to that of a torus (or a sphere), the bisector of the surface of revolution and the torus (or the sphere) is a rational surface of revolution. This is because the bisector of a circle and a planar rational curve with a rational offset is the same as the bisector of the center of the circle and the rational offset curve; therefore the latter curve is also rational. Peternell [16] showed that the bisector of a line and a rational curve with a rational offset is also a rational curve. Similar arguments also apply to the cylinder, cone, and plane, when the axis of rotation is shared with the surface of revolution.

Dutta and Hoffmann [2] also considered the bisector of two cylinders of the same radius, and the bisector of two parallel cylinders. The bisector of two cylinders of the same radius is the same as the bisector of their axes, which is a hyperbolic paraboloid and therefore rational. Moreover, the bisector of two parallel cylinders is a cylindrical surface which is obtained by linearly extruding the bisector of two circles. Thus, the bisector of two parallel cylinders is an elliptic or hyperbolic cylinder, which is also rational.

Again, we can slightly extend this result. Consider two rational canal surfaces obtained by sweeping a sphere (of a fixed radius) along two rational space curves. The bisector of these canal surfaces is the same as that of their skeleton space curves, which is a rational surface. Moreover, two parallel cylindrical rational surfaces have a rational bisector surface if their cross-sectional curves have a rational bisector curve. In particular, when one cross-section is a circle and the other cross-section is a planar rational curve with a rational offset, the bisector must be a rational cylindrical surface.

2.3.3. Bisectors on the Unit Sphere S^2

Consider two conic surfaces that share the same apex. Their bisector surface is another conic surface with the same apex, which we may assume to be located at the origin. Thus the conic surfaces are ruled surfaces with their directrix curves fixed at the origin. The intersection of these conic surfaces with the unit sphere S^2 generates spherical curves; the curve corresponding to the bisector surface is indeed the bisector of the two spherical curves obtained from the original conic surfaces. Thus, the bisector curve construction on S^2 is equivalent to the bisector surface construction for two conic surfaces sharing the same apex. In the present section we consider the construction of bisector curves on S^2.

Given two points P and Q on S^2, let their spherical (geodesic) distance $\rho(P, Q)$ on S^2 be the angle between P and Q: $\rho(P, Q) = \arccos\langle P, Q\rangle$, where P and Q are two unit vectors. Consequently, for three points $P, Q, R \in S^2$, we have $\rho(P, Q) = \rho(P, R)$, if and only if $\langle P, Q \rangle = \langle P, R \rangle$. Let $Q \in S^2$ be a point and $C(t) \in S^2$ be a regular C^1 rational spherical curve. Their spherical bisector curve $\mathscr{B}(t) \in S^2$ must satisfy the following three constraints:

$$\langle \mathscr{B}(t), Q \rangle = \langle \mathscr{B}(t), C(t) \rangle, \tag{12}$$

$$\left\langle \mathscr{B}(t) - C(t), \frac{dC(t)}{dt} \right\rangle = 0, \tag{13}$$

$$\langle \mathscr{B}(t), \mathscr{B}(t) \rangle = 1. \tag{14}$$

Equation (12) locates the bisector curve $\mathscr{B}(t)$ at an equal spherical geodesic distance from Q and $C(t)$. Since the normal plane $\mathscr{P}_n(t)$ of a spherical curve $C(t) \in S^2$ contains the origin, it intersects S^2 in a great circle that is orthogonal to $C(t)$. Equation (13) implies that the bisector point is contained in the normal plane $\mathscr{P}_n(t)$. Finally, Eq. (14) constrains the bisector curve to the unit sphere S^2.

Unfortunately, Eq. (14) is quadratic in $\mathscr{B}(t)$; thus the spherical curve is, in general, non-rational. Fortunately, the ruling directions of conic surfaces may be represented by nonunit vectors. Thus, for the construction of rational direction curves, we replace the unitary condition of Equation (14) by the following *linear* equation:

$$\langle \mathscr{B}(t), (0, 0, 1) \rangle = 1. \tag{15}$$

Equation (15) constrains the bisector curve to the plane $Z = 1$. Equations (12), (13), and (15) form a system of three linear equations in $\mathscr{B}(t)$, whose solution is a rational curve on the plane $Z = 1$, which we denote as $\bar{\mathscr{B}}(t)$. Normalizing $\bar{\mathscr{B}}(t)$, we obtain a spherical bisector curve: $\mathscr{B}(t) = \bar{\mathscr{B}}(t)/\|\bar{\mathscr{B}}(t)\| \in S^2$. Because of the square root in the denominator, the bisector curve $\mathscr{B}(t) \in S^2$ will be, in general, non-rational.

Given two regular C^1 rational curves $C_1(u)$ and $C_2(v)$ on S^2, their bisector curve $\mathscr{B}(u(v)) \in S^2$ must satisfy the following three conditions:

$$\langle \mathscr{B}(u(v)), C_1(u) - C_2(v) \rangle = 0, \tag{16}$$

$$\langle \mathscr{B}(u(v)) - C_1(u), C_1'(u) \rangle = 0, \tag{17}$$

$$\langle \mathscr{B}(u(v)) - C_2(v), C_2'(v) \rangle = 0. \tag{18}$$

Equation (16) is the constraint of equal distance. Equations (17) and (18) imply that the bisector is simultaneously on the normal planes of the two curves. All three planes pass through the origin and they intersect, in general, only at the origin. However, there is a singular case where the three planes intersect in a line and their normal vectors are coplanar:

$$\lambda(u, v) = \begin{vmatrix} C_1(u) - C_2(v) \\ C_1'(u) \\ C_2'(v) \end{vmatrix} = 0. \tag{19}$$

In fact, it is a necessary and sufficient condition for a bisector point $\mathscr{B}(u(v)) \in S^2$ to have its foot points at $C_1(u)$ and $C_2(v)$ [7]. The bisector point $\mathscr{B}(u(v)) \in S^2$ is then computed as one of the intersection points between the line and the unit sphere. Because of this extra constraint $\lambda(u, v) = 0$, the spherical bisector curve is, in general, non-rational (see also Elber and Kim [5]). However, the spherical bisector curve of two circles on S^2 is an interesting special case which allows a rational bisector.

In a slightly more general case, let us assume that one curve $C_1(u)$ is a circle and the other curve $C_2(v)$ has a rational spherical offset (e.g. a circle on the sphere). Then the curve-curve bisector on the unit sphere is the same as the bisector of a point and an offset curve on S^2. To obtain this bisector, we first offset both curves on S^2 until the circular offset degenerates to a point, and then solve this simplified system of equations for the spherical point-curve bisector. Using this technique, we can reduce the spherical circle-circle bisectors to the spherical point-circle bisectors.

2.3.4. Line-Plane and Cone-Plane Bisectors

A plane is a special case of a circular cone with $\frac{\pi}{2}$ as its spanning angle. Moreover, the set of all planes is closed under offsetting. Based on these two properties, and by combining results discussed in Sections 2.3.2 and 2.3.3, we can construct the line-plane and cone-plane bisectors.

Consider the bisector of a line $\mathscr{L}$ and a plane $\mathscr{P}$. Without loss of generality, we may assume that $\mathscr{P}$ is the XY-plane and $\mathscr{L}$ intersects $\mathscr{P}$ at the origin. (We assume that $\mathscr{P}$ and $\mathscr{L}$ are not parallel, since the parallel case reduces to the point-line bisector.) Let $Q = \mathscr{L} \cap S^2$ and $C(t) = \mathscr{P} \cap S^2$ be a point and a great circle, respectively, both on S^2. Moreover, let $\bar{\mathscr{B}}(t)$ be the bisector of Q and $C(t)$ on the plane $Z = 1$. Then, the bisector surface of $\mathscr{L}$ and $\mathscr{P}$ is given by

$$\mathscr{B}(t, r) = r\bar{\mathscr{B}}(t), \quad r \in \mathbb{R}.$$

Next we consider the bisector of a circular cone $\mathscr{C}$ and a plane $\mathscr{P}$. Without loss of generality, we may assume that $\mathscr{P}$ is the XY-plane and that the apex of the circular cone $\mathscr{C}$ is located at the origin. Let $C_1(u) = \mathscr{C} \cap S^2$ and $C_2(t) = \mathscr{P} \cap S^2$ be a circle and a great circle, respectively, both on S^2. Moreover, let $\bar{\mathscr{B}}(t)$ be the bisector of $C_1(u)$ and $C_2(t)$ on the plane $Z = 1$. (Note that the bisector curve is constructed by the spherical offset technique discussed at the end of Section 2.3.3.) Then, the bisector surface of $\mathscr{C}$ and $\mathscr{P}$ is again given by

$$\mathscr{B}(t, r) = r\bar{\mathscr{B}}(t), \quad r \in \mathbb{R}.$$

If the apex of the cone $\mathscr{C}$ is not contained in $\mathscr{P}$, we can offset both the cone and the plane until the apex is contained in $\mathscr{P}$. A translation moves both varieties so that the new apex is now located at the origin. All cone-plane bisectors can thus be reduced to the standard form discussed above. Note that the same technique can be applied to non-circular cones $\mathscr{C}$ as well if their spherical curves $\mathscr{C} \cap S^2$ have rational spherical offsets.

3. Bisectors in Higher Dimensions

We now examine the existence of rational bisectors in higher dimensions. Let $\mathscr{V}_1$ and $\mathscr{V}_2$ be two varieties of dimensions d_1 and d_2, respectively, both in $\mathbb{R}^d$. The bisector $\mathscr{B}$ of $\mathscr{V}_1$ and $\mathscr{V}_2$ must be located in the normal subspaces of the two varieties. Hence, there are $d_1 + d_2$ orthogonality constraints to be considered. The bisector must, of course, also be at an equal distance from the two varieties, so there are in total $d_1 + d_2 + 1$ linear constraints. When the two varieties $\mathscr{V}_1$ and $\mathscr{V}_2$ are in general position, their bisector $\mathscr{B}$ has a rational representation if

$$d_1 + d_2 + 1 \leq d.$$

For example, consider two curves in $\mathbb{R}^3$. Each curve contributes one orthogonality constraint; that is, the bisector must be contained in the normal plane of each curve. Together with the requirement of equidistance from two input curves, the total number of constraints is three, which is equal to the dimension of the space. Thus, the bisector has a rational representation.

In contrast, a bivariate surface imposes two orthogonality constraints; namely that the bisector of two surfaces must be contained in the normal line of each. Including equidistance, the total number of constraints is therefore five. Hence the bisector of two bivariate surfaces has a rational representation in $\mathbb{R}^d$, for $d \geq 5$, but not in $\mathbb{R}^3$. Similarly, the bisector of a bivariate surface and a univariate curve has a rational representation in $\mathbb{R}^d$, for $d \geq 4$, but not in $\mathbb{R}^3$.

The bisector curve of two curves in $\mathbb{R}^2$, the bisector surface of a curve and a surface in $\mathbb{R}^3$, and the bisector of two surfaces in $\mathbb{R}^3$ are all, in general, non-rational; therefore we need to approximate them numerically. Methods for approximating the bisectors of two curves were presented by Farouki and Ramamurthy [11] and by Elber and Kim [5]. Additionally, methods for approximating the bisector of two surfaces or that of a curve and a surface in $\mathbb{R}^3$ were recently proposed by the latter authors [8].

4. α-Sectors

By definition, the shortest distances from a bisector point to the two varieties being bisected are always equal. Consider an intermediate surface with weighted distances from the two varieties,

$$\alpha\|\mathscr{B} - \mathscr{V}_1\| = (1 - \alpha)\|\mathscr{B} - \mathscr{V}_2\|, \tag{20}$$

where $0 \leq \alpha \leq 1$. We denote the locus of points that are at relative distances α and $(1-\alpha)$ from the two varieties as the *α-sector*. Unfortunately, the square of Eq. (20) is linear in $\mathscr{B}$ only for $\alpha = \frac{1}{2}$. Nevertheless, there is a nice property that the two special α-sectors are identical with the original varieties when $\alpha = 0$ or $\alpha = 1$. Note that the α-sector reduces to the bisector when $\alpha = \frac{1}{2}$.

The ability to change α continuously could be a useful tool in a range of applications, e.g., to produce metamorphosis between two freeform shapes. In the next sections we consider a few simple examples of the α-sectors of two varieties. While Eq. (20) is quadratic, we later 'linearize' this constraint and introduce the *pseudo α-sector* which is simple to represent as a rational function.

4.1. The Point-Line α-Sector in $\mathbb{R}^2$

We may assume without loss of generality that the line is the Y-axis, that is, the parametric line $C(t) = (0, t)$, and that the point is $Q = (1, 0)$. We choose α so that $\alpha = 0$ corresponds to the line and $\alpha = 1$ corresponds to the point.

The α-sector $\mathscr{B} = (b_x, b_y)$ between the Y-axis and the point Q satisfies the line-orthogonality constraint

$$0 = \left\langle \mathscr{B} - C(t), \frac{dC(t)}{dt} \right\rangle = \langle (b_x, b_y) - (0, t), (0, 1) \rangle = b_y - t, \tag{21}$$

and the distance constraint

$$\alpha^2 \left((b_x - 1)^2 + b_y^2 \right) = (1-\alpha)^2 (b_x^2 + (b_y - t)^2). \tag{22}$$

Solving Eqs. (21) and (22) and replacing (b_x, b_y) with (x, y), we obtain the quadratic curve

$$\left(\frac{2\alpha - 1}{\alpha^2} \right) x^2 + y^2 - 2x + 1 = 0. \tag{23}$$

Figure 3 shows the α-sectors of the line $(0, t)$ and the point (1, 0) for various different values of α. When $\alpha < \frac{1}{2}$, the coefficients of x^2 and y^2 have opposite signs, and so the α-sector is a hyperbola. When $\alpha = \frac{1}{2}$, the coefficient of x^2 vanishes, and so the bisector is a parabola. When $\alpha > \frac{1}{2}$, the coefficients of x^2 and y^2 have the same sign, and so the α-sector is an ellipse.

4.2. The Point-Plane α-Sector in $\mathbb{R}^3$

A similar α-sector exists for a point and a plane in three dimensions. We may assume without loss of generality that the plane is the YZ-plane, that is, the parametric plane $S(u, v) = (0, u, v)$, and that the point is $Q = (1, 0, 0)$. We choose α such that $\alpha = 0$ corresponds to the plane and $\alpha = 1$ corresponds to the point.

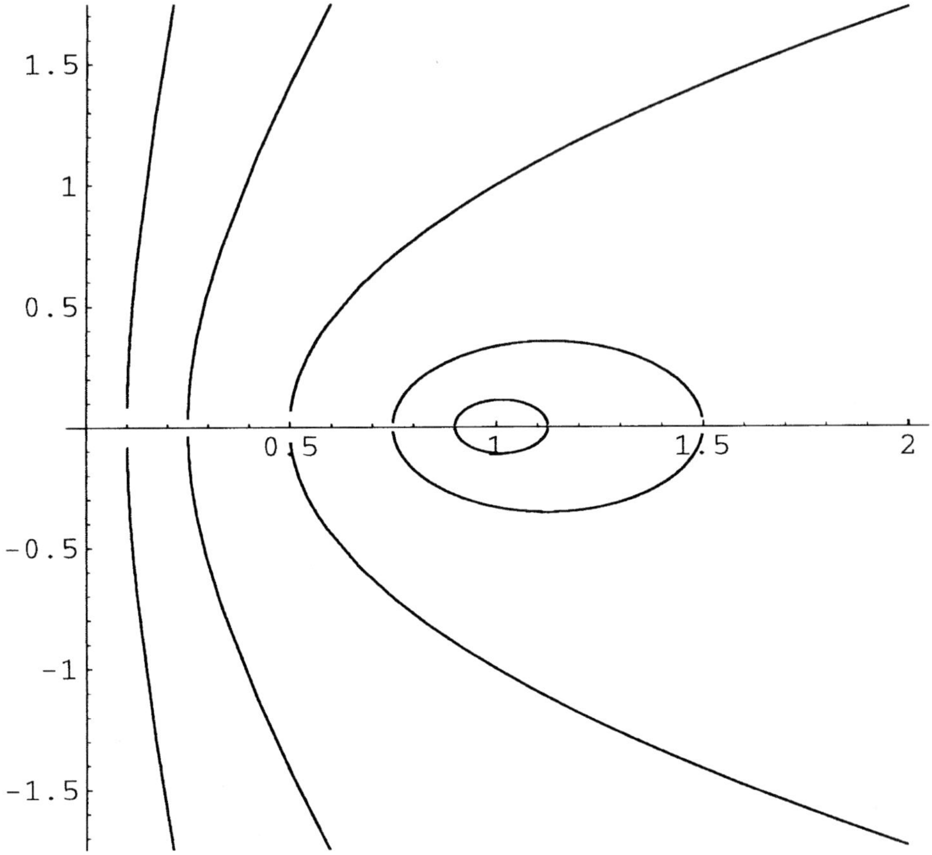

Figure 3. The α-sectors of the point (1, 0) and the line (0, t) for $\alpha = 0.10, 0.25, 0.50, 0.75, 0.90$

Let $\mathcal{B} = (b_x, b_y, b_z)$ be the α-sector of $S(u, v)$ and Q. As in the two-dimensional case we have the two plane-orthogonality constraints

$$0 = \left\langle \mathcal{B} - S(u, v), \frac{\partial S(u, v)}{\partial u} \right\rangle = \langle (b_x, b_y, b_z) - (0, u, v), (0, 1, 0) \rangle = b_y - u, \qquad (24)$$

$$0 = \left\langle \mathcal{B} - S(u, v), \frac{\partial S(u, v)}{\partial v} \right\rangle = \langle (b_x, b_y, b_z) - (0, u, v), (0, 0, 1) \rangle = b_z - v, \qquad (25)$$

and the distance constraint

$$\alpha^2 \left((b_x - 1)^2 + b_y^2 + b_z^2 \right) = (1 - \alpha)^2 (b_x^2 + (b_y - u)^2 + (b_z - v)^2). \qquad (26)$$

Solving Eqs. (24)–(26) and replacing (b_x, b_y, b_z) with (x, y, z), we obtain the quadratic surface

$$\left(\frac{2\alpha-1}{\alpha^2}\right)x^2 + y^2 + z^2 - 2x + 1 = 0. \tag{27}$$

This is a hyperboloid of two sheets for $0 < \alpha < \frac{1}{2}$, an elliptic (circular) paraboloid for $\alpha = \frac{1}{2}$, and an ellipsoid for $\frac{1}{2} < \alpha < 1$.

4.3. The Line-Line α-Sector in $\mathbb{R}^3$

Yet another simple example is the α-sector of two straight lines $C_1(u) = (1, u, 0)$ and $C_2(v) = (0, 0, v)$. We choose α such that $\alpha = 0$ corresponds to $C_2(v)$ and $\alpha = 1$ corresponds to $C_1(u)$. Now let $\mathscr{B} = (b_x, b_y, b_z)$ be the α-sector of $C_1(u)$ and $C_2(v)$, and we have the two line-orthogonality constraints

$$0 = \left\langle \mathscr{B} - C_1(u), \frac{dC_1(u)}{du} \right\rangle = \langle (b_x, b_y, b_z) - (1, u, 0), (0, 1, 0) \rangle = b_y - u, \tag{28}$$

$$0 = \left\langle \mathscr{B} - C_2(v), \frac{dC_2(v)}{dv} \right\rangle = \langle (b_x, b_y, b_z) - (0, 0, v), (0, 0, 1) \rangle = b_z - v, \tag{29}$$

and the distance constraint

$$\alpha^2\left((b_x - 1)^2 + (b_y - u)^2 + b_z^2\right) = (1-\alpha)^2(b_x^2 + b_y^2 + (b_z - v)^2). \tag{30}$$

The solution of Eqs. (28)–(30) is the quadratic surface

$$\left(\frac{2\alpha-1}{\alpha^2}\right)x^2 - \left(\frac{1-\alpha}{\alpha}\right)^2 y^2 + z^2 - 2x + 1 = 0. \tag{31}$$

Thus the $\frac{1}{2}$-sector (bisector) of $C_1(u)$ and $C_2(v)$ is the surface

$$y^2 - z^2 + 2x - 1 = 0,$$

whose parametric form is given as $\left(\frac{1-u^2+v^2}{2}, u, v\right)$. This confirms the result of [4, §2.2].

When $\alpha = \frac{1}{2}$, Eq. (31) yields a hyperbolic paraboloid. Otherwise, when $0 < \alpha < 1$, but $\alpha \neq \frac{1}{2}$, it yields a hyperboloid of one sheet, which reduces to a line for $\alpha = 0$ or 1. However, the α-sector of two general rational curves in $\mathbb{R}^3$ is usually a non-rational surface.

4.4. The Pseudo α-Sector

In the case of a spherical bisector, we resorted to the linear constraint $Z = 1$. Similarly, we now seek a linear constraint that replaces the quadratic L_2-norm of

Eq. (20) while yielding similar properties to the α-sector in constraining the relative distances to the two given varieties. We choose the plane that is at relative distances of α and $(1-\alpha)$ from the closest point on each variety.

For example, for the pseudo α-sector of a curve $C(t)$ and a point Q in $\mathbb{R}^2$, we impose the two linear constraints

$$\left\langle \mathscr{B}(t) - C(t), \frac{dC(t)}{dt} \right\rangle = 0, \tag{32}$$

$$\langle \mathscr{B}(t) - (\alpha Q + (1-\alpha)C(t)), C(t) - Q \rangle = 0. \tag{33}$$

Equation (32) is the regular orthogonality constraint, and Eq. (33) ensures that the bisector is on the plane containing the point $\alpha Q + (1-\alpha)C(t)$ and orthogonal to the vector $C(t) - Q$. If $C(t)$ has a rational representation, we can easily use Cramer's rule to obtain a rational representation for $\mathscr{B}(t) = (b_x(t), b_y(t))$.

Figure 4 shows three examples of planar pseudo α-sectors of: (i) a point and a line (Fig. 4a), (ii) a point and a cubic curve (Fig. 4b), and (iii) a point and a circle (Fig. 4c). These examples were all created using the IRIT solid-modeling environment [12].

The extension to $\mathbb{R}^3$ follows the same guidelines. The pseudo α-sector of two curves $C_1(u)$ and $C_2(v)$ in $\mathbb{R}^3$ imposes the three linear constraints

$$\left\langle \mathscr{B}(t) - C_1(u), \frac{dC_1(u)}{du} \right\rangle = 0, \tag{34}$$

$$\left\langle \mathscr{B}(t) - C_2(v), \frac{dC_2(v)}{dv} \right\rangle = 0, \tag{35}$$

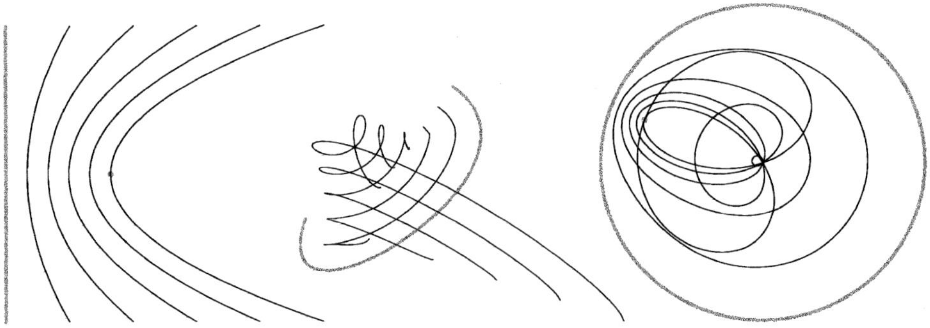

(a) (b) (c)

Figure 4. a The pseudo α-sectors of a point and a line in $\mathbb{R}^2$ for $\alpha = 0.10, 0.25, 0.50, 0.75, 0.90$ (cf. Fig. 3). **b** The pseudo α-sectors of a point and a cubic curve in $\mathbb{R}^2$ for $\alpha = 0.2, 0.4, 0.6, 0.8, 1.0$. **c** The pseudo α-sectors of a point and a circle in $\mathbb{R}^2$ for $\alpha = 0.2, 0.4, 0.6, 0.8, 1.0$. The original curves and points are shown in gray

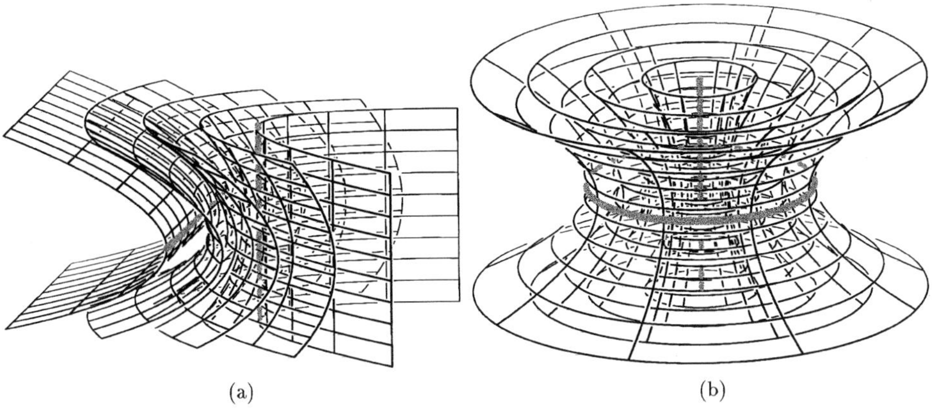

Figure 5. **a** The pseudo α-sectors of two lines $\mathbb{R}^3$ for $\alpha = 0.0, 0.25, 0.5, 0.75, 1.0$. **b** The pseudo α-sectors of a line and a circle in $\mathbb{R}^3$ for $\alpha = 0.0, 0.25, 0.5, 0.75, 1.0$. The original curves are shown in gray

$$\langle \mathscr{B}(t) - (\alpha C_1(u) + (1-\alpha)C_2(v)), C_1(u) - C_2(v)\rangle = 0. \tag{36}$$

Again, if $C_1(u)$ and $C_2(v)$ have rational representations, we can use Cramer's rule to obtain a rational representation for $\mathscr{B}(t)$. Figure 5 shows two such pseudo α-sectors in $\mathbb{R}^3$, for (i) two lines (Fig. 5a), and (ii) a line and a circle (Fig. 5b).

The pseudo α-sector is identical to the α-sector only when $\alpha = \frac{1}{2}$; in that case, they are both equivalent to the bisector. Note also that the pseudo 0- and 1-sectors are only approximations to the original varieties. This is because of the approximate distance constraint: points on the pseudo α-sector do not satisfy the $\alpha{:}(1-\alpha)$ distance ratio; instead, this property constrains only their projections on the lines joining the respective points on the varieties.

5. Conclusions

In this paper we have examined various special cases for which rational bisectors exist. We showed constructively that the point-curve bisectors in $\mathbb{R}^2$, and all point-curve, point-surface, and curve-curve bisectors in $\mathbb{R}^3$, have rational representations. We have also considered some special cases where the surface-surface bisectors are rational.

Further, we describe the exact and pseudo α-sectors, extensions of the bisector that should be useful in various applications, such as metamorphosis between the pseudo α-sector.

Acknowledgements

The authors are grateful for the anonymous reviewer who pointed us at the classification of line-line α-sectors and bisectors; Chasles, Journal de Math 1, 1836; Schoenflies, Zeitschrift für Mathematik und Physik **23**, 1878. This research was supported in part by the Fund for Promotion of Research at The

Technion, Haifa, Israel, by the Abraham and Jennie Failkow Academic Lectureship, and by the Korean Ministry of Science and Technology (MOST) under the National Research Laboratory Project.

References

[1] Choi, H. I., Han, C. Y., Moon, H. P., Roh, K. H., Wee, N.-S.: Medial axis transform and offset curves by Minkowski Pythagorean hodograph curves. Comput. Aided Des. *31*, 59–72 (1999).
[2] Dutta, D., Hoffmann, C.: On the skeleton of simple CSG objects. ASME J. Mech. Des. *115*, 87–94 (1993).
[3] Dutta, D., Martin, R., Pratt, M.: Cyclides in surface and solid modeling. IEEE Comput. Graphics Appl. *13*, 53–59 (1993).
[4] Elber, G., Kim, M.-S.: The bisector surface of freeform rational space curves. ACM Trans. Graphics *17*, 32–49 (1998).
[5] Elber, G., Kim, M.-S.: Bisector curves of planar rational curves. Comput. Aided Des. *30*, 1089–1096 (1998).
[6] Elber, G., Kim, M.-S.: Computing rational bisectors. IEEE Comput. Graph. Appl. *19*, 76–81 (1999).
[7] Elber, G., Kim, M.-S.: Rational bisectors of CSG primitives. Proc. 5th ACM/IEEE Symposium on Solid Modeling and Applications, Ann Arbor, Michigan, pp. 246–257, June 1999.
[8] Elber, G., Kim, M.-S.: A computational model for non-rational bisector surfaces: curve-surface and surface-surface bisectors. Proc. Geometric Modeling and Processing 2000, Hong Kong, April 2000, pp. 364–372.
[9] Farouki, R., Sakkalis, T.: Pythagorean hodographs. IBM J Res. Dev. *34*, 736–752 (1990).
[10] Farouki, R., Johnstone, J.: The bisector of a point and a plane parametric curve. Comput. Aided Geom. Des. *11*, 117–151 (1994).
[11] Farouki, R., Ramamurthy, R.: Specified-precision computation of curve/curve bisectors. Int. J. Comput. Geom. Appl. *8*, 599–617 (1998).
[12] IRIT 7.0 User's Manual. The Technion–IIT, Haifa, Israel, 1997. Available at http://www.cs.technion.ac.il/irit.
[13] O'Dunlaing, C., Yap, C. K.: A "retraction" method for planning the motion of a disk. J. Algorithms *6*, 104–111 (1985).
[14] Persson, H.: NC machining of arbitrary shaped pockets. Comput. Aided Des. *10*, 169–174 (1978).
[15] Peternell, M., Pottmann, H.: Computing rational parameterizations of canal surfaces. J Symb. Comput. *23*, 255–266 (1997).
[16] Peternell, M.: Geometric properties of bisector surfaces. Graph. Models Image Proc. *62*, 202–236 (2000).
[17] Pottmann, H.: Rational curves and surfaces with rational offsets. Comput. Aided Geom. Des. *12*, 175–192 (1995).
[18] Pottmann, H., Lü, W., Ravani, B.: Rational ruled surfaces and their offsets. Graph. Models Image Proc. *58*, 544–552 (1996).
[19] Sheehy, D., Armstrong, C., Robinson, D.: Shape description by medial surface construction. IEEE Trans. Visual. Comput. Graph. *2*, 42–72 (1996).
[20] Sherbrooke, E., Patrikalakis, N., Brisson, E.: An algorithm for the medial axis transform of 3D polyhedral solids. IEEE Trans. Visual. Comput. Graph. *2*, 44–61 (1996).

G. Elber
G. Barequet
Department of Computer Science
Technion, Israel Institute of Technology
Haifa 32000, Israel
e-mails: gershon@cs.technion.ac.il
barequet@cs.technion.ac.il

M.-S. Kim
Department of Computer Engineering
Seoul National University Seoul 151-742
South Korea
e-mail: mskim@comp.snu.ac.kr

Computing [Suppl] 14, 89–103 (2001)

Computing

Piecewise Linear Wavelets over Type-2 Triangulations

M. S. Floater and **E. G. Quak**, Oslo

Abstract

The idea of summing pairs of so-called *semi-wavelets* has been found to be very useful for constructing piecewise linear wavelets over refinements of arbitrary triangulations. In this paper we demonstrate the versatility of the semi-wavelet approach by using it to construct bases for the piecewise linear wavelet spaces induced by uniform refinements of four-directional box-spline grids.

AMS Subject Classifications: 41A15, 41A63, 65D07.

Key Words: Wavelets, prewavelets, piecewise linear splines, triangulations, local support.

1. Introduction

In a recent paper [2], piecewise linear (pre-) wavelets over uniformly refined triangulations were constructed. The construction was later simplified in [3], [4] by recognizing these wavelets as the sum of two so-called semi-wavelets. Though the main emphasis in all three papers was on triangulations of arbitrary topology, an important special case is a triangulation of Type-1, formed by adding diagonal lines in a single direction to a rectangular grid. This can also be viewed as a three-directional box spline grid. The (interior) wavelets in [2] reduce in this case to the elements previously found in [6].

However, Type-I triangulations are asymmetric in the sense that one of the two possible diagonal directions is favoured over the other. In view of the fact that this might lead to asymmetric wavelet decompositions of symmetric data, we construct in this paper piecewise linear wavelets over Type-2 triangulations, or four-directional box spline grids. Bivariate splines on Type-2 triangulations have been studied as an alternative to three-directional and tensor-product splines; see Chapter 3 of [1] and [7] and the references therein.

In this paper, we will see how the semi-wavelet approach of [4] again turns out to be a useful tool for constructing wavelets. We derive a complete set of wavelet functions, including special elements at the (rectangular) boundary of the triangulation and we show that the whole set forms a basis for the wavelet space.

2. Multiresolution for Type-2 Triangulations

The two diagonals of each square $S_{ij} = [i, i+1] \times [j, j+1], i, j \in \mathbb{Z}$, in the plane, divide the square into four congruent triangles. Following convention, we will refer to the set of all such triangles as a *Type-2 triangulation*. We will also refer to any subtriangulation as a Type-2 triangulation and we will be concerned with the bounded subtriangulation $\mathcal{T}^0$, generated by the squares S_{ij} for $i = 0, 1, \ldots, m-1$ and $j = 0, 1, \ldots, n-1$, for some arbitrary m, n, see Fig. 1. Throughout the paper we will assume, for the sake of simplicity, that $m \geq 2$ and $n \geq 2$, though wavelet constructions can be made in a similar way when either $m = 1$ or $n = 1$ (or both). We let V^0 and E^0 denote the vertices and edges respectively in $\mathcal{T}^0$, so that $V^0 = \{(i,j)\}_{i=0,\ldots,m,j=0,\ldots,n} \cup \{(i+1/2, j+1/2)\}_{i=0,\ldots,m-1,j=0,\ldots,n-1}$.

Let $S^0 = S_1^0(\mathcal{T}^0)$ be the linear space of continuous functions over $\mathcal{T}^0$ which are linear over every triangle. A basis for S^0 is given by the nodal functions ϕ_v^0 in S^0, for $v \in V^0$, satisfying $\phi_v^0(w) = \delta_{vw}$. The support of $\phi_{i+1/2,j+1/2}^0$ is the square S_{ij}, while the support of ϕ_{ij}^0 is the diamond enclosed by the polygon with vertices $(i-1, j), (i, j-1), (i+1, j), (i, j+1)$, suitably truncated if the point (i, j) lies on the boundary of the domain $D = [0, m] \times [0, n]$.

Next consider the refined triangulation $\mathcal{T}^1$, also of Type-2, formed by adding lines in the four directions halfway between each pair of existing parallel lines, as in Fig. 2, and define V^1, E^1, the linear space S^1, and the basis $\phi_u^1, u \in V^1$ accordingly. Then S^0 is a subspace of S^1 and a refinement equation relates the coarse nodal functions ϕ_v^0 to the fine ones ϕ_v^1. In order to formulate this equation we define

$$V_v^0 = \{w \in V^0 : w \text{ and } v \text{ are neighbours in } V^0\},$$

and

$$V_v^1 = \{u = (w+v)/2 \in V^1 : w \in V_v^0\}.$$

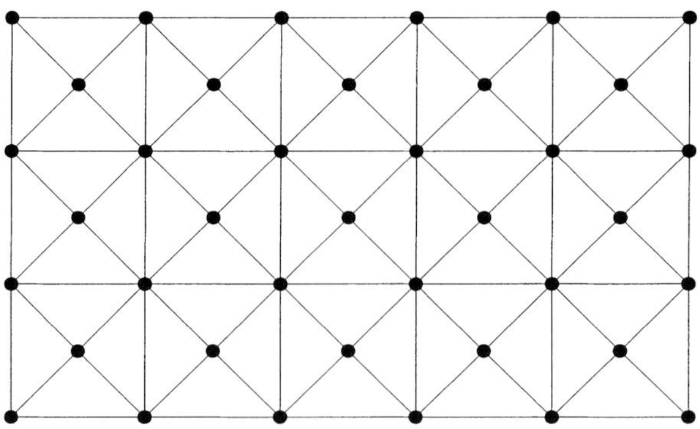

Figure 1. A Type-2 triangulation

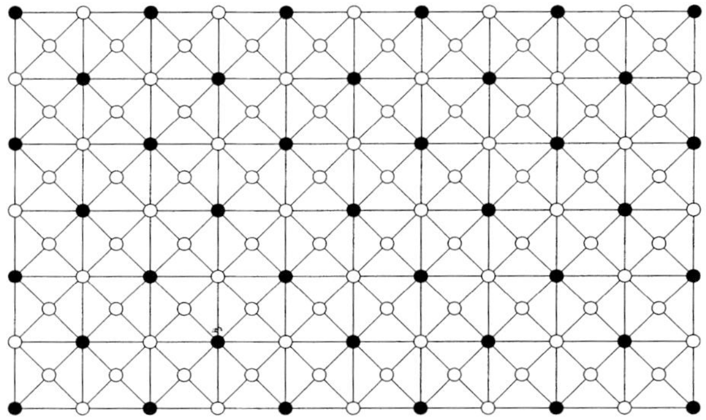

Figure 2. The first refinement

Thus V_v^0 is the set of neighbours of v in V^0 while V_v^1 is the set of midpoints between v and its coarse neighbours. For example when v is an interior vertex, there are two cases:

$$V^1_{i+1/2,j+1/2} = \{(i+1/4, j+1/4), (i+3/4, j+1/4),\\ (i+3/4, j+3/4), (i+1/4, j+3/4)\},$$

and

$$V^1_{i,j} = \{(i+1/2, j), (i+1/4, j+1/4), (i, j+1/2), (i-1/4, j+1/4),\\ (i-1/2, j), (i-1/4, j-1/4), (i, j-1/2), (i+1/4, j-1/4)\}.$$

Then the refinement equation is easily seen to be

$$\phi_v^0(x) = \phi_v^1(x) + \frac{1}{2}\sum_{u \in V_v^1} \phi_u^1(x).$$

The main aim of this paper is to build a basis for the unique orthogonal complement W^0 of S^0 in S^1, treating S^0 and S^1 as Hilbert spaces equipped with the inner product

$$\langle f, g\rangle = \int_D f(x)g(x)dx, \quad f, g \in L_2(D).$$

Ideally we would like a basis of functions with small support for the purpose of conveniently representing the *decomposition* of a given function f^1 in S^1 into its two unique components $f^0 \in S^0$ and $g^0 \in W^0$;

$$f^1 = f^0 \oplus g^0.$$

We will call any basis functions *wavelets*. Clearly the refinement of $\mathcal{T}^0$ can be continued indefinitely, generating a nested sequence

$$S^0 \subset S^1 \subset \cdots S^k \subset \cdots,$$

and if we define the wavelet space W^{k-1} to be the orthogonal complement at every refinement level k,

$$S^k = S^{k-1} \oplus W^{k-1},$$

we obtain the decomposition

$$S^n = S^0 \oplus W^0 \oplus W^1 \oplus \cdots \oplus W^{n-1},$$

for any $n \geq 1$. By combining wavelet bases for the spaces W^k with the nodal bases for the spaces S^k, we obtain the framework for a multiresolution analysis (MRA). We refer the reader to [5] for a discussion of the corresponding filter bank algorithms and the approximation of functions by thresholding wavelet coefficients. Note that the basis elements of any W^k can simply be taken to be dilations of the basis elements for W^0 and therefore we restrict our study purely to W^0.

3. Semi-Wavelets and Wavelets

Our approach to constructing wavelets for the wavelet space W^0 is to sum pairs of *semi-wavelets*, elements of the fine space which have smaller support and are close to being in the wavelet space, in the sense that they are orthogonal to all but two of the nodal functions in the coarse space.

Letting v_1 and v_2 be two neighbouring vertices in V^0, and denoting by $u \in V^1 \backslash V^0$ their midpoint, we define the *semi-wavelet* $\sigma_{v_1,u} \in S^1$ as the element with support contained in the support of $\phi^0_{v_1}$ and having the property that, for all $v \in V^0$,

$$\langle \sigma_{v_1,u}, \phi^0_v \rangle = \begin{cases} -1 & \text{if } v = v_1; \\ 1 & \text{if } v = v_2; \\ 0 & \text{otherwise.} \end{cases} \tag{3.1}$$

Thus $\sigma_{v_1,u}$ has the form

$$\sigma_{v_1,u}(x) = \sum_{v \in N^1_{v_1}} a_v \phi^1_v(x),$$

where

$$N^1_{v_1} = \{v_1\} \cup V^1_{v_1}$$

denotes the *fine neighbourhood* of v_1. The only non-trivial inner products between $\sigma_{v_1,u}$ and coarse nodal functions ϕ_v^0 occur when v belongs to the *coarse neighbourhood* of v_1,

$$N_{v_1}^0 = \{v_1\} \cup V_{v_1}^0.$$

Thus the number of coefficients and conditions are the same and, as we will subsequently establish, the element $\sigma_{v_1,u}$ is unique.

Since the dimension of W^0 is equal to the number of fine vertices in V', i.e. $|V^1| - |V^0|$, it is natural to associate one wavelet ψ_u per fine vertex $u \in V^1 \backslash V^0$. Since each u is the midpoint of some edge in E^0 connecting two coarse vertices v_1 and v_2 in V^0, the element of S^1,

$$\psi_u = \sigma_{v_1,u} + \sigma_{v_2,u} \tag{3.2}$$

is a wavelet since it is orthogonal to all nodal functions $\phi_v^0, v \in V^0$.

Thus in the remainder of this section we turn our attention to establishing the uniqueness of all the semi-wavelets with regard to (3.1) and to finding their coefficients. Initially we consider only interior vertices v_1 and there are two cases: (i) $v_1 = (i+1/2, j+1/2)$ and (ii) $v_1 = (i,j)$. Firstly, if $v_1 = (i+1/2, j+1/2)$, then $\sigma_{v_1,u}$ has support contained in S_{ij} and its fine and coarse neighbourhoods are

$$\begin{aligned} N_{v_1}^1 = \{&(i+1/2, j+1/2), (i+1/4, j+1/4), (i+3/4, j+1/4), \\ &(i+3/4, j+3/4), (i+1/4, j+3/4)\}, \end{aligned} \tag{3.3}$$

and

$$N_{v_1}^0 = \{(i+1/2, j+1/2), (i,j), (i+1,j), (i+1,j+1), (i,j+1)\}. \tag{3.4}$$

Thus there are five coefficients and five constraints imposed by (3.1) and we must solve the linear system

$$Ax = b, \tag{3.5}$$

where

$$A = (\langle \phi_v^0, \phi_w^1 \rangle)_{v \in N_{v_1}^0, w \in N_{v_1}^1}$$

and

$$b = \begin{cases} (-1,1,0,0,0)^T & \text{if } v_2 = (i,j); \\ (-1,0,1,0,0)^T & \text{if } v_2 = (i+1,j); \\ (-1,0,0,1,0)^T & \text{if } v_2 = (i+1,j+1); \\ (-1,0,0,0,1)^T & \text{if } v_2 = (i,j+1), \end{cases}$$

and the ordering of the vertices in $N_{v_1}^0$ and $N_{v_1}^1$ is the same as in (3.3) and (3.4). Due to symmetry, we can simply assume that $b = (-1,1,0,0,0)^T$ and the coefficients of the remaining three semi-wavelets are the same but rotated appropriately around v_1. In order to compute the entries in the 5×5 matrix A, we apply the following standard lemma.

Lemma 1. *Let $T = [x_1, x_2, x_3]$ be a triangle and let $f, g : T \to \mathbb{R}$ be two linear functions. If $f_i = f(x_i)$ and $g_i = g(x_i)$ for $i = 1, 2, 3$, and $a(T)$ is the area of the triangle T, then*

$$\int_T f(x)g(x)dx = \frac{a(T)}{12}(f_1g_1 + f_2g_2 + f_3g_3 + (f_1 + f_2 + f_3)(g_1 + g_2 + g_3)).$$

Using this lemma, and the fact that

$$\langle f, g \rangle = \sum_{T \in \mathcal{T}^1} \int_T f(x)g(x)dx$$

for any f and g in S^1, one can compute the entries $\langle \phi_v^0, \phi_w^1 \rangle$ of A and one finds that

$$A = \frac{1}{192}\begin{pmatrix} 20 & 6 & 6 & 6 & 6 \\ 3 & 8 & 1 & 0 & 1 \\ 3 & 1 & 8 & 1 & 0 \\ 3 & 0 & 1 & 8 & 1 \\ 3 & 1 & 0 & 1 & 8 \end{pmatrix},$$

which is non-singular with inverse

$$B = A^{-1} = \frac{1}{2}\begin{pmatrix} 30 & -18 & -18 & -18 & -18 \\ -9 & 55 & -1 & 7 & -1 \\ -9 & -1 & 55 & -1 & 7 \\ -9 & 7 & -1 & 55 & -1 \\ -9 & -1 & 7 & -1 & 55 \end{pmatrix}.$$

Thus the vector x of coefficients of $\sigma_{v_1,u}$ is given by

$$x = Bb = (1/2)(-48, 64, 8, 16, 8)^T.$$

The coefficients are shown in Fig. 3a after multiplying them by a factor of 2 (the same scaling will be applied to all later semi-wavelet coefficients). The vertex v_1 is in the centre of the figure (the only coarse vertex where $\sigma_{v_1,u}$ is non-zero) and the fine vertex $u = (v_1 + v_2)/2$ is circled.

In case (ii), we suppose that $v_1 = (i, j)$, whose fine neighbourhood is

$$N_{v_1}^1 = \{(i, j), (i + 1/2, j), (i + 1/4, j + 1/4), (i, j + 1/2), (i - 1/4, j + 1/4),$$
$$(i - 1/2, j), (i - 1/4, j - 1/4), (i, j - 1/2), (i + 1/4, j - 1/4)\},$$

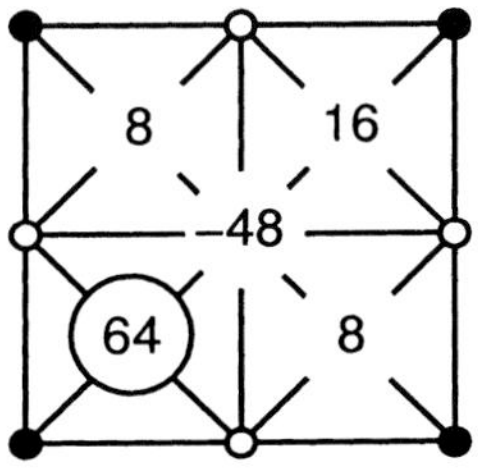

Figure 3a. First interior semi-wavelet

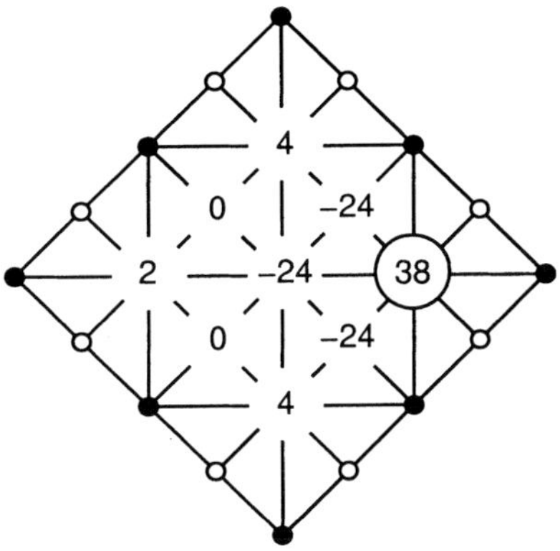

Figure 3b. Second interior semi-wavelet

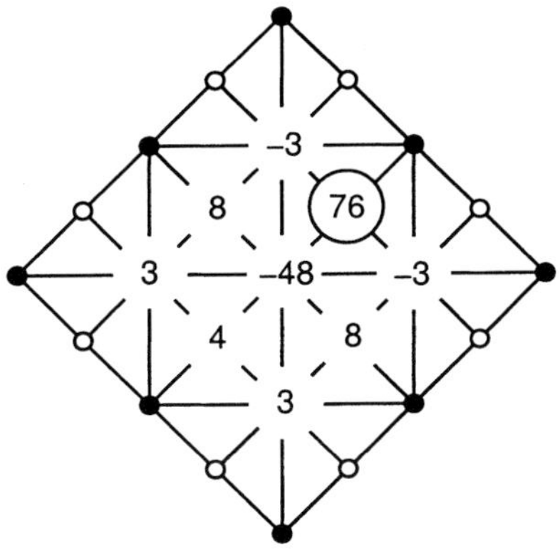

Figure 3c. Third interior semi-wavelet

and whose coarse neighbourhood is

$$N_{v_1}^0 = \{(i,j),(i+1,j),(i+1/2,j+1/2),(i,j+1),(i-1/2,j+1/2),\\ (i-1,j),(i-1/2,j-1/2),(i,j-1),(i+1/2,j-1/2)\}.$$

Thus we again solve the linear system (3.5) where A is this time a 9×9 matrix and b is either

$$(-1,1,0,0,0,0,0,0,0)^T \text{ or } (-1,0,1,0,0,0,0,0,0)^T,$$

depending on whether $v_2 = (i + 1/2, j)$ or $v_2 = (i + 1/4, j + 1/4)$ and the remaining six possible coarse neighbours v_2 lead to the same coefficients, only rotated. Applying Lemma 1 we find after some straightforward calculation that

$$A = \frac{1}{192}\begin{pmatrix} 24 & 12 & 8 & 12 & 8 & 12 & 8 & 12 & 8 \\ 1 & 12 & 1 & 0 & 0 & 0 & 0 & 0 & 1 \\ 1 & 4 & 6 & 4 & 0 & 0 & 0 & 0 & 0 \\ 1 & 0 & 1 & 12 & 1 & 0 & 0 & 0 & 0 \\ 1 & 0 & 0 & 4 & 6 & 4 & 0 & 0 & 0 \\ 1 & 0 & 0 & 0 & 1 & 12 & 1 & 0 & 0 \\ 1 & 0 & 0 & 0 & 0 & 4 & 6 & 4 & 0 \\ 1 & 0 & 0 & 0 & 0 & 0 & 1 & 12 & 1 \\ 1 & 4 & 0 & 0 & 0 & 0 & 0 & 4 & 6 \end{pmatrix}$$

which is invertible with inverse

$$B = A^{-1} = \frac{1}{4}\begin{pmatrix} 42 & -6 & -54 & -6 & -54 & -6 & -54 & -6 & -54 \\ -3 & 73 & -9 & 5 & 3 & 1 & 3 & 5 & -9 \\ -3 & -51 & 149 & -51 & 13 & -3 & 5 & -3 & 13 \\ -3 & 5 & -9 & 73 & -9 & 5 & 3 & 1 & 3 \\ -3 & -3 & 13 & -51 & 149 & -51 & 13 & -3 & 5 \\ -3 & 1 & 3 & 5 & -9 & 73 & -9 & 5 & 3 \\ -3 & -3 & 5 & -3 & 13 & -51 & 149 & -51 & 13 \\ -3 & 5 & 3 & 1 & 3 & 5 & -9 & 73 & -9 \\ -3 & -51 & 13 & -3 & 5 & -3 & 13 & -51 & 149 \end{pmatrix}.$$

Thus if $b = (-1, 1, 0, 0, 0, 0, 0, 0)^T$, we find that

$$x = Bb = \frac{1}{2}(-24, 38, -24, 4, 0, 2, 0, 4, -24)^T,$$

and if $b = (-1, 0, 1, 0, 0, 0, 0, 0)^T$, we have

$$x = Bb = \frac{1}{2}(-48, -3, 76, -3, 8, 3, 4, 3, 8)^T.$$

These two semi-wavelets are illustrated in Fig. 3b and 3c. Using the three interior semi-wavelets of Fig. 3 provides us with two wavelets ψ_u, from (3.2). The first of these, in Fig. 4a, is the sum of two semi-wavelets from Fig. 3b and the second, in Fig. 4b, the sum of the semi-wavelets in Fig. 3a and 3c. Symmetries and rotations of these two give us all interior wavelets ψ_u in the sense that v_1 and v_2 are both interior vertices of $\mathcal{T}^0$.

Now consider the case where v_1 is a boundary vertex, which means that $v_1 = (i, j)$. Let us suppose first that v_1 lies on an edge of the domain, but not at one of the

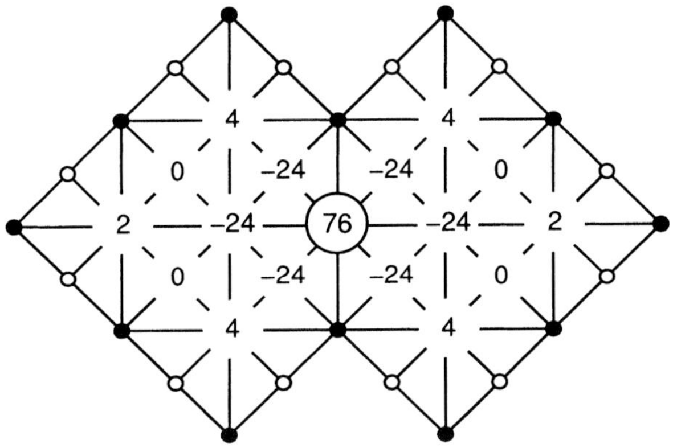

Figure 4a. First interior wavelet

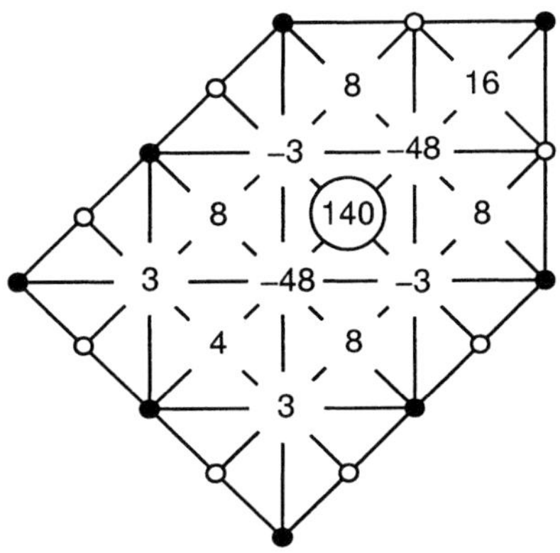

Figure 4b. Second interior wavelet

four corners, thus we assume without loss of generality that $j = 0$ and $0 < i < m$. The coarse and fine neighbourhoods of v_1 are then

$$N_{v_1}^0 = \{(i, 0), (i + 1, 0), (i + 1/2, 1/2), (i, 1), (i - 1/2, 1/2), (i - 1, 0)\},$$

and

$$N_{v_1}^1 = \{(i, 0), (i + 1/2, 0), (i + 1/4, 1/4), (i, 1/2), (i - 1/4, 1/4), (i - 1/2, 0)\},$$

respectively, and the matrix A has dimension 6×6. From Lemma 1 we find that

$$A = \frac{1}{192}\begin{pmatrix} 12 & 6 & 8 & 12 & 8 & 6 \\ 1/2 & 6 & 1 & 0 & 0 & 0 \\ 1 & 4 & 6 & 4 & 0 & 0 \\ 1 & 0 & 1 & 12 & 1 & 0 \\ 1 & 0 & 0 & 4 & 6 & 4 \\ 1/2 & 0 & 0 & 0 & 1 & 6 \end{pmatrix}$$

which is invertible and its inverse is

$$B = A^{-1} = \frac{1}{2}\begin{pmatrix} 42 & -6 & -54 & -6 & -54 & -6 \\ -3 & 73 & -9 & 5 & 3 & 1 \\ -3 & -51 & 81 & -27 & 9 & -3 \\ -3 & 5 & -3 & 37 & -3 & 5 \\ -3 & -3 & 9 & -27 & 81 & -51 \\ -3 & 1 & 3 & 5 & -9 & 73 \end{pmatrix}.$$

There are three solutions of interest, depending on v_2. If $v_2 = (i+1, j)$ then $b = (-1, 1, 0, 0, 0, 0)^T$ and $x = Bb = (1/2)(-48, 76, -48, 8, 0, 4)^T$. If $v_2 = (i + 1/2, j + 1/2)$ then $b = (-1, 0, 1, 0, 0, 0)^T$ and $x = Bb = (1/2)(-96, -6, 84, 0, 12, 6)^T$. If $v_2 = (i, j+1)$ then $b = (-1, 0, 0, 1, 0, 0)^T$ and $x = Bb = (1/2)(-48, 8, -24, 40, -24, 8)^T$. These three elements are shown in Fig. 5a, 5b, and 5c. Summing two of the first edge semi-wavelets gives us the edge wavelet ψ_u shown in Fig. 6a. Summing the second edge semi-wavelet and the first interior semi-wavelet gives us the edge wavelet ψ_u shown in Fig. 6b. Finally, summing the third edge semi-wavelet and the second interior semi-wavelet gives us the edge wavelet ψ_u shown in Fig. 6c. Up to rotation and symmetries these elements provide all wavelets ψ_u for which one of v_1 and v_2 is an interior vertex while the other one lies on the boundary but not at a corner.

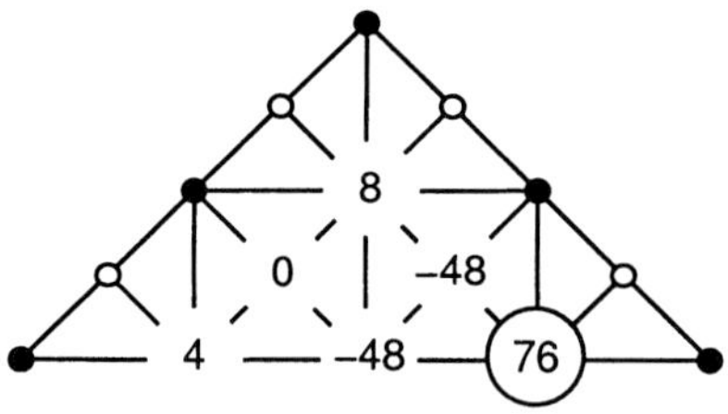

Figure 5a. First edge semi-wavelet

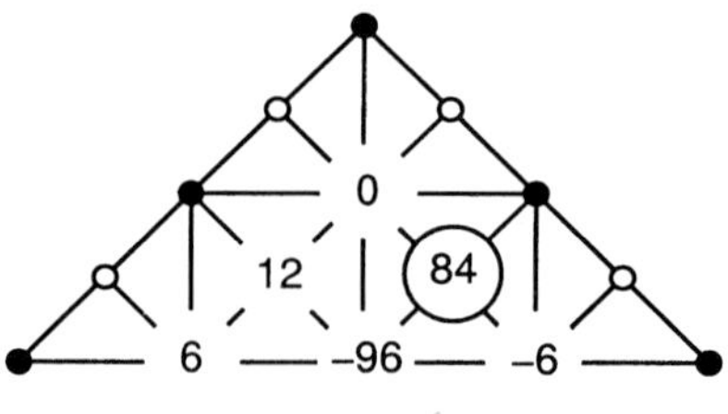

Figure 5b. Second edge semi-wavelet

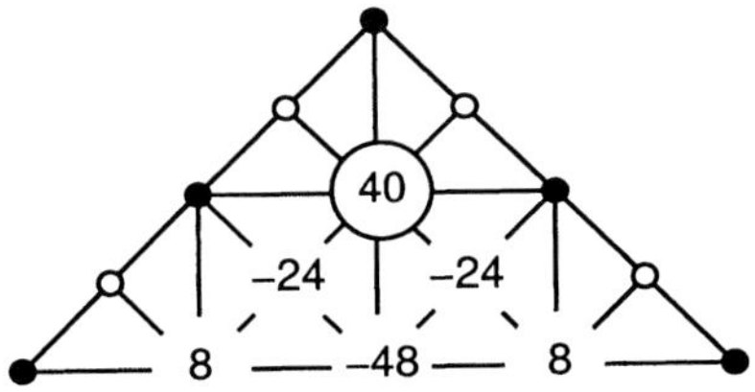

Figure 5c. Third edge semi-wavelet

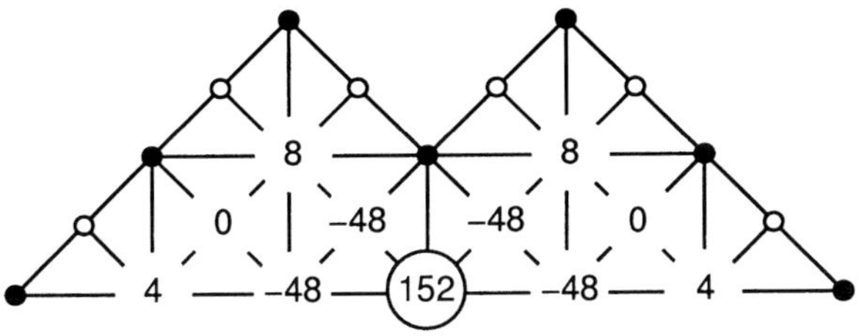

Figure 6a. First edge wavelet

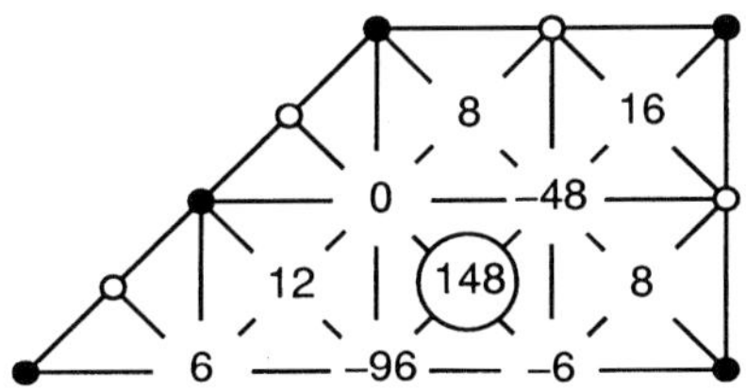

Figure 6b. Second edge wavelet

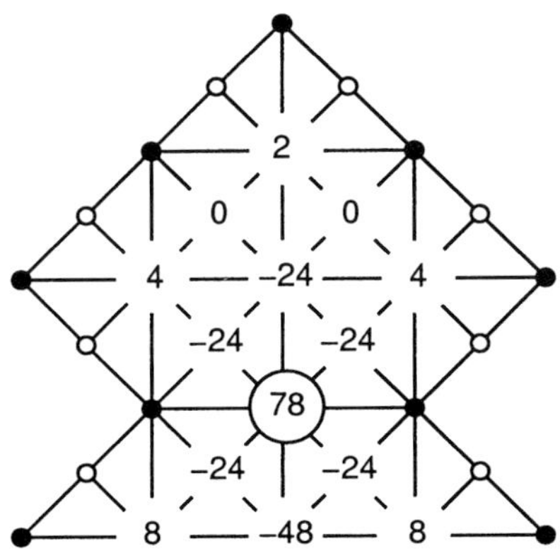

Figure 6c. Third edge wavelet

In the case that v_1 is one of the four corners of the domain, we may suppose without loss of generality that $v_1 = (0, 0)$. The coarse and fine neighbourhoods of v_1 are then

$$N_{v_1}^0 = \{(0,0),(1,0),(1/2,1/2),(0,1)\},$$

and

$$N_{v_1}^1 = \{(0,0),(1/2,0),(1/4,1/4),(0,1/2),\},$$

and the matrix A has dimension 4×4, specifically,

$$A = \frac{1}{192}\begin{pmatrix} 6 & 6 & 8 & 6 \\ 1/2 & 6 & 1 & 0 \\ 1 & 4 & 6 & 4 \\ 1/2 & 0 & 1 & 6 \end{pmatrix}$$

which is invertible with inverse

$$B = A^{-1} = \begin{pmatrix} 42 & -6 & -54 & -6 \\ -3 & 37 & -3 & 5 \\ -3 & -27 & 45 & -27 \\ -3 & 5 & -3 & 37 \end{pmatrix}.$$

There are only two cases, up to symmetry: if $v_2 = (1,0)$ then $b = (-1,1,0,0)^T$ and $x = Bb = (1/2)(-96,80,-48,16)^T$; while if $v_2 = (1/2,1/2)$ then $b = (-1,0,1,0)^T$ and $x = Bb = (1/2)(-192,0,96,0)^T$. These two semi-wavelets are shown in Fig. 7a and 7b. Summing the first corner semi-wavelet and the first edge semi-wavelet yields the wavelet in Fig. 8a and summing the second corner semi-wavelet

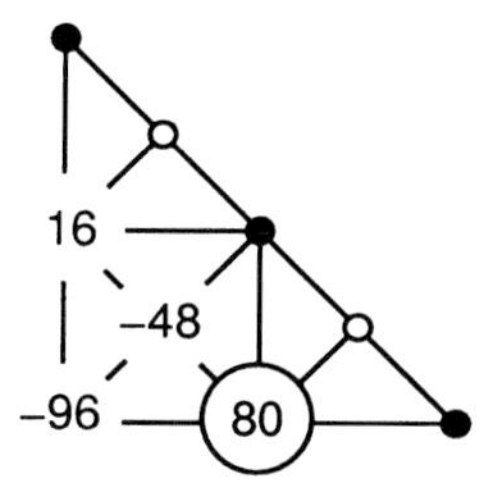

Figure 7a. First corner semi-wavelet

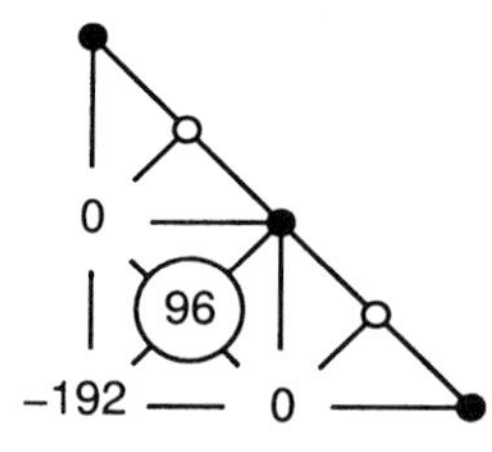

Figure 7b. Second corner semi-wavelet

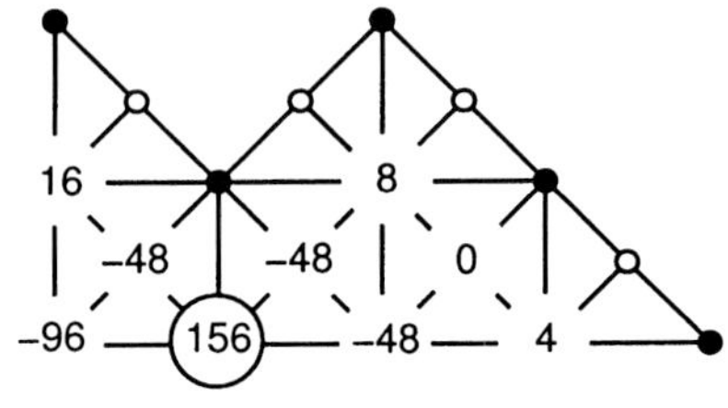

Figure 8a. First corner wavelet

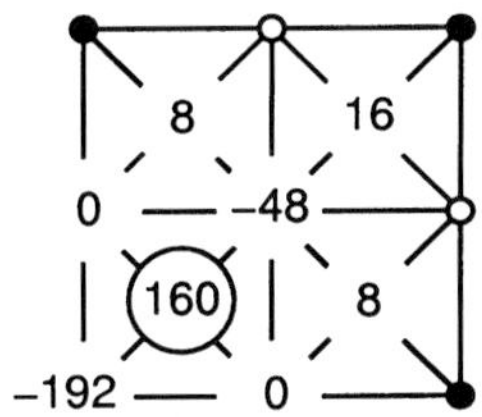

Figure 8b. Second corner wavelet

and the first interior semi-wavelet yields the wavelet in Fig. 8b. Symmetries and rotations of these give us all remaining wavelets ψ_u.

We complete the paper by proving the following theorem.

Theorem 1. *The set of wavelets* $\{\psi_u\}_{u \in V^1 \setminus V^0}$ *defined by* (3.2) *is a basis for the wavelet space* W^0.

Proof: It is sufficient to show that the wavelets ψ_u are linearly independent. We demonstrate this by showing that the square matrix

$$Q = (\psi_v(u))_{u,v \in V^1 \setminus V^0}$$

is diagonally dominant and therefore non-singular. Diagonal dominance is clearly equivalent to the condition that

$$\psi_v(v) > \sum_{u \in V^1 \setminus V^0, u \neq v} |\psi_u(v)|, \quad \text{for all } v \in V^1 \setminus V^0.$$

Thus for each v in $V^1 \setminus V^0$ we need to show that the sum of the absolute values of coefficients at v of wavelets other than ψ_v is less than the coefficient at v of ψ_v itself. It turns out that this condition does indeed hold in every topological case. In Fig. 9 each distinct topological case of $v \in V^1 \setminus V^0$ is illustrated by placing the

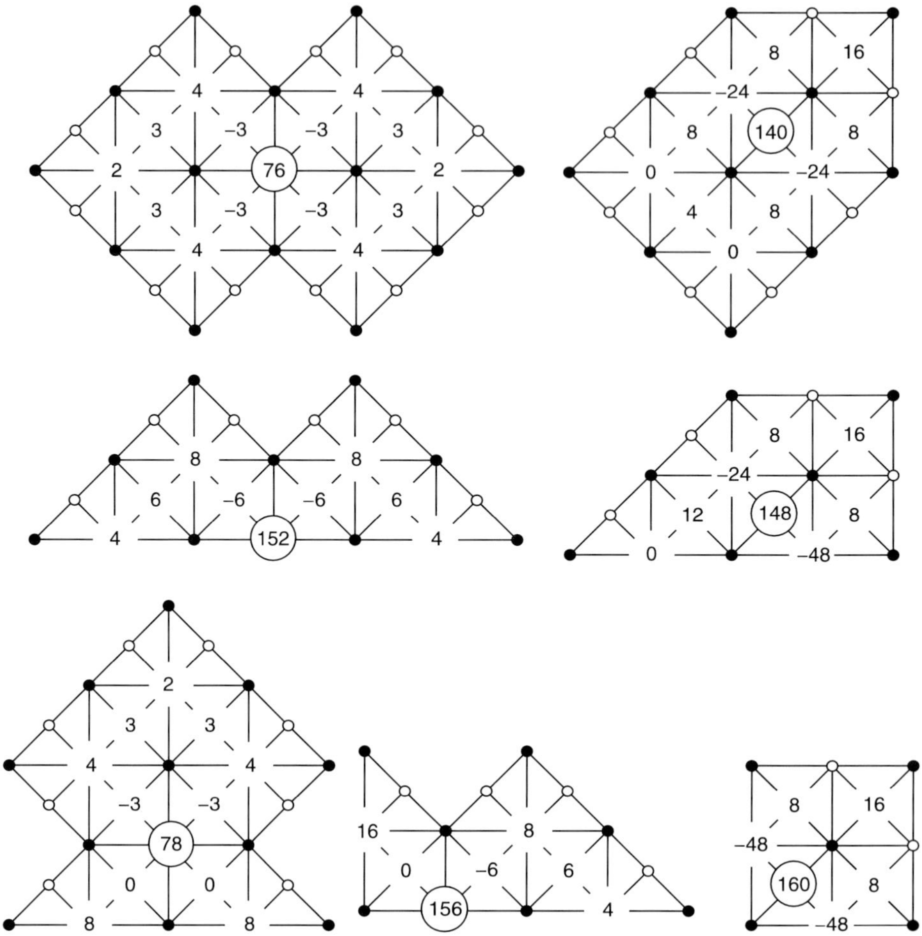

Figure 9. Wavelet evaluations

value $\psi_u(v)$ at u for each relevant u. The vertex v is circled in each case. Thus the coefficients in each figure are the non-zero elements of the row v of the matrix Q. □

References

[1] Chui, C. K.: Multivariate splines. Philadelphia: SIAM, 1988.

[2] Floater, M. S., Quak, E. G.: Piecewise linear prewavelets on arbitrary triangulations. Numer. Math. *82*, 221–252 (1999).

[3] Floater, M. S., Quak, E. G.: A semi-prewavelet approach to piecewise linear pre-wavelets on triangulations. In: Approximation theory IX, vol. 2: computational aspects (Chui, C. K., Schumaker, L. L., eds.), pp. 63–70. Nashville: Vanderbilt University Press, 1998.

[4] Floater, M. S., Quak, E. G.: Linear independence and stability of piecewise linear prewavelets on arbitrary triangulations. SIAM J. Numer. Anal. *38*, 58–79 (2001).

[5] Floater, M. S., Quak, E. G., Reimers, M.: Filter bank algorithms for piecewise linear prewavelets on arbitrary triangulations. J. Comp. Appl. Math. *119*, 185–207 (2001).

[6] Kotyczka, U., Oswald P.: Piecewise linear prewavelets of small support. In: Approximation theory VIII, vol. 2 (Chui, C. K., Schumaker, L. L., eds.), pp. 235–242. World Scientific: Singapore, 1995.
[7] Nürnberger, G., Walz, G.: Error analysis in interpolation by bivariate C^1-splines. IMA J. Numer. Anal. *18*, 485–508 (1998).

M. S. Floater
E. G. Quak
SINTEF, Applied, Mathematics
Post Box 124,
Blindern, N-0314 Oslo
Norway
e-mails: mif@math.sintef.no
ewq@math.sintef.no

Computing [Suppl] 14, 105–118 (2001)

Computing

Feature-Based Matching of Triangular Meshes

M. Fröhlich, H. Müller, C. Pillokat, and **F. Weller,** Dortmund

Abstract

Given two triangular surface meshes M and N in space and an error criterion, we want to find a rigid motion A so that the deviation of $A(M)$ from N minimizes the error criterion. We present a solution to this problem for the case that the surface represented by M is known to be part of the surface represented by N. The solution consists of two steps: coarse matching and refined matching. Coarse matching is performed by first selecting a limited number of mesh vertices with special properties for which suitable numerical feature values are defined. From the selected characteristic vertices, labeled by their feature values, well-matching triples of vertices are selected which are additionally filtered by checking for whether they define an acceptable matching of the given meshes. For refined matching the iterated closest point approach is used which is speeded up by using an "nearest-neighbor-octree" for search space reduction. The solution aims at meshes with a high number of vertices.

1. Introduction

The problem treated in this contribution is

Matching of triangular surface meshes.

Input. Two triangular surface meshes M and N in space with the property that M represents a part of the surface represented by N, and an error criterion.

Output. A rigid motion A so that the deviation of $A(M)$ from N minimizes the error criterion.

The similarity of the two surfaces represented by the mesh is assumed to be of geometric nature, that is, the same geometry is generally approximated by meshes of different connectivity. This means that approaches based on finding similar patterns in the two meshes cannot be applied.

The problem of surface matching occurs in computer-aided engineering, for example with optimization of milling programs. The workpiece to be produced is constructed in a CAD-system. From the CAD-model, a milling program is derived, for instance by a path planning module of a CAD-system. With the resulting milling program, a prototype workpieces is produced. Usually the prototype workpiece will not perfectly match with the CAD-model. For computer-based detection of the deviations, the prototype workpiece is digitized. Then the resulting data are matched with the original CAD-model data, in order to check for inaccuracies of the milling program. At locations of high deviation the milling program is adapted.

Surface matching belongs to the class of geometric pattern matching problems. It is known that the computational complexity of many of these problems is high, in particular under worst-case-considerations, and even in case where polynomial solutions are known [2]. Since our goal is to treat meshes of several thousands of vertices, we will follow a heuristic approach. Our solution consists of two main steps: coarse matching and refined matching.

Coarse matching is performed by first selecting a hopefully small number of vertices of the meshes with special properties. These properties are defined by numerical values called feature values. The result consists of two point sets, one for each mesh, every point labeled by feature values. The next step is to move one point set so that at least three points match approximately. This approach is related to those of [6, 9] in that matching of labeled point sets is considered there, too. However, the difference is that we are satisfied with a coarse matching, that is a matching which need not to be quite close to the optimum. For that reason we can simplify this step somewhat. The crucial point of this approach is to find suitable features, for which we make a suggestion in this paper.

For the second step, refined matching, we use the iterated closest point approach of [3]. This approach needs careful implementation because otherwise it may be quite time-consuming for the large number of vertices we want to treat. We propose a closest point octree subdivision for restriction of search space which turns out to yield a considerable speed-up.

In Section 2 we describe our solution to coarse matching. Section 3 considers refined matching. In Section 4, experimental results obtained with an implementation of the suggested algorithms are presented. The experiments show that the method is applicable in practice.

2. Coarse Matching

Coarse feature-based matching consists of two basic steps: selection of vertices on both meshes which have characteristic properties with respect to some feature, and point matching based on the selected characteristic points. Section 2.1 is devoted to the definition of suitable features. In Section 2.2 we describe the procedure of selection of characteristic points. The matching algorithm is presented in Section 2.3.

2.1. Feature Definition

The goal of feature definition is to find feature values that describe a limited number of vertices of a mesh as having a particular property. In our application we consider a point as important if the curvature in its environment is high. High curvature occurs for example at edges or corners of the geometric shape described by the mesh. As already noted at the beginning, we cannot make use of combinatorial properties of the meshes because the shape geometry may be approximated differently by the two meshes that have to be matched. For this reason, our goal is to define the features independent from the given mesh combinatorics.

Several possibilities of estimating the local curvature of surfaces from approximating meshes have been invented in the past, see e.g. [8] and further references therein. Difficulties which we have realized with different definitions of curvature are their sensitivity with respect to the sampling strategy by which the mesh is obtained, and to the noise typically present in sampled data sets. The feature defined in the following tries to remedy these troubles by two actions. First, additional points are interpolated on the surface around a vertex. Then the additional points are used for curvature calculation at the vertex. We take the average of the distances of these points from a fitting a plane as a simple curvature estimate. Second, we do not just use one curvature value as a feature, but a feature vector of d of these values, d a fixed dimension. The components of the feature vector are basically calculated in the same manner and represent estimates of the curvature calculated from the surface behavior at different distances from the considered vertex.

2.1.1. Calculation of a Single Feature Value

Our basic feature value at a mesh vertex $\mathbf{p}$ is calculated by the following algorithm.

Calculation of a feature value at a mesh vertex p.

1. *Calculate their intersection points of the edges of the mesh with a sphere with center* $\mathbf{p}$, *arranged on the intersection curves between the mesh and the sphere.*

The desired points of the intersection curves are calculated by a marching triangle approach on the mesh [1]. In order to find all intersection curves in the case that more than one exist, the set of edges is calculated which intersect the sphere and have one of its endpoints in the interior of the sphere, by a depth-first search on the mesh starting at the center $\mathbf{p}$ of the sphere. These edges are used as starting points for the marching algorithm. Those found during marching are removed because they are no longer useful for finding another intersection curve.

2. *Insert points into the intersection curves.*

The reason for this step is to diminish the dependency of the feature on the combinatorics of the mesh caused by different sampling densities. Point insertion is performed in the following steps,

(a) Choose a number k.

(b) Connect consecutive points of every intersection curve by the shortest path on the sphere, that is a great circular arc.

(c) Calculate the sum l of the arc lengths of all intersection curves.

(d) Determine k points with distance l/k, measure in arc length, on the circular arc chains.

3. *Calculate a fitting plane E of the curves.*

The plane can be found with the least-square method, according to [10] using the code of [5] in our implementation.

4. *Report the average distance of the points to the plane E as the feature value.*

In step 1, more than one curve may occur if the vertex **p** is close to the surface boundary if the boundary multiply enters and leaves the sphere around **p**. The case that other parts of the surface enter the sphere is treated by considering only points that can be reached on the mesh on a path completely in the interior of the sphere.

$k = 200$ is a typical choice in step 2.

As an alternative in step 4, we have also checked the distance of **p** to the plane E for a possible feature. This feature, however, has turned out in our experiments to be less effective in identifying characteristic points.

2.1.2. Calculation of a Feature Vector

The feature vector $v(\mathbf{p})$ of a mesh vertex **p** is composed of a constant number d of feature values. The feature values are calculated for different spheres, with increasing radii. The radii are chosen by defining a minimum and a maximum radius, and subdividing the interval equidistantly, in order to obtain d radii. The minimum and the maximum radius is defined heuristically by using the location of the first local maximum of the edge length histogram for the minimum radius, and one third of the length of the bounding box diagonal for the maximum radius.

For later comparison of feature vectors, we additionally need a so-called significant index $i_s(\mathbf{p})$ of the feature vector $v(\mathbf{p})$. The significant index is defined as the smallest index i of v so that the sphere radius belonging to the entry $v_i(\mathbf{p})$ of $v(\mathbf{p})$ is greater than the length of the shortest incident edge of **p**. If this is impossible, we take $i_s(\mathbf{p}) := d$. The reason for this definition of $i_s(\mathbf{p})$ is that for a radius less than $v_{i_s}(\mathbf{p})$ the feature values are very similar and hence not useful.

2.2. Selection of Characteristic Points

Characteristic points are points with extraordinary feature vectors. For our definition of feature vectors we consider a feature vector as extraordinary if it has high feature values. The reason is that high feature values indicate regions of high curvature on the surface which are of particular interest as characteristic points.

A point is selected as a characteristic point if it has higher relevant features than its adjacent points. The relevant features are those with index i_m and $i_m + 1$, if $i_m < d$, and i_m and $i_m - 1$, if $i_m = d$, respectively, of the feature vectors of the two points **p** and **q** to be compared, where $i_m := \max\{i_s(\mathbf{p}), i_s(\mathbf{q})\}$.

2.3. *Algorithm for Coarse Matching*

The first step of the algorithm for coarse matching is to calculate the list of characteristic points for each of the two meshes, L_M and L_N, according to the previous subsection. In order to explain the further steps, we need to define different measures of similarity:

Feature similarity. The feature similarity of a characteristic point $\mathbf{p}$ of M and a characteristic point $\mathbf{q}$ of N is defined as the average of the relative error of corresponding relevant entries of the feature vector,

$$s_f(\mathbf{p},\mathbf{q}) := \frac{1}{d_s} \sum_{i=i_m}^{i_m+d_s-1} \frac{|v_i(\mathbf{p}) - v_i(\mathbf{q})|}{|v_i(\mathbf{p})| + |v_i(\mathbf{q})|}$$

if $i_m + d_s - 1 \leq d$, and

$$s_f(\mathbf{p},\mathbf{q}) := \frac{1}{d - i_m + 1} \sum_{i=i_m}^{d} \frac{|v_i(\mathbf{p}) - v_i(\mathbf{q})|}{|v_i(\mathbf{p})| + |v_i(\mathbf{q})|}$$

if $i_m + d_s - 1 > d$, where $i_m := \max\{i_s(\mathbf{p}), i_s(\mathbf{q})\}$, and d_s the number of relevant dimensions which is a free parameter of the algorithm.

Two characteristic points are accepted as ε_f-feature-similar if $s_f(\mathbf{p},\mathbf{q}) < \varepsilon_f$.

Distance similarity. Let $e = (\mathbf{p},\mathbf{p}')$ be a line segment defined by two characteristic points $\mathbf{p}$ and $\mathbf{p}'$ of M, and $f = (\mathbf{q},\mathbf{q}')$ be a line segment defined by two characterisitc points $\mathbf{q}$ and $\mathbf{q}'$ of N. The distance similarity of e and f is defined as the relative difference of the lengths $l(e)$, $l(f)$ of the line segments,

$$s_d(e,f) := \frac{|l(e) - l(f)|}{l(e) + l(f)}.$$

Two edges are accepted as ε_d-distance-similar if $s_d(e,f) < \varepsilon_d$.

Matching similarity. For every vertex $\mathbf{p}$ of the (possibly moved) mesh M, let $\mathbf{r}(\mathbf{p})$ be a point on mesh N which is closest to $\mathbf{p}$. The matching similarity of M and N is defined as

$$s_m(M,N) := \max_{\mathbf{p} \in M} ||\mathbf{p} - \mathbf{r}(\mathbf{p})||.$$

M and N are accepted as ε_m-matching if $s_m(M,N) < \varepsilon_m$.

With these measures, the algorithm works as follows.

Algorithm of coarse matching.

1. *Characteristic point calculation.*

Determine the characteristic points as described in Section 2.2.

2. *Characterisitic point assignment.*

Assign to each characteristic point of M all characteristic points of N which are ε_f-feature-similar. The result is stored in a list of lists.

3. *Line segment assignment.*

Using the result of step 1, assign to each line segment $e = (\mathbf{p}, \mathbf{p}')$ of M all line segments $f = (\mathbf{q}, \mathbf{q}')$ of N so that $\mathbf{p}$, $\mathbf{q}$ and $\mathbf{p}'$, $\mathbf{q}'$, respectively, are ε_f-feature-similar, and e and f are ε_d-distance-similar. The result is stored in a list of lists.

4. *Triangle assignment and matching quality test.*

Take all triples of characteristic points $\mathbf{p}$, $\mathbf{p}'$, $\mathbf{p}''$ of M so that there exist characteristic points $\mathbf{q}$, $\mathbf{q}'$, $\mathbf{q}''$ of N so that the line segment $(\mathbf{q}, \mathbf{q}')$ was assigned to $(\mathbf{p}, \mathbf{p}')$, $(\mathbf{q}', \mathbf{q}'')$ to $(\mathbf{p}', \mathbf{p}'')$, and $(\mathbf{q}'', \mathbf{q})$ to $(\mathbf{p}'', \mathbf{p})$ in step 2. Let $s = (\mathbf{p}, \mathbf{p}', \mathbf{p}'')$ be the triangle corresponding to M, and $t = (\mathbf{q}, \mathbf{q}', \mathbf{q}'')$ the triangle corresponding to N.

Calculate a rigid motion A which matches s with t. The rigid motion consists of three steps: translation of s so that its center point coincides with that of t, rotation around the axis in direction of the cross-product of the normal vectors of both triangles so that the triangles are co-planar, and rotation around the common normals of the triangles so that the sum of squared distances of corresponding vertices is minimized [7].

If $s_m(A(M), N) < \varepsilon_m$, insert A into a list of matching transforms, together with $s_m(A(M), N)$, sorted according to this value.

3. Refined Matching

Refined matching is performed by the iterated closest point approach (ICP) of [3]. In each step of iteration the algorithm determines for each vertex $\mathbf{p}$ of the first mesh M the closest point $\mathbf{r}(\mathbf{p})$ on a triangle of the second mesh N. Then a rigid motion of M is determined which minimizes the sum of squared distances between points $\mathbf{p}$ and $\mathbf{r}(\mathbf{p})$. With the mesh M in its new location, this procedure is iterated until a (local) minimum is reached.

From the point of view of computational efficiency, the main problem is to find the closest points $\mathbf{r}(\mathbf{p})$. In order to reduce this effort, the triangles of the mesh N are inserted into an octree. The octree covers the axis-parallel bounding box of N plus a margin of 10 percent on each side. To each cell of the octree, the set of triangles of N is assigned which are possibly the closest to one of the points of the cell.

Initially, all triangles of N are assigned to the root cell. From a given cell to which its triangles are already assigned, we calculate the triangle assignment for its eight successors by testing its triangles. For each of the eight cells c and every triangle t, we calculate the extremal distances from points in c to t,

$$\underline{d}(c, t) := \min_{\mathbf{s} \in c} \min_{\mathbf{r} \in t} ||\mathbf{s} - \mathbf{r}||,$$

$$\bar{d}(c, t) := \max_{\mathbf{s} \in c} \min_{\mathbf{r} \in t} ||\mathbf{s} - \mathbf{r}||.$$

Let

$$d^*(c) := \min_t \bar{d}(c,t)$$

where the minimum is taken over all triangles of the parent cell. Those triangles t for which $\underline{d}(c,t) \leq d^*(c)$ are assigned to c. If some triangle t has $\underline{d}(c,t) > d^*(c)$, there exists a closer triangle for each point of the cell, so t needs not be assigned to c.

The iteration of subdivision is stopped if the number of triangles either falls below a given threshold, or the depth of the octree would exceed a given bound.

Using the octree, the closest point $\mathbf{r}(\mathbf{p})$ of a point $\mathbf{p}$ is calculated by first finding the leaf of the octree into which $\mathbf{p}$ falls. For each of the triangles found at the leaf, the closest point is calculated, and the minimum over these points is taken as $\mathbf{r}(\mathbf{p})$.

4. Results

In the following we show the results of an experimental evaluation of our algorithms on two data sets.

Example 1 is based on the mesh of a part of a rim. The mesh N, in the terminology of the previous sections, has 4787 vertices (Fig. 1, left). Mesh M is obtained from sample points of part of the same shape, with 4393 vertices (Fig. 1, right). The samplings of both meshes are different. The bounding box diagonals of the data sets have length 226 and 189, respectively.

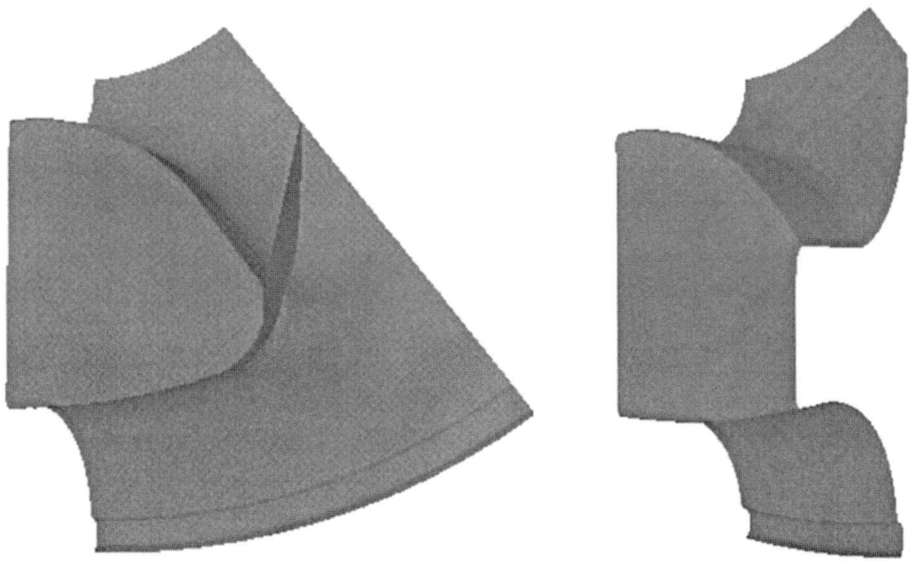

Figure 1. Shaded segment of a part of a rim, with 4787 vertices (left), and a submesh of it, differently sampled, with 4393 vertices (right)

We have first applied coarse matching with minimum radius 1 and maximum radius 50, according to the rule of Section 2.1.2. The matching similarity bound was chosen as $\varepsilon_m = 5$. With these parameters, coarse matching reported four matchings, with matching similarities between 3.23 and 4.45. The matching with the lower value is shown in Fig. 2, left.

Afterwards we have applied refined matching to this result. It reduces the error from initially 0.47 to a final value of 0.02 after 24 iterations (Fig. 2, right).

A refined matching with an unfavorable initial coarse matching is shown in Fig. 3. It reduces the error from 177.8 to 0.66 in 50 iterations. This example shows the importance of a not too bad initial coarse matching.

In the following experiments we have varied the radii and the matching similarity bound of coarse matching. The first setting has a minimum radius 1.5, a maximum radius 15, and $\varepsilon_m = 10$. It yields two matchings with matching similarities between 3.42 and 4.44. In the second example we have chosen quite extreme values: minimum radius 10, maximum radius 150, $\varepsilon_m = 50$. The result is very bad. It consists of two matchings with matching similarities of about 40. Figure 4, left, confirms that the matching is also visually bad. Nevertheless, refined matching for this example is surprisingly successful, cf. Fig. 4, right, which, however is not typical. It reduces the error from 294.8 to 0.005 in 31 iterations.

Example 2 uses a hook. Mesh N covers the whole hook, together with a part of the plane environment on which it is located. It has 4501 vertices (Fig. 5, left). Mesh M has 1470 vertices and represents a differently sampled part of the same

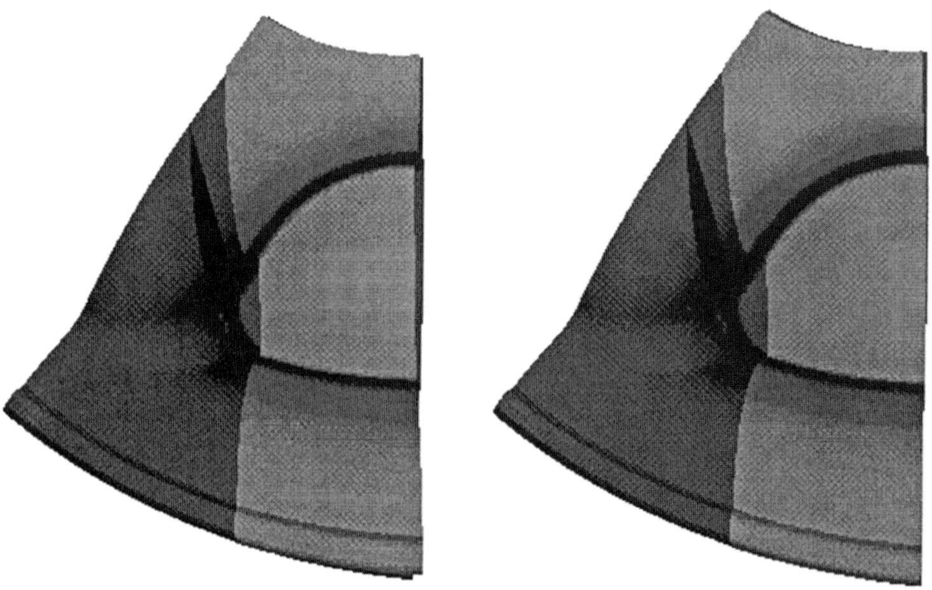

Figure 2. Coarse matching (left), with a corresponding refined matching (right)

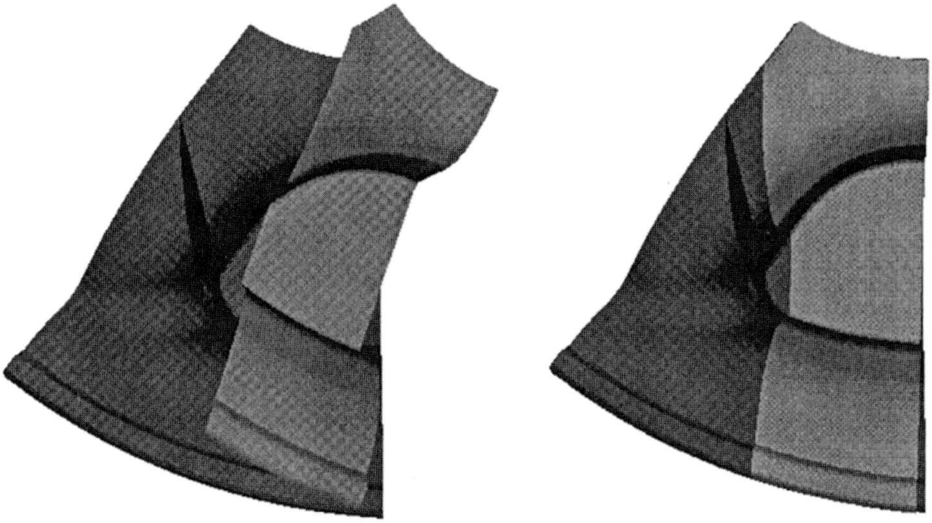

Figure 3. A bad coarse matching (left) which is successfully corrected by refined matching (right)

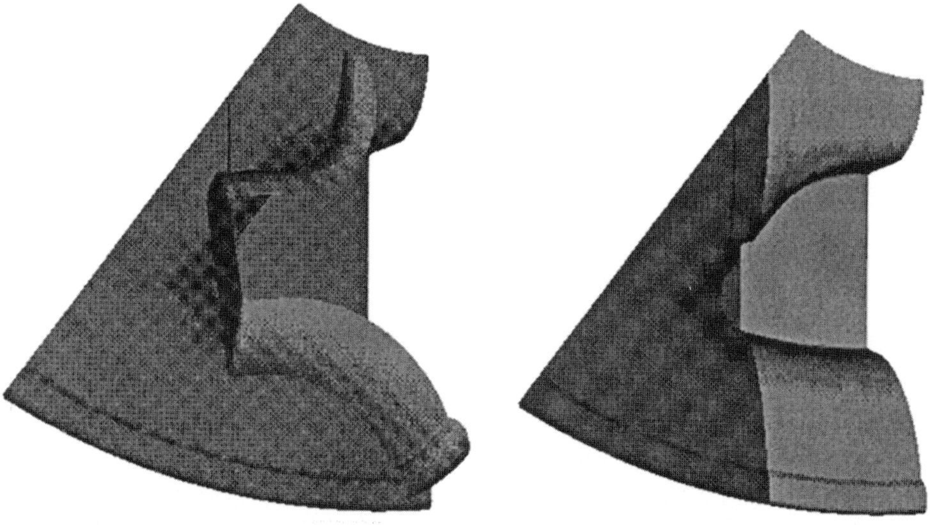

Figure 4. A very bad coarse matching (left) which, surprisingly, is successfully corrected by refined matching (right)

object (Fig. 5, right). Figure 6 displays the characteristic points for minimum radius 2 and maximum radius 40. $\varepsilon_m = 20$ yields ten coarse matchings with matching similarities of about 14. Smaller values of ε_m did not help to reduce the number of matchings. Figure 7, left, shows that reasonable solutions are among the reported matchings. Surprisingly, the matching similarity of this example is slightly worse than that of example of Fig. 8, left, which is the matching with best

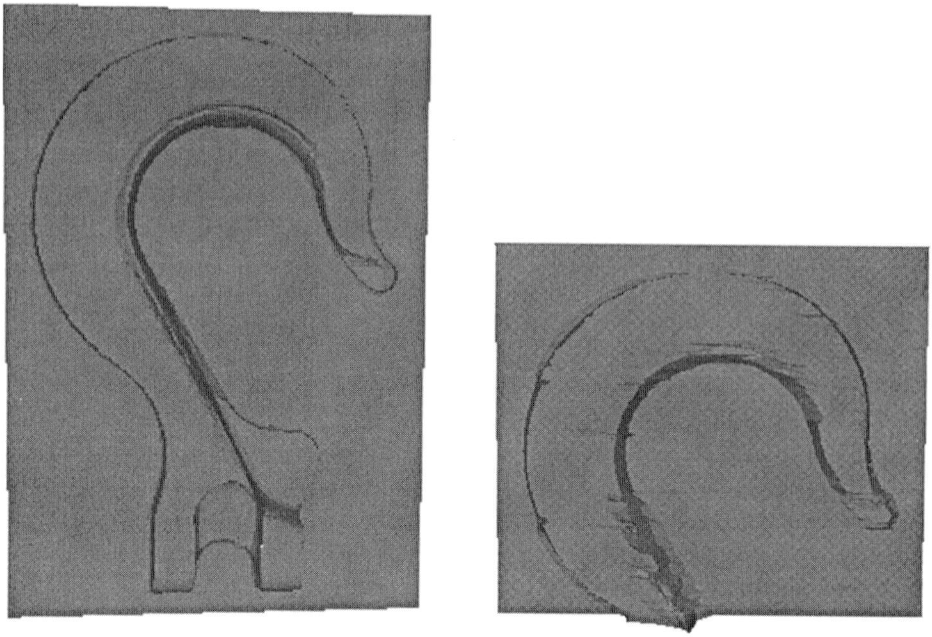

Figure 5. The mesh of a hook, including a plane environment, with 4501 vertices (left), and a differently sampled mesh of a part of it with 1470 vertices (right)

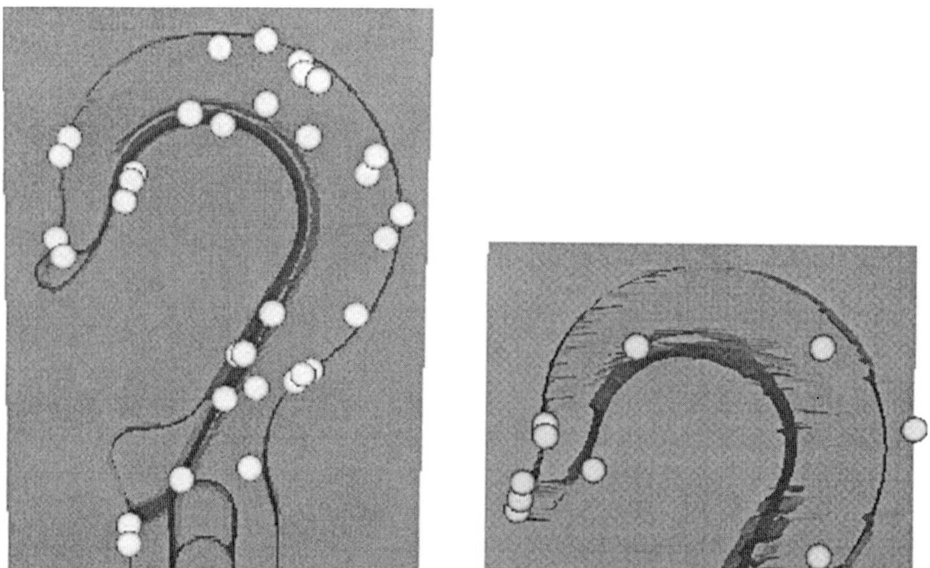

Figure 6. Characteristic points found on the two meshes of the previous figure

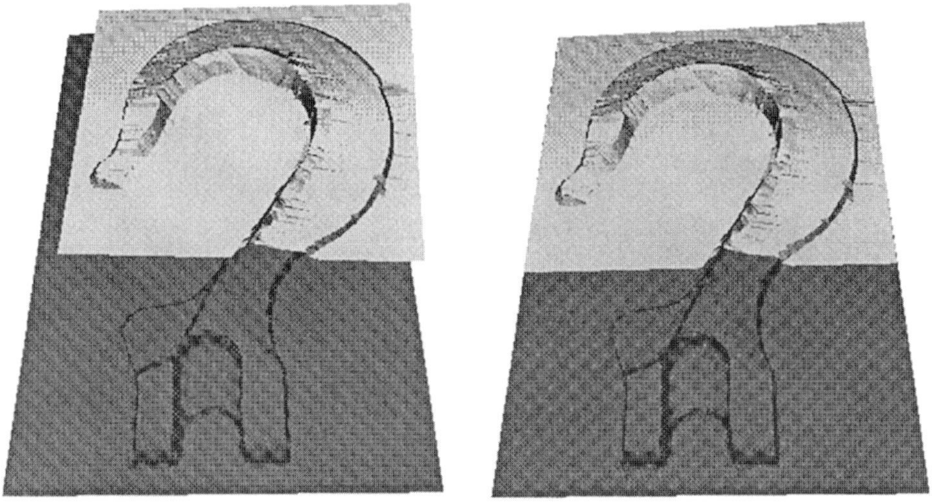

Figure 7. A favorable coarse matching (left) and its successful improvement by refined matching (right)

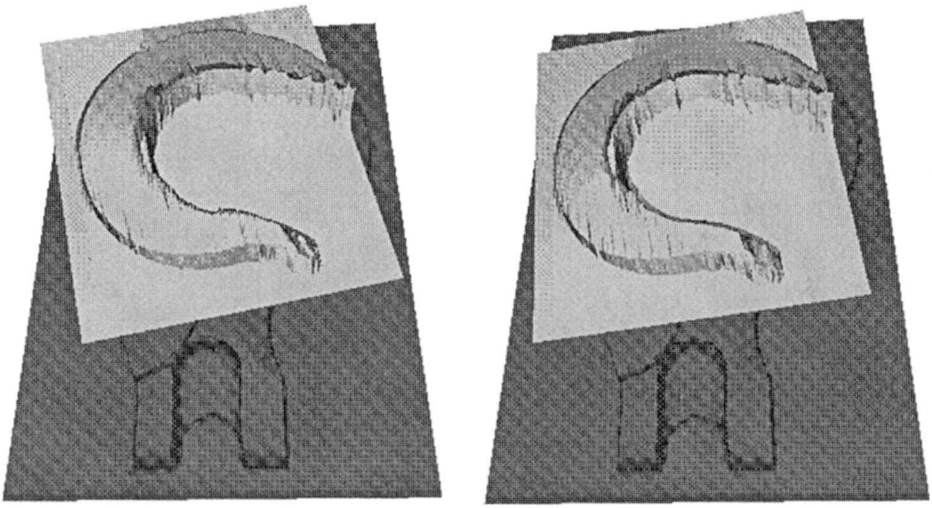

Figure 8. A bad coarse matching (left) which could not be corrected by refined matching (right)

matchings similarity. Refined matching improves the error from 22.9 to 0.02 in 34 iterations (Fig. 7, right). Refined matching applied to the solution of coarse matching of Fig. 8 reduces the error from 30.35 to 13.63, but evidently sticks in a local minimum which is unequal to the desired match.

The experiments show that the algorithm can be applied successful, but that no deterministic rule seems to exist which guarantees that it is successful. For that reason we have embedded the algorithm in an interactive environment in which the user can select those of the coarse matchings offered by that algorithm, for refined matching, which are reasonable for him. The role of coarse matching can be seen as to avoid that the refined matching falls into an unfavorable local minimum.

With respect to computational efficiency, we have experimentally analyzed the calculation of the characteristic points, which is the most time consuming part of coarse matching. We have used the meshes of the two examples. The dimension of the feature vector was $d = 6$. Table 1 shows the results of measurements for different ranges of sphere radii, 1–4, 1–25, 1–50, and 1–75. We have measured the total number of edges that were visited for calculation of the intersection curves (first line), the average number of edges visited for one feature value (second line), and the total time required in hours:minutes:seconds (third line). The times were measured on a Pentium 100 PC with 32 MB main memory. Evidently, calculation time increases more than linearly as a function of the radius.

Furthermore, we have analyzed the behavior of octree calculation. Depending on the depth limit of the tree, we have measured the time of computation in hours, the number of leaves, the rounded average number of triangles per leaf, and the time required for error calculation which needs a closest distance calculation for all vertices, for the mesh of the rim of Fig. 5, matched to itself based on the octree (Table 2). On the used computer with 32 MB RAM, depth four was the maximum

Table 1. Experimental analysis of the calculation of characteristic points, for feature vector dimension $d = 6$

Shape	1–5	1–25	1–50	1–75
Hook mesh N (27 006 edges)				
#edges	1 231 574	5434 033	8 687 369	9 661 803
#edges/feature	46	201	322	358
comp. time	0:04:34 h	0:20:17 h	0:45:44 h	1:11:17 h
Rim mesh N (28 622 edges)				
#edges	1 228 576	4 577 822	7 411 953	8 953 471
#edges/feature	43	159	258	311
comp. time	0:04:46 h	0:16:03 h	0:37:37 h	1:03:02 h

We have used the meshes N of the two examples (main rows). The columns correspond to different ranges of sphere radii. Within the main rows, the total number of edges that were visited for calculation of the intersection curves (first line), the average number of edges visited for one feature value (second line), and the total time required in hours:minutes:seconds (third line) are compiled

Table 2. Behavior of octree calculation

Max. depth	Comp. time	#Leaves	#Triangles/leaf	Error calc.
1	0:03 h	8	7277	994 934 ms
2	0:20 h	64	2754	446 687 ms
3	1:13 h	512	849	168 926 ms
4	3:24 h	4096	260	47 192 ms

Dependent on the depth limit of the tree, the columns show the time of computation in hours, the number of leaves, the average number of triangles per leaf, and the time required for error calculation which needs a closest distance calculation for all vertices, for the mesh against itself based on the octree that could be treated. The computation times show that the time reduction of error calculation is significant. In order to minimize the total time of computation, the level should be chosen so that the sum of octree preprocessing time and the time of iteration of refinement matching is minimized.

In summary, the computation times on a relatively slow computer with little main memory show that the algorithms are applicable for meshes with isolated extraordinary points that can be used as characteristic points.

We close this section with some ideas of possible improvements. Concerning computational efficiency of the phase of refined matching, one possibility might be to replace the octree, at least during the iteration, by links to triangles and local search on the triangles. This may save memory in that phase.

If a relatively good coarse matching can be expected, the octree possibly needs not to be stored. Instead, the octree subdivision strategy may be used to define an initial assignment of vertices of one mesh to triangles on the other mesh, for local minimum search. Only one path down the octree with the starting nodes of not yet investigated branches has to be stored. Based on the initial assignment, iteration may be performed as outlined in the preceding paragraph.

For dense meshes, replacing the closest neighbor which is currently a point in a triangle, by vertices of a mesh which may be found more quickly, might be feasible.

Feature calculation in the phase of coarse matching is a crucial topic. We have suggested to use a vector of features based on a simple curvature estimate in order to cope with varying sample strategies and noise. An alternative approach might be to consider more advanced curvature estimates [8] in combination with non-shrinking mesh-smoothing filters [11].

We have assumed that the surface represented by M is part of the surface represented by N. This assumption simplifies the formulation of the approach, but basically the method should be extensible to the case of overlapping surfaces. The reason is that pairs of similar triangles can be found for coarse matching in the set of feature points of the two vertices analogously to the subset case. For refined matching, we currently use all vertices of M. In the overlapping case, only vertices of M and N potentially located in the overlapping zone should be considered in the goal function of optimization. One simple possibility in this direction could be to consider only those pairs of vertices whose distance in the current matching does not exceed a given threshold.

In our examples we have used very dense meshes. Possible it is feasible and useful to thin them out, in particular in regions or low curvature, in order to speed-up the calculation. A survey of mesh simplification is given in [4].

Acknowledgement

The authors would like to thank the referees for their helpful hints.

References

[1] Allgower, E. L., Schmidt, P. H.: An algorithm for piecewise linear approximation of an implicitly defined manifold. SIAM J. Numer. Anal. *22*, 322–346 (1985).
[2] Alt, H., Guibas, L.: Discrete geometric shapes: matching, interpolation, and approximation. In: Handbook of computational geometry (Urrutia, J., Sack, J.-R., eds.) Amsterdam: North-Holland.
[3] Besl, P. J., McKay, N. D.: A method of registration of 3-D shapes. IEEE Trans. Pattern Anal. Mach. Intell. *14*, 239–256 (1992).
[4] Cignoni, P., Montani, C., Scopigno, R.: A comparison of mesh simplification algorithms. Comput. Graphics *22*, 37–54 (1998).
[5] Eberly, D.: Magic – may alternate graphics and image code. Department of Computer Science, University of North Carolina at Chapel Hill, ftp://ftp.cs.unc.edu/pub/packages/magic, 1998.
[6] Hoffmann, F., Kriegel, K., Wenk, C.: Matching 2D patterns of proteins spots. In: Proc. 14th ACM Symposium on Computational Geometry, pp. 231–239 (1998).
[7] Horn, B. K. P.: Closed-form solution of absolute orientation using unit quaternions. J. Optical Soc. Am. *4*, 629–642 (1987).
[8] Krsek, P., Lucács, G., Martin, R. R.: Algorithms for computing curvature from range data. In: The Mathematics of Surfaces VIII (Cripps, R., ed.), pp. 1–16. Winchester: Information Geometers. 1998.
[9] Ogawa, H.: Labeled point pattern matching by Delaunay triangulations and maximal cliques. Pattern Rec. *19*, 35–40 (1986).
[10] Press, W. H., Teukoldy, S. A., Vetterling, W. T., Flannery, B. P.: Numerical recipes in C – the state of scientific computing, 2nd edn. Cambridge: CUP, 1992.
[11] Taubin, G.: A signal processing approach to fair surface design. In: Proceedings SIGGRAPH'95, pp. 351–358 (1995).

M. Fröhlich
H. Müller
C. Pillokat
F. Weller
Informatik VII
Universität Dortmund
D-44221 Dortmund
Germany
e-mail: froehliz@ls7.informatik.uni-dortmund.de

Computing [Suppl] 14, 119–154 (2001)

Computing

C^4 Interpolatory Shape-Preserving Polynomial Splines of Variable Degree

N. C. Gabrielides and **P. D. Kaklis,** Athens

Abstract

This paper introduces a new family of C^4-continuous interpolatory variable-degree polynomial splines and investigates their interpolation and asymptotic properties as the segment degrees increase. The basic outcome of this investigation is an iterative algorithm for constructing C^4 interpolants, which conform with the *discrete* convexity and torsion information contained in the associated polygonal interpolant. The performance of the algorithm, in particular the *fairness* effect of the achieved high parametric continuity, is tested and discussed for a planar and a spatial data set.

1. Introduction

The problem of shape-preserving curve interpolation can be regarded as a topic that is well studied in the planar case (see, e.g., the references in Hoschek and Lasser ([9], §§3.6, 3.8) and Messac and Sivanandan [14]), while it receives constantly increasing attention in the case of three-dimensional space; see, e.g., Asaturyan et al. [1], Goodman and Ong ([7], [8]), Kaklis and Karavelas [11].

Despite the diversity of techniques employed for handling the various versions of this problem, one may dare stating that, in general, the parametric continuity, achieved by the proposed schemes, is restricted to order two for the planar and order three for the spatial case. Obviously, these orders seem to be sufficient from the Differential-Geometry point of view, ensuring the continuity of the basic invariant quantities: curvature, torsion, etc. Nevertheless, the authors of the present paper consider that, improving further the continuity-order of a shape-preserving interpolation scheme, may be a worthwhile task if it is anticipated that additional continuity may improve the *fairness* profile of a shape-preserving curve by, e.g., lowering absolute curvature maxima. This is especially true for schemes that, by their very nature, tend to sacrifice *fairness* in favour of *shape* with staying as close as necessary to a readily available shape-preserving but non-smooth interpolant, e.g., the associated polygonal interpolant.

The present paper attempts to materialize the above task for the so-called family of *variable degree polynomial splines*. These splines have been successfully employed by various researchers for handling the shape-preserving interpolation problem not only for curves, but for surfaces as well; see, e.g., Costantini [2], [3], [4], Kaklis and Sapidis [13], Kaklis and Ginnis [10], Ginnis et al. [6]. The specific aim of this paper is to develop a new family of variable degree polynomial splines, that offer fourth-order parametric continuity, and test their performance in the context of both the planar and the spatial shape-preserving-interpolation problem.

The rest of the paper is structured in eight sections and an appendix. In Section 2, we introduce the basic representation of the new spline family Γ_4 (§2.1), formulate and prove the well posedness of the associated interpolation problem (§2.2), and investigate the structure of the Bézier control polygon of the polynomial segments of an element in Γ_4 (§2.3). Section 3 studies the asymptotic behaviour of an interpolant $\mathbf{Q}(u) \in \Gamma_4(\mathscr{K})$ as the segment degrees $\mathscr{K}$ increase *locally, semi-locally or globally*. In Section 4 we adopt from the pertinent literature a shape-preserving-interpolation notion (see Def. 4.1) consisting of two parts, the so-called *convexity* and *torsion* criteria. In Section 5, Theorems 5.1 and 5.2 establish that, if the degrees increase appropriately, then $\Gamma_4(\mathscr{K})$ is able to conform with both parts of the *convexity criterion* of the adopted shape-preservation notion. On the contrary, $\Gamma_4(\mathscr{K})$ is able to satisfy the *torsion criterion* only in the interior of each parametric segment (Th. 6.1), for the nodal torsion of an element in $\Gamma_4(\mathscr{K})$ always vanishes. The obtained asymptotic results of Sections 5 and 6, rely heavily on the use of a lemma that is stated and proved in the Appendix; see Lemma A.1.

Exploiting the outcome of the two previous sections, §7 formulates an iterative algorithm for the automatic construction of C^4 shape-preserving interpolants in $\Gamma_k(\mathscr{K})$. The numerical performance of this algorithm is presented and discussed in §8, for two data sets; see Table 8.1 and Figs. 8.1–8.3 for the 2D point-set, and Table 8.2 and Figs. 8.4–8.8 for the 3D point-set.

The paper ends with Section 9, containing comparative remarks between the performance of the herein proposed algorithm and that proposed in Kaklis and Karavelas [11] for shape-preserving interpolation with C^2 variable degree splines. On the basis of these remarks, we can legitimately assert that increasing the parametric continuity of variable degree splines leads to fairer curvature distributions, which justifies the undertaken task at least in the area of fair shape-preserving interpolation in the plane. As for the torsion distribution, larger parametric continuity seems to yield larger torsion values in the interior of the parametric intervals, apparently due to the intrinsic property that not only torsion but its arc-length derivative as well vanish at the parametrization nodes.

2. A C^4 Family of Variable Degree Splines

Let be given a point-set

$$\mathscr{D} = \{\mathbf{I}_m \in \mathbb{E}^3,\ m = 1, \ldots, N\}, \quad \mathbf{I}_m \neq \mathbf{I}_{m+1},\ m = 1, \ldots, N-1,$$

along with a strictly increasing sequence $\mathscr{U} = \{u_m,\ m = 1, \ldots N\}$ of parametric nodes. We aim to construct a C^4-continuous polynomial spline $\mathbf{Q}(u)$, $u \in [u_1, u_N]$, that interpolates $\mathscr{D}$ over the nodes of $\mathscr{U}$, while its degrees may vary from segment to segment. Let $\mathscr{K} = \{k_m + \ell, \ m = 1, \ldots, N-1\}$ be the associated degree sequence, with k_m being the variable part of the polynomial degree in the segment $[u_m, u_{m+1}]$, ℓ being its constant part. The restriction of $\mathbf{Q}(u)$ in $[u_m, u_{m+1}]$ is chosen to have the following representation:

$$\begin{aligned}\mathbf{Q}(u) &= \mathbf{L}(u) + h_m^2 \mathbf{Q}_m^{(2)} \Theta_m(1-t) + h_m^2 \mathbf{Q}_{m+1}^{(2)} \Theta_m(t) \\ &\quad + h_m^4 \mathbf{Q}_m^{(4)} \Phi_m(1-t) + h_m^4 \mathbf{Q}_{m+1}^{(4)} \Phi_m(t), \quad t \in [0,1], \end{aligned} \tag{2.1}$$

with $t = \frac{u - u_m}{h_m}$ denoting the local variable and $h_m = u_{m+1} - u_m > 0$. Here $\mathbf{L}(u)$ is the linear interpolant of $\mathbf{I}_m$, $\mathbf{I}_{m+1}$ and $\mathbf{Q}_m^{(2)}$, $\mathbf{Q}_m^{(4)}$ denote the second and fourth order nodal derivatives of $\mathbf{Q}(u)$ at $u = u_m$, respectively. Finally, $\Theta_m(t)$ and $\Phi_m(t)$ are auxiliary polynomials that should satisfy the following boundary conditions at $t = 0, 1$:

$$\Theta_m^{(2q)}(0) = 0, \quad \Theta_m^{(2q)}(1) = \delta_{1q}, \ \ q = 0, 1, 2, \tag{2.2a}$$

$$\Phi_m^{(2q)}(0) = 0, \quad \Phi_m^{(2q)}(1) = \delta_{2q}, \tag{2.2b}$$

where the superscript q denotes the qth derivative of the underlying function and δ_{ij} is the Kronecker's delta. Once equalities (2.2a) and (2.2b) hold true, it can be readily shown that the family of polynomial splines, defined by (2.1), interpolates $\mathscr{D}$ and its one-sided derivatives of order $2q$, $q = 1, 2$, are continuous at the internal nodes of $\mathscr{U}$, i.e.,

$$\mathbf{Q}^{(2q)}(u_m^-) = \mathbf{Q}^{(2q)}(u_m^+) = \mathbf{Q}_m^{(2q)}, \quad q = 1, 2, \ \ m = 2, \ldots, N-1.$$

Obviously, in order to achieve C^4-continuity on $[u_1, u_N]$, one has further to ensure continuity of the one-sided derivatives of odd order $2q - 1$, $q = 1, 2$. Towards this aim, we have first to appropriately construct the auxiliary polynomials $\Theta_m(t)$ and $\Phi_m(t)$.

2.1. Constructing the Auxiliary Polynomials $\Theta_m(t)$ and $\Phi_m(t)$

Attempting to inherit the asymptotic behaviour of the C^2 variable degree splines, constructed in Kaklis and Pandelis [12], we choose to express $\Theta_m(t)$ and $\Phi_m(t)$ in terms of the lacunary polynomial

$$F(t;p) = \frac{t^p - t}{p(p-1)}, \quad t \in [0,1]. \tag{2.3}$$

More specifically, we set:

$$\Theta_m(t) = a_\Theta F(t; k_m + \ell) + b_\Theta F(t; k_m), \tag{2.4a}$$

$$\Phi_m(t) = a_\Phi F(t; k_m + \ell) + b_\Phi F(t; k_m), \tag{2.4b}$$

where $\{a_\Theta, b_\Theta\}$ and $\{a_\Phi, b_\Phi\}$ will be specified via conditions (2.2a) and (2.2b), respectively. Now, if $k_m \geq 5$, then all boundary conditions at $t = 0$ and the boundary conditions for $q = 0$ at $t = 1$ ($\Theta_m(1) = \Phi_m(1) = 0$) are obviously satisfied. Then we are left with a pair of conditions ($q = 1, 2, t = 1$) for each auxiliary polynomial, which leads to a 2×2 linear system. Taking $l \geq 1$, these linear systems can be readily solved, yielding:

$$a_\Theta = \frac{-(k_m - 2)(k_m - 3)}{\ell^2 + \ell(2k_m - 5)}, \quad b_\Theta = \frac{(k_m + \ell - 2)(k_m + \ell - 3)}{\ell^2 + \ell(2k_m - 5)}. \tag{2.5a}$$

$$a_\Phi = -b_\Phi = \frac{1}{\ell^2 + \ell(2k_m - 5)}. \tag{2.5b}$$

The below given asymptotic estimates quantify the behaviour of the so-constructed auxiliary polynomials for large values of $k_m (k_m \geq 5)$ with $\ell (\ell \geq 1)$ being kept fixed. These estimates can be readily derived with the aid of the defining formulae (2.4a) and (2.4b) and the asymptotic estimate:

$$t^k(1 - t^\ell) = o(1), \quad t \in [0, 1]. \tag{2.6}$$

Furthermore, they are divided in two groups: the so-called *interval estimates*, holding uniformly with respect to t in [0, 1] or in an arbitrary, but fixed, closed subinterval of [0, 1), denoted by $[0, 1)_c$, and the *boundary estimates*, holding for $t = 0, 1$.

Auxiliary polynomial: $\Theta_m(t)$

(i) **Interval estimates**:

$$\Theta_m^{(q)}(t) = o(k_m^{q-1}), \quad t \in [0, 1], \quad q = 0, 1, \ldots, 4, \tag{2.7a}$$

$$\Theta_m^{(q)}(t) = O(k_m^{q-1} t^{k_m}), \quad t \in [0, 1)_c. \tag{2.7b}$$

(ii) **Boundary estimates**:

$$\Theta_m^{(1)}(0) \approx k_m^{-2}, \quad \Theta_m^{(q)}(0) = 0, \quad q = 2, 3, 4, \tag{2.7c}$$

$$\Theta_m^{(1)}(1) \approx k_m^{-1}, \quad \Theta_m^{(2)}(1) = 1, \quad \Theta_m^{(3)}(1) \approx k_m, \quad \Theta_m^{(4)}(1) = 0, \tag{2.7d}$$

where $\approx$ is an asymptotic equivalence symbol that should be understood as: $f \approx g$ if and only if $f = O(g)$ and $g = O(f)$; see Eckhaus ([5], Def. 1.1.3).

Auxiliary polynomial: $\Phi_m(t)$

(i) **Interval estimates**:

$$\Phi_m^{(q)}(t) = o\left(k_m^{q-3}\right), \ t \in [0,1], \ q = 0,1,\ldots,4, \tag{2.8a}$$

$$\Phi_m^{(q)}(t) = O\left(k_m^{q-3} t^{k_m}\right), \ t \in [0,1)_c. \tag{2.8b}$$

(ii) **Boundary estimates**:

$$\Phi_m^{(1)}(0) \approx k_m^{-4}, \ \Phi_m^{(q)}(0) = 0, \ q = 2,3,4, \tag{2.8c}$$

$$\Phi_m^{(1)}(1) \approx k_m^{-3}, \ \Phi_m^{(2)}(1) = 0, \ \Phi_m^{(3)}(1) \approx k_m^{-1}, \ \Phi_m^{(4)}(1) = 1. \tag{2.8d}$$

Table 2.1 collects information regarding the sign of the boundary derivatives of $\Theta_m(t)$ and $\Phi_m(t)$. Furthermore, it can be shown that:

$$\Theta_m^{(q)}(t) > 0, \ q = 2,3, \ \Phi_m^{(2)}(t) < 0, \ t \in (0,1), \tag{2.9}$$

which, along with the contents of Table 2.1, will be of intensive use in the next sections.

Table 2.1. Signs of the boundary derivatives of the auxiliary polynomials $\Theta_m(t)$ and $\Phi_m(t)$

	$t=0$	$t=1$		$t=0$	$t=1$
$\Theta_m^{(1)}(t)$	<0	>0	$\Phi_m^{(1)}(t)$	>0	<0
$\Theta_m^{(2)}(t)$	$=0$	$=1$	$\Phi_m^{(2)}(t)$	$=0$	$=0$
$\Theta_m^{(3)}(t)$	$=0$	>0	$\Phi_m^{(3)}(t)$	$=0$	>0
$\Theta_m^{(4)}(t)$	$=0$	$=0$	$\Phi_m^{(4)}(t)$	$=0$	$=1$

2.2. Solving the Interpolation Problem

The family of splines $\mathbf{Q}(u)$, defined by (2.1) and (2.4a)–(2.5b), will be henceforth denoted by $\Gamma_4(\mathcal{K})$. In addition, and for the sake of notational conformity, the family of C^2 variable degree splines, introduced in Kaklis and Karavelas [11] will be denoted in this paper by $\Gamma_2(\mathcal{K})$. We now proceed to investigate the well-posedness of the following:

Interpolation Problem: Let $\mathcal{D}$, $\mathcal{U}$ and $\mathcal{K}$ be given. Find an element $\mathbf{Q}(u) \in \Gamma_4(\mathcal{K})$ that is C^4-continuous on $[u_1, u_N]$, interpolates $\mathcal{D}$ at the nodes $\mathcal{U}$ and satisfies one of the following two types of boundary conditions:

a) *Type-I boundary conditions*: $\mathbf{Q}_n^{(1)} = \mathbf{v}_n$, $\mathbf{Q}_n^{(3)} = \mathbf{0}$, $n = 1, N$,

b) *Periodic boundary conditions* ($\mathbf{I}_1 = \mathbf{I}_N$): $\mathbf{Q}^{(q)}(u_1) = \mathbf{Q}^{(q)}(u_N)$, $q = 1,2,3,4$.

As previously noted, for a spline $\mathbf{Q}(u)$ to be C^4-continuous, it suffices to ensure continuity of the first- and third-order parametric derivative of $\mathbf{Q}(u)$ at the internal nodes u_m, $m = 2, \ldots, N-1$, of $\mathcal{U}$. The first-order continuity conditions yield, along with the first two of the *type-I* boundary conditions ($\mathbf{Q}^{(1)}(u_n) = \mathbf{v}_n$, $n = 1, N$), the following set of linear equations for the even-order nodal derivatives $\mathbf{Q}_m^{(2q)}$, $q = 1, 2$, $m = 1, \ldots, N$ ($\mathbf{Q}_0^{(2q)} := \mathbf{0}$, $\mathbf{Q}_{N+1}^{(2q)} := \mathbf{0}$):

$$\left\{ \begin{array}{l} a_{m,m-1}\mathbf{Q}_{m-1}^{(2)} + a_{mm}\mathbf{Q}_m^{(2)} + a_{m,m+1}\mathbf{Q}_{m+1}^{(2)} + \\ b_{m,m-1}\mathbf{Q}_{m-1}^{(4)} + b_{mm}\mathbf{Q}_m^{(4)} + b_{m,m+1}\mathbf{Q}_{m+1}^{(4)} = \mathbf{R}_m, \ m = 1, \ldots, N, \end{array} \right\} \tag{2.10a}$$

with

$$\left\{ \begin{array}{rcl} a_{11} & = & h_1 \Theta_1^{(1)}(1), \\ a_{mm} & = & h_{m-1}\Theta_{m-1}^{(1)}(1) + h_m \Theta_m^{(1)}(1), \ m = 2, \ldots, N-1, \\ a_{NN} & = & h_{N-1}\Theta_{N-1}^{(1)}(1), \\ a_{m,m+1} & = & -h_m \Theta_m^{(1)}(0), \ m = 1, \ldots, N-1, \\ a_{m,m-1} & = & a_{m-1,m}, \ m = 2, \ldots, N, \end{array} \right\} \tag{2.10b}$$

$$\left\{ \begin{array}{rcl} b_{11} & = & h_1^3 \Phi_1^{(1)}(1), \\ b_{mm} & = & h_{m-1}^3\Phi_{m-1}^{(1)}(1) + h_m^3 \Phi_m^{(1)}(1), \ m = 2, \ldots, N-1, \\ b_{NN} & = & h_{N-1}^3\Phi_{N-1}^{(1)}(1), \\ b_{m,m+1} & = & -h_m^3 \Phi_m^{(1)}(0), \ m = 1, \ldots, N-1, \\ b_{m,m-1} & = & b_{m-1,m}, \ m = 2, \ldots, N, \end{array} \right\} \tag{2.10c}$$

and

$$\left\{ \begin{array}{rcl} \mathbf{R}_1 & = & \Delta\mathbf{I}_1 - \mathbf{v}_1, \\ \mathbf{R}_m & = & \Delta\mathbf{I}_m - \Delta\mathbf{I}_{m-1}, \ \Delta\mathbf{I}_m = \frac{\mathbf{I}_{m+1} - \mathbf{I}_m}{h_m}, \ m = 2, \ldots, N-1, \\ \mathbf{R}_N & = & \mathbf{v}_N - \Delta\mathbf{I}_{N-1}. \end{array} \right\} \tag{2.10d}$$

The following lemma summarizes the properties of the matrices $\mathbb{A} = \{a_{ij}\}$ and $\mathbb{B} = \{b_{ij}\}$, appearing in the linear system (2.10a).

Lemma 2.1. The matrix $\mathbb{A}(\mathbb{B})$ is $N \times N$ tridiagonal, symmetric and strictly diagonally dominant with positive (negative) elements.

Proof: As it is readily seen from (2.10b) and (2.10c), both $\mathbb{A}$ and $\mathbb{B}$ are tridiagonal and symmetric. Now, using (2.4a) in conjunction with (2.5a), one gets after some straightforward calculus, the inequality:

$$-\Theta_m^{(1)}(0) \leq \frac{4}{13}\Theta_m^{(1)}(1), \tag{2.11}$$

which, in view of (2.10b) and Table 2.1, implies that $\mathbb{A}$ is strictly diagonally dominant with positive elements. Working analogously with (2.4b) and (2.5b), we arrive at

$$\Phi_m^{(1)}(0) \leq -\frac{1}{2}\Phi_m^{(1)}(1), \tag{2.12}$$

implying, with the aid of (2.10c) and Table 2.1, that the elements of $\mathbb{B}$ are negative and $\mathbb{B}$ is strictly diagonally dominant too. The validity of the Lemma then follows readily. □

Next, we turn to impose continuity of the third-order parametric derivative of $\mathbf{Q}(u)$ at the internal nodes of $\mathscr{U}$. Combining these conditions with the last two of type-I boundary conditions ($\mathbf{Q}^{(3)}(u_n) = \mathbf{0},\ n = 1, N$), we are lead to the set of equations:

$$\mathbf{Q}_m^{(4)} = c_m \mathbf{Q}_m^{(2)},\ m = 1, \dots, N, \tag{2.13a}$$

where

$$\left\{ \begin{array}{ll} c_1 = -h_1^{-2} \dfrac{\Theta_1^{(3)}(1)}{\Phi_1^{(3)}(1)}, & \\ c_m = -\dfrac{h_{m-1}^{-1}\Theta_{m-1}^{(3)}(1) + h_m^{-1}\Theta_m^{(3)}(1)}{h_{m-1}\Phi_{m-1}^{(3)}(1) + h_m\Phi_m^{(3)}(1)}, & m = 2, \dots, N-1, \\ c_N = -h_{N-1}^{-2} \dfrac{\Theta_{N-1}^{(3)}(1)}{\Phi_{N-1}^{(3)}(1)}. & \end{array} \right\} \tag{2.13b}$$

Now, noting that $\Theta_m^{(3)}(1)$ and $\Phi_m^{(3)}(1)$ are both positive(see Table 2.1), formula (2.13b) yields readily.

Lemma 2.2. The diagonal elements of matrix $\mathbb{C} = diag\{c_m\}$ are negative.

Summarizing the hitherto obtained results, we can say that the interpolation problem in $\Gamma_4(\mathscr{K})$ leads to a pair of linear systems for $\mathbb{Q}^{(2q)} = (\mathbf{Q}_1^{(2q)}, \dots, \mathbf{Q}_N^{(2q)})^T$, $q = 1, 2$. This pair can be written in matrix form as below:

$$\mathbb{A}\mathbb{Q}^{(2)} + \mathbb{B}\mathbb{Q}^{(4)} = \mathbb{R} \tag{2.14a}$$

$$\mathbb{Q}^{(4)} = \mathbb{C}\mathbb{Q}^{(2)}, \tag{2.14b}$$

where the matrices $\mathbb{A}$, $\mathbb{B}$, $\mathbb{R} = (\mathbf{R}_1, \dots, \mathbf{R}_N)^T$ and $\mathbb{C}$ are defined by (2.10b)–(2.10d) and (2.13b), respectively. Substituting (2.14b) into (2.14a), we arrive at a single matrix equation for $\mathbb{Q}^{(2)}$, namely:

$$\mathbb{D}\mathbb{Q}^{(2)} = \mathbb{R},\ \mathbb{D} = \mathbb{A} + \mathbb{B}\mathbb{C}. \tag{2.15}$$

The well-posedness of the linear system (2.15) stems from

Lemma 2.3. The matrix $\mathbb{D} = \mathbb{A} + \mathbb{B}\mathbb{C}$ is tridiagonal with positive elements. Furthermore, $\mathbb{D}$ is strictly diagonally dominant columnwise.

Proof: The first part of the Lemma follows readily from Lemmata 2.1 and 2.2. Next, since $\mathbb{B}$ is symmetric and strictly diagonally dominant (Lemma 2.1), its right-hand side multiplication by the diagonal matrix $\mathbb{C}$ (Lemma 2.2) preserves diagonal dominance along columns only. On the other hand, $\mathbb{A}$ is symmetric and strictly diagonally dominant with positive elements; see again Lemma 2.1. Then, by virtue of the previous remarks we conclude that the second part of the Lemma holds true as well. □

We thus can state:

Theorem 2.1. Let $k_m \geq 5$, $m = 1, \ldots, N-1$, and $\ell \geq 1$. Then there exists a unique element $\mathbf{Q}(u)$ in $\Gamma_4(\mathcal{K})$ that is C^4-continuous on $[u_1, u_N]$, interpolates $\mathcal{D}$ at the nodes of $\mathcal{U}$ and satisfies *type-I* boundary conditions.[1]

It can easily be proved that, for $\ell = 2$ and $k_m = 3$, $m = 1, \ldots, N-1$, $\Gamma_4(\mathcal{K})$ recovers the standard C^4 quintic interpolation spline, the basic difference being that the second equation of the interpolation system (2.14b) is altered from $\mathbb{Q}^{(4)} = \mathbb{C}\mathbb{Q}^{(2)}$, where $\mathbb{C}$ is a diagonal matrix, to $\mathbb{A}\mathbb{Q}^{(4)} = \mathbb{C}_q\mathbb{Q}^{(2)}$, with $\mathbb{C}_q$ being now a tridiagonal matrix. Nevertheless, one cannot continuously attach $\Gamma_4(k_m = 3; \ell = 2)$ to the family of Theorem 2.1, for the construction process of the auxilliary functions $\Theta_m(t)$ and $\Phi_m(t)$, described in §2.1, fails for $k_m = 4$ independently of the value attributed to ℓ; more accurately the first of (2.2a) and (2.2b) cannot be fulfilled for $q = 2$.

2.3. The Bézier Control Net of an Element in $\Gamma_4(\mathcal{K})$

We conclude this section with a subsection devoted on the investigation of the structure of the Bézier control polygon of an element $\mathbf{Q}(u) \in \Gamma_4(\mathcal{K})$. To start with, the Bernstein-Bézier representation of the restriction of $\mathbf{Q}(u)$ on $[u_m, u_{m+1}]$ has as follows:

$$\mathbf{Q}(u) = \sum_{j=0}^{k_m+\ell} \mathbf{b}_j^{(m)} B_j^{k_m+\ell}(t), \;\; u \in [u_m, u_{m+1}], \tag{2.16}$$

where $\mathbf{b}_j^{(m)}$ are the Bézier control vertices and $B_j^{k_m+\ell}(t)$ are the Bernstein polynomials of degree $k_m + \ell$.

Substituting (2.4a) and (2.4b) into (2.1), $\mathbf{Q}(u)$ can be alternatively represented as:

$$\mathbf{Q}(u) = \mathbf{Q}_1(u) + \mathbf{Q}_2(u) - \mathbf{L}(u), \tag{2.17}$$

where:

$$\begin{aligned}\mathbf{Q}_1(u) = \mathbf{L}(u) &+ h_m^2\left[\mathbf{Q}_m^{(2)} a_\Theta + h_m^2 \mathbf{Q}_m^{(4)} a_\Phi\right] F(1-t; k_m+\ell) \\ &+ h_m^2\left[\mathbf{Q}_{m+1}^{(2)} a_\Theta + h_m^2 \mathbf{Q}_{m+1}^{(4)} a_\Phi\right] F(t; k_m+\ell),\end{aligned} \tag{2.18a}$$

$$\begin{aligned}\mathbf{Q}_2(u) = \mathbf{L}(u) &+ h_m^2\left[\mathbf{Q}_m^{(2)} b_\Theta + h_m^2 \mathbf{Q}_m^{(4)} b_\Phi\right] F(1-t; k_m) \\ &+ h_m^2\left[\mathbf{Q}_{m+1}^{(2)} b_\Theta + h_m^2 \mathbf{Q}_{m+1}^{(4)} b_\Phi\right] F(t; k_m).\end{aligned} \tag{2.18b}$$

The polynomial segments $\mathbf{Q}_i(u)$, $i = 1, 2$, admit of the same representation with the polynomial segments of an element in family $\Gamma_2(\mathcal{K})$, whose Bézier control

[1] A directly analogous result can be drawn for periodic boundary conditions. The only difference with the case of type-I boundary conditions is that $\mathbb{A}$ and $\mathbb{B}$ are, now, $(N-1) \times (N-1)$ cyclic matrices.

polygon is well studied in Sapidis and Kaklis [15]. More specifically, the following result holds true (ibid. Th. 3.1):

Proposition 2.1. The Bézier control vertices $\{\mathbf{b}_j^{(m)},\ j = 0, \ldots, k_m\}$ of the restriction of $[u_m, u_{m+1}]$ of an element $\mathbf{Q}(u) \in \Gamma_2(\mathscr{K})$ are given by:

$$
\begin{aligned}
\mathbf{b}_0^{(m)} &= \mathbf{I}_m, \\
\mathbf{b}_j^{(m)} &= \mathbf{I}_m + j\frac{h_m}{k_m}\mathbf{Q}_m^{(1)} + (j-1)\frac{h_m^2}{k_m(k_m-1)}\mathbf{Q}_m^{(2)},\ j = 1, \ldots, k_m - 1, \\
\mathbf{b}_{k_m}^{(m)} &= \mathbf{I}_{m+1}.
\end{aligned}
\tag{2.19}
$$

Differentiating twice (2.18a) and (2.18b) and setting $u = u_m$, one can readily determine the nodal values $\mathbf{Q}_{i,m}^{(1)}$ and $\mathbf{Q}_{i,m}^{(2)}$, $i = 1, 2$. Substituting these expressions into (2.19), we derive the Bézier control points $\{\mathbf{b}_{1j}^{(m)},\ j = 0, \ldots, k_m + \ell\}$ and $\{\mathbf{b}_{2j}^{(m)},\ j = 0, \ldots, k_m\}$ of $\mathbf{Q}_1$ and $\mathbf{Q}_2$, respectively. Then (2.17) becomes:

$$
\mathbf{Q}(u) = \sum_{j=0}^{k_m+\ell} \mathbf{b}_{1j}^{(m)} B_j^{k_m+\ell}(t) + \sum_{j=0}^{k_m} \mathbf{b}_{2j}^{(m)} B_j^{k_m}(t) - \sum_{j=0}^{1} \mathbf{b}_{3j}^{(m)} B_j^1(t), \tag{2.20}
$$

with $\mathbf{b}_{30}^{(m)} = \mathbf{I}_m$ and $\mathbf{b}_{31}^{(m)} = \mathbf{I}_{m+1}$. Now, if we raise the degree of $\mathbf{Q}_2(u)$ ℓ times and the degree of $\mathbf{L}(u)$ $(k_m + \ell - 1)$ times, we get:

$$
\mathbf{Q}(u) = \sum_{j=0}^{k_m+\ell} \left(b_{1j}^{(m)} + b_{2j}^{\prime(m)} - b_{3j}^{\prime(m)} \right) B_j^{k_m+\ell}(t), \tag{2.21}
$$

where $\mathbf{b}_{2j}^{\prime(m)}$ and $\mathbf{b}_{3j}^{\prime(m)}$ are the control points of the degree-elevated curves $\mathbf{Q}_2(u)$ and $\mathbf{L}(u)$, respectively. Comparing (2.16) with (2.21), we get the control points of $\mathbf{Q}(u)$:

$$
\mathbf{b}_j^{(m)} = \mathbf{b}_{1j}^{(m)} + \mathbf{b}_{2j}^{\prime(m)} - \mathbf{b}_{3j}^{\prime(m)},\ j = 0, 1, \ldots, k_m + \ell. \tag{2.22}
$$

Let us now turn back to the second of formulae (2.19) and observe that the intermediate control points of $\mathbf{Q}_i(u)$, $i = 1, 2$, are collinear, i.e., the shape of the Bézier control polygon of the splines in $\Gamma_2(\mathscr{K})$ can be fully described, in each segment, by only four control points, just like that of C^2 cubic splines. It is now easy to prove that an analogous result holds true for the splines in $\Gamma_4(\mathscr{K})$, in reference with the standard C^4 quintic spline. During the afore-mentioned degree elevations, the collinearity property of the intermediate control points of $\mathbf{Q}_2$ is partially destroyed, due to the corner cutting procedure. More accurately, ℓ degree elevations generate $\ell(\ell+1)/2$ corner cuttings over the left-hand side portion of $\mathbf{Q}_2(u)$, thus inserting $(\ell + 1)$ new control points that are not, in general, collinear. The very same procedure produces another $(\ell + 1)$ non-collinear control points over the right-hand side portion of $\mathbf{Q}_2(u)$. Nevertheless, the remaining control points, indexed from $\mathbf{b}_{2,\ell+1}^{\prime(m)}$ up to $\mathbf{b}_{2,k_m-1}^{\prime(m)}$, are still collinear. On the other hand, Proposition 2.1 implies that the control points of $\mathbf{Q}_1(u) \in \Gamma_2(\mathcal{K}\{k_m+\ell,\ m=1,\dots,N-1\})$, indexed from $\ell+1$ to k_m-1 are collinear. This is also true for the linear interpolant $\mathbf{L}(u)$, all control points of which are collinear. Then noting that (2.22) preserves collinearity, as affine transformation of the involved control points, we arrive at:

Theorem 2.2. The control points of the Bézier curve $\mathbf{Q}(u) \in \Gamma_4(\mathcal{K})$, $u \in [u_m, u_{m+1}]$, indexed from $\ell+1$ up to k_m-1, are collinear.

For $\ell = 1$, the above theorem establishes a readily seen similarity between the control polygon of $\mathbf{Q}(u) \in \Gamma_4(\mathcal{K})$ and that of the standard C^4 quintic spline.

Corollary 2.1. If $\ell = 1$, the shape of the control polygon of $\mathbf{Q}(u) \in \Gamma_4(\mathcal{K})$, $u \in [u_m, u_{m+1}]$ can be fully determined only by six control points, namely $\mathbf{b}_0^{(m)}$, $\mathbf{b}_1^{(m)}$, $\mathbf{b}_2^{(m)}$, $\mathbf{b}_{k_m-1}^{(m)}$, $\mathbf{b}_{k_m}^{(m)}$ and $\mathbf{b}_{k_m+1}^{(m)}$, the remaining lying equidistantly on the line segment joining $\mathbf{b}_2^{(m)}$ and $\mathbf{b}_{k_m-1}^{(m)}$.

3. Asymptotic Behaviour for Large Segment Degrees

In this section we investigate the asymptotic behaviour of a C^4-continuous interpolant $\mathbf{Q}(u) \in \Gamma_4(\mathcal{K})$ as the segment degrees increase according to one of the following three ways:

(i) **local increase**: $k_m \to \infty$, while $k_n, \ell < M, n \neq m$,

(ii) **semi-local increase**: $k_{m-1}, k_m, k_{m+1} \to \infty$, while $k_n, \ell < M, n \neq m-1, m, m+1$, or

(iii) **global increase**: $k_m \to \infty$, $m = 1, \dots, N-1$, with $\ell < M$,

M being a fixed positive constant.

In view of the asymptotic properties of the auxiliary functions $\Theta_m(t)$ and $\Phi_m(t)$, summarized in estimates (2.7a)–(2.8d), the sought-for asymptotic behaviour of $\mathbf{Q}(u)$ can be derived once the analogous asymptotic behaviour of the even-order nodal derivatives $\mathbf{Q}_m^{(2)}$ and $\mathbf{Q}_m^{(4)}$ is available. For this purpose, we first recall formulae (2.14b) and (2.15):

$$\mathbb{D}\mathbb{Q}^{(2)} = \mathbb{R},\ \mathbb{Q}^{(4)} = \mathbb{C}\mathbb{Q}^{(2)}, \tag{3.1}$$

where the non-zero elements of $\mathbb{C} = diag\{c_m\}$ and $\mathbb{D} = \{d_{mn}\}$ are negative and positive, respectively; see Lemmata 2.2 and 2.3. Next, we scale $\mathbb{Q}^{(2)}$ by $\mathbb{F} = diag\{d_{mm}\}$ and rewrite the first of the matrix equations (3.1) in the following form:

$$\mathbb{E}\hat{\mathbb{Q}}^{(2)} = \mathbb{R},\ \mathbb{E} = \mathbb{D}\mathbb{F}^{-1},\ \hat{\mathbb{Q}}^{(2)} = \mathbb{F}\mathbb{Q}^{(2)}. \tag{3.2}$$

$\mathbb{E}$ can now be decomposed as

$$\mathbb{E} = \hat{\mathbb{E}} + \mathbb{I}, \tag{3.3}$$

where $\hat{\mathbb{E}}$ is a tridiagonal matrix, whose non-zero elements on the m-th column have as follows:

$$\text{row } m-1 : \frac{d_{m-1,m}}{d_{mm}}, \qquad \text{row } m+1 : \frac{d_{m+1,m}}{d_{mm}}.$$

Since $\mathbb{D}$ is strictly diagonally dominant columnwise, it is easy to see that

$$\frac{d_{m-1,m}}{d_{mm}} + \frac{d_{m+1,m}}{d_{mm}} < 1,$$

leading to

$$\|\hat{\mathbb{E}}\|_1 = \max_{1 \le m \le N} \left\{ \frac{d_{m-1,m}}{d_{mm}} + \frac{d_{m+1,m}}{d_{mm}} \right\} < 1. \tag{3.4}$$

We shall prove, however, a stronger result, namely $\|\hat{\mathbb{E}}\|_1 < \delta < 1$, where δ is a constant not depending on the degree distribution $\mathscr{K}$. Since $d_{mn} = a_{mn} + b_{mn}c_n$ (see Eq. (2.15)), formulae (2.10b), (2.10c) and (2.13b) give:

$$\frac{d_{m-1,m}}{d_{mm}} + \frac{d_{m+1,m}}{d_{mm}} = -\frac{\left[h_{m-1}\Theta_{m-1}^{(1)}(0) + h_{m-1}^3 c_m \Phi_{m-1}^{(1)}(0)\right] + \left[h_m \Theta_1^{(m)}(0) + h_m^3 c_m \Phi_1^{(m)}(0)\right]}{\left[h_{m-1}\Theta_{m-1}^{(1)}(1) + h_{m-1}^3 c_m \Phi_{m-1}^{(1)}(1)\right] + \left[h_m \Theta_m^{(1)}(1) + h_m^3 c_m \Phi_m^{(1)}(1)\right]}$$

Then, if we weaken inequality (2.11) by taking $-\Theta_m^{(1)}(0) < \frac{1}{2}\Theta_m^{(1)}(1)$ and use the inequality (2.12), it can easily be shown that:

$$\frac{d_{m-1,m}}{d_{mm}} + \frac{d_{m+1,m}}{d_{mm}} < \frac{1}{2} = \delta,$$

which, in conjunction with (3.4), gives:

$$\|\hat{\mathbb{E}}\|_1 < \frac{1}{2}. \tag{3.5}$$

Next, by virtue of (3.2), (3.3) and (3.5), Neumann's lemma yields the following inequality:

$$\|\hat{\mathbb{Q}}^{(2)}\|_1 \le \frac{\|\mathbb{R}\|_1}{1-\delta} = 2\|\mathbb{R}\|_1. \tag{3.6}$$

Since $\mathbb{R}$ depends only on the data set $\mathscr{D}$ and the parametrization $\mathscr{U}$, one may write:

$$\|\mathbb{R}\|_1 \le \frac{\mu}{2}. \tag{3.7}$$

where μ is a positive constant depending on $\mathscr{D}$ and $\mathscr{U}$, exclusively. Combining now (3.6) with (3.7), the former can be strengthened as

$$\|\hat{\mathbb{Q}}^{(2)}\|_1 \leq \mu. \tag{3.8}$$

Then, recalling the defining relation $\mathbb{Q}^{(2q)} = (\mathbf{Q}_1^{(2q)}, \mathbf{Q}_2^{(2q)}, \ldots, \mathbf{Q}_N^{(2q)})^T$, $q = 1, 2$, and appealing to (3.8), (3.2) and (3.1), we are led to the following basic result:

Lemma 3.1. There exists a positive constant $\mu_1 (= 3\mu)$, depending exclusively on the data set $\mathscr{D}$ and the parametrization $\mathscr{U}$, such that:

$$\|\mathbf{Q}_m^{(2)}\|_1 \leq \frac{1}{d_{mm}} \mu_1, \quad \|\mathbf{Q}_m^{(4)}\|_1 \leq \frac{|c_m|}{d_{mm}} \mu_1, \quad m = 1, \ldots, N. \tag{3.9}$$

As it is readily implied by the above lemma, the asymptotic behaviour of the second-order nodal derivatives is dominated by that of the factor d_{mm}. In this connection, we state and prove the following key result:

Lemma 3.2. (i) If the degrees increase locally, then $d_{mm}^{-1} = O(k_m^{-1})$. (ii) If $k_n \to \infty$ with $n = m-1, m$, then $d_{mm}^{-1} = O(k_n)$.

Proof: Using (2.10b) and (2.10c), the inverse of $d_{mm} = a_{mm} + c_m b_{mm}$ is given by the formula:

$$\frac{1}{d_{mm}} = \frac{1}{\left[h_{m-1}\Theta_{m-1}^{(1)}(1) + h_{m-1}^3 c_m \Phi_{m-1}^{(1)}(1)\right] + \left[h_m \Theta_m^{(1)}(1) + h_m^3 c_m \Phi_m^{(1)}(1)\right]}. \tag{3.10}$$

On the basis of the sign information contained in Table 2.1, and the defining relations (2.13b) of c_m, it is readily seen that all four terms in the denominator of (3.10) are non-negative. We proceed by distinguishing between the following cases:

(i) If the degrees increase locally, then k_m tends to infinity, while the remaining degrees are kept fixed. Appealing to the defining relations (2.13b) of c_m and the second of the sharp asymptotic estimates (2.7d) and (2.8d) of $\Theta_m^{(3)}(1)$ and $\Phi_m^{(3)}(1)$, respectively, we arrive at:

$$c_m \approx k_m. \tag{3.11}$$

Using the above asymptotic equivalence relation and recalling the non-negativity of the denominator terms in (3.10), we then get:

$$\frac{1}{d_{mm}} \leq \frac{1}{h_{m-1}^3 c_m \Phi_{m-1}^{(1)}(1)} = O(k_m^{-1}),$$

which proves part (i) of the Lemma.

(ii) Suppose now that both k_{m-1} and k_m tend to infinity. Appealing once again the non-negativity argument, we can write:

$$\frac{1}{d_{mm}} \leq \frac{1}{h_{m-1}\Theta^{(1)}_{m-1}(1) + h_m\Theta^{(1)}_m(1)}. \tag{3.12}$$

Combining the above inequality with the first of the sharp asymptotic estimates (2.7d), part (ii) of the Lemma follows readily. □

The quantification of the asymptotic behaviour of the fourth-order nodal derivatives $\mathbf{Q}^{(4)}_m$ pre-assumes the asymptotic evaluation of the ratio c_m/d_{mm}; see the second of inequalities (3.9). Recalling that a_{mm} is positive, while b_{mm} and c_m are both negative, we can write, with the aid of (2.10c),

$$\frac{|c_m|}{d_{mm}} = \left|\frac{b_{mm}c_m}{a_{mm} + b_{mm}c_m} \cdot \frac{1}{b_{mm}}\right| \leq \frac{1}{h^3_{m-1}|\Phi^{(1)}_{m-1}(1)| + h^3_m|\Phi^{(1)}_m(1)|}. \tag{3.13}$$

Then, exploiting the first of the sharp asymptotic estimates (2.8d), we are lead to:

Lemma 3.3. (i) If the degrees increase locally, then $c_m/d_{mm} = O(1)$. (ii) If $k_n \to \infty$ with $n = m-1,\, m$, then $c_m/d_{mm} = O(k_n^3)$.

We are now ready to asymptotically evaluate both $\|\mathbf{Q}^{(2)}_m\|$ and $\|\mathbf{Q}^{(4)}_m\|$ as the degrees increase locally, semi-locally or globally. Exploiting Lemmata 3.1, 3.2 and 3.3, we arrive, after some simple asymptotic algebra, at the following result:

Theorem 3.1. (i) If the degrees increase locally, then:

$$\|\mathbf{Q}^{(2)}_n\| = O(1/k_m), \quad n = m, m+1, \tag{3.14a}$$

$$\|\mathbf{Q}^{(2)}_n\| = O(1), \quad n \neq m, m+1, \tag{3.14b}$$

$$\|\mathbf{Q}^{(4)}_n\| = O(1), \quad n = 1, \ldots, N-1. \tag{3.14c}$$

(ii) If the degrees increase semi-locally, then:

$$\|\mathbf{Q}^{(2)}_n\| = O(k_\upsilon), \quad n = m, m+1, \;\; \upsilon = n-1, n, \tag{3.15a}$$

$$\|\mathbf{Q}^{(2)}_{m-1}\| = O(1/k_{m-1}), \quad \|\mathbf{Q}^{(2)}_{m+2}\| = O(1/k_m), \tag{3.15b}$$

$$\|\mathbf{Q}^{(2)}_n\| = O(1), \quad n \neq m-1, \ldots, m+2, \tag{3.15c}$$

$$\|\mathbf{Q}^{(4)}_n\| = O(k_\upsilon^3), \quad n = m, m+1, \;\; \upsilon = n-1, n, \tag{3.15d}$$

$$\|\mathbf{Q}_n^{(4)}\| = O(1), \quad n \neq m, m+1. \tag{3.15e}$$

(iii) If the degrees increase globally, then:

$$\|\mathbf{Q}_n^{(2)}\| = O(k_v), \quad \|\mathbf{Q}_n^{(4)}\| = O(k_v^3), \qquad n = 2, \ldots, N-1, \ v = n-1, n. \tag{3.16a}$$

Combining the above theorem, with the internal asymptotic estimates (2.7a)–(2.7b) and (2.8a)–(2.8b) of the auxiliary polynomials $\Theta_m(t)$ and $\Phi_m(t)$, respectively, we can materilize the main task of this section, namely to investigate the asymptotic behaviour of a C^4 element $\mathbf{Q}(u) \in \Gamma_4(\mathcal{K})$ as the degrees increase locally, semi-locally and globally. More accurately, the deviation between $\mathbf{Q}(u)$ and the associated linear interpolant $\mathbf{L}(u)$ (see equ. (2.1)) behaves as follows:

Theorem 3.2. (i) If the degrees increase locally, then:

$$\|\mathbf{Q}^{(q)}(u) - \mathbf{L}^{(q)}(u)\| = o(k_m^{q-2}), \quad q = 0, 1, \ldots, 4, \ u \in [u_m, u_{m+1}], \tag{3.17a}$$

$$\|\mathbf{Q}^{(q)}(u) - \mathbf{L}^{(q)}(u)\| = o(1), \quad q = 3, 4, \ u \in (u_m, u_{m+1})_c, \tag{3.17b}$$

where $(u_m, u_{m+1})_c$ denotes an arbitrary, but fixed, closed subinterval of $[u_m, u_{m+1}]$.

(ii) If the degrees increase semi-locally of globally, then:

$$\|\mathbf{Q}^{(q)}(u) - \mathbf{L}^{(q)}(u)\| = o(k_m^q), \quad q = 0, 1, \ldots, 4, \ u \in [u_n, u_{n+1}], \tag{3.18a}$$

$$\|\mathbf{Q}^{(q)}(u) - \mathbf{L}^{(q)}(u)\| = o(1), \quad q = 1, \ldots 4, \ u \in (u_n, u_{n+1})_c \tag{3.18b}$$

with $n = m-1, m, m+1$ for semi-local or $n = 1, \ldots, N-1$ for global increase, respectively.

4. The Adopted Notion of Shape Preservation

The last theorem of the preceding section demonstrates the tension-parameter role of the degree distribution $\mathcal{K}$ by showing that, as the degrees increase, the spline interpolant $\mathbf{Q}(u) \in \Gamma_4(\mathcal{K})$ will tend, in an accurately defined sense, to the linear interpolant $\mathbf{L}(u)$ that connects the interpolation point-set $\mathcal{D}$.

This nice asymptotic property does not, however, ensure that, for sufficiently large degrees, the spline curve will not exhibit unwanted oscillations or twists. In other words, the usefulness of the spline family $\Gamma_4(\mathcal{K})$ depends on the ability of its elements $\mathbf{Q}(u)$, not only to lie closely to the corresponding polygonal interpolant, but to conform with the convexity and torsion information contained in it, as well. In this connection, we adopt the following notion of shape-preserving interpolation, grounded on the notion introduced by Asaturian et al. [1]; see also Kaklis and Karavelas [11] and Goodman and Ong [7].

Definition 4.1. Let $\mathbf{Q}(u), u \in [u_1, u_N]$, be a C^3 continuous parametric curve that interpolates the point-set $\mathscr{D}$ over the nodes of the parametrization $\mathscr{U}$ and obeys type I or periodic boundary conditions. $\mathbf{Q}(u)$ will be called *shape preserving* provided that:

(i) (*Convexity criterion*) Let

$$\mathbf{P}_m = (\mathbf{I}_m - \mathbf{I}_{m-1}) \times (\mathbf{I}_{m+1} - \mathbf{I}_m), \quad m = 1, \ldots, N, \tag{4.1}$$

be the so-called *convexity indicator* of the polygonal interpolant at $\mathbf{I}_m$.

(i.1) If $\mathbf{P}_m \cdot \mathbf{P}_{m+1} > 0$, then

$$\mathbf{P}_n \cdot \mathbf{w}(u) > 0, \quad n = m, m+1, \ u \in [u_m, u_{m+1}], \tag{4.2}$$

where

$$\mathbf{w}(u) = \mathbf{Q}^{(1)}(u) \times \mathbf{Q}^{(2)}(u) \tag{4.3}$$

is the vector appearing in the numerator of the rational expression for the curvature $\kappa(u)$ of $\mathbf{Q}(u)$ and sharing the same direction with the binormal of $\mathbf{Q}(u)$.

(i.2) If $\mathbf{P}_m \cdot \mathbf{P}_{m+1} < 0$, then $\mathbf{P}_n \cdot \mathbf{w}(u_n) > 0, n = m, m+1$,and $\mathbf{P}_n \cdot \mathbf{w}(u)$ changes sign only once in $[u_m, u_{m+1}]$.

(ii) (*Torsion criterion*) Let

$$\Delta_m = [\mathbf{I}_m - \mathbf{I}_{m-1}, \mathbf{I}_{m+1} - \mathbf{I}_m, \mathbf{I}_{m+2} - \mathbf{I}_{m+1}], \quad m = 1, \ldots, N-1, \tag{4.4}$$

be the so-called *torsion indicator* for the segment of the polygonal interpolant that connects $\mathbf{I}_m$ with $\mathbf{I}_{m+1}$.

(ii.1) If $\Delta_m \neq 0$, then

$$\Delta_m \sigma(u) > 0, \quad u \in (u_m, U_{m+1}), \tag{4.5}$$

where

$$\sigma(u) = det[\mathbf{Q}^{(1)}(u), \mathbf{Q}^{(2)}(u), \mathbf{Q}^{(3)}(u)] \tag{4.6}$$

is the numerator of the rational expression of the torsion $\tau(u)$ of $\mathbf{Q}(u)$ that determines its sign.

(ii.2) If $\Delta_m \Delta_{m+1} > 0$, then $\Delta_m \sigma(u_m) > 0$.

According to the type of the imposed boundary conditions, the above definition obeys, respectively, the following conventions for type I (periodic) boundary conditions: $\mathbf{I}_0 = \mathbf{I}_1 - h_0 \mathbf{v}_1 (\mathbf{I}_0 = \mathbf{I}_{N-1})$ and $\mathbf{I}_{N+1} = \mathbf{I}_N + h_N \mathbf{v}_N (\mathbf{I}_{N+1} = \mathbf{I}_1)$ with $h_0, h_N > 0$.

5. Asymptotic Validity of the Convexity Criterion

The aim of this section is to establish that, for appropriately large values of the segment degrees, the new spline family $\Gamma_4(\mathcal{K})$ is able to conform with the convexity criterion of Definition 4.1. Before we proceed on with the consideration of the asymptotic behaviour of the curvature numerator $\mathbf{w}(u)$, we rewrite $\mathbf{Q}(u)$ in a more compact form, exploiting the collinearity of the second and the fourth-order nodal derivatives (see Eq. (2.13a)):

$$\mathbf{Q}(u) = \mathbf{L}(u) + h_m^2 \mathbf{Q}_m^{(2)} H_m(1-t; c_m) + h_m^2 \mathbf{Q}_{m+1}^{(2)} H_m(t; c_{m+1}), \tag{5.1}$$

where

$$H_m(t; c) = \Theta_m(t) + c h_m^2 \Phi_m(t), \tag{5.2}$$

c being one of the coefficients $c_n, n = 1, \ldots, N$. Using (5.1) we get, after some straight-forward calculus, the following expression for the curvature numerator:

$$\mathbf{w}(u) = \mathbf{w}_m H_m^{(2)}(1-t; c_m) + \mathbf{w}_{m+1} H_m^{(2)}(t; c_{m+1}) + h_m \mathbf{Q}_{m+1}^{(2)} \times \mathbf{Q}_m^{(2)} \psi(t), \tag{5.3}$$

where $\mathbf{w}_m$ denotes the nodal value of $\mathbf{w}(u)$ at $u = u_m$ and

$$\begin{aligned} \psi(t) = {} & H_m^{(2)}(t; c_{m+1}) \left[H_m^{(1)}(1-t; c_m) - H_m^{(1)}(0; c_m) \right] \\ & + H_m^{(2)}(1-t; c_m) \left[H_m^{(1)}(t; c_{m+1}) - H_m^{(1)}(0; c_{m+1}) \right]. \end{aligned} \tag{5.4}$$

The ensuing lemma is a basic result that marks out the asymptotic behaviour of $\mathbf{w}_m$ as the neighbouring segment degrees tend to infinity.

Lemma 5.1. The following limiting relation holds true:

$$\lim_{k_{m-1}, k_m \to \infty} d_{mm} \mathbf{w}_m = \frac{\mathbf{P}_m}{h_{m-1} h_m}.$$

Proof: After differentiating twice (5.1) and setting $u = u_m$, the quantity $d_{mm}\mathbf{w}_m$ can be written as:

$$d_{mm}\mathbf{w}_m = \Delta \mathbf{I}_m \times (d_{mm}\mathbf{Q}_m^{(2)}) + h_m H_m^{(1)}(0; c_{m+1}) \mathbf{Q}_{m+1}^{(2)} \times (d_{mm}\mathbf{Q}_m^{(2)}). \tag{5.5}$$

Let us first deal with the product $d_{mm}\mathbf{Q}_m^{(2)}$, appearing in both terms of the right-hand side of (5.5). Appealing to the *m-th* row of the linear system (2.15) and recalling Lemma 3.1, we get the following inequalities:

$$\begin{aligned}\|d_{mm}\mathbf{Q}_m^{(2)} - \mathbf{R}_m\| &\leq d_{m,m-1}\|\mathbf{Q}_{m-1}^{(2)}\| + d_{m,m+1}\|\mathbf{Q}_{m+1}^{(2)}\| \\ &\leq \left(\frac{d_{m,m-1}}{d_{m-1,m-1}} + \frac{d_{m,m+1}}{d_{m+1,m+1}}\right)\mu_1.\end{aligned} \tag{5.6}$$

Rewriting the first of the above fractions as

$$\begin{aligned}\frac{d_{m,m-1}}{d_{m-1,m-1}} &= \frac{-h_{m-1}\Theta_{m-1}^{(1)}(0) - h_{m-1}^3 c_{m-1}\Phi_{m-1}^{(1)}(0)}{\left[h_{m-2}\Theta_{m-2}^{(1)}(1) + h_{m-2}^3 c_{m-1}\Phi_{m-1}^{(1)}(1)\right] + \left[h_{m-1}\Theta_{m-1}^{(1)}(1) + h_{m-1}^3 c_{m-1}\Phi_{m-1}^{(1)}(1)\right]} \\ &= -\frac{\Theta_{m-1}^{(1)}(0) + h_{m-1}^2 c_{m-1}\Phi_{m-1}^{(1)}(0)}{\Theta_{m-1}^{(1)}(1) + h_{m-1}^2 c_{m-1}\Phi_{m-1}^{(1)}(1)}\left(\frac{1}{1 + \frac{h_{m-2}\Theta_{m-2}^{(1)}(1) + h_{m-2}^3 c_{m-1}\Phi_{m-1}^{(1)}(1)}{h_{m-1}\Theta_{m-1}^{(1)}(1) + h_{m-1}^3 c_{m-1}\Phi_{m-1}^{(1)}(1)}}\right),\end{aligned}$$

and using the sign information of Table 2.1, we obtain the following bound for

$$\frac{d_{m,m-1}}{d_{m-1,m-1}} \leq \left| -\frac{\Theta_{m-1}^{(1)}(0) + h_{m-1}^2 c_{m-1}\Phi_{m-1}^{(1)}(0)}{\Theta_{m-1}^{(1)}(1) + h_{m-1}^2 c_{m-1}\Phi_{m-1}^{(1)}(1)} \right|. \tag{5.7}$$

Relying, once again, on Table 2.1, inequality (5.7) can be strengthened further as:

$$\frac{d_{m,m-1}}{d_{m-1,m-1}} \leq \left|\frac{\Theta_m^{(1)}(0)}{\Theta_m^{(1)}(1)}\right| + \left|\frac{\Theta_m^{(1)}(0)}{\Theta_m^{(1)}(1)}\right|.$$

Assuming now that k_{m-1} tends to infinity and recalling the sharp asymptotic estimates (2.7c), (2.7d) and (2.8c), (2.8d), the above inequality leads to the following limiting relation:

$$\lim_{k_{m-1}\to\infty} \frac{d_{m,m-1}}{d_{m-1,m-1}} = 0. \tag{5.8}$$

Working analogously for the second fraction in the right-hand side of (5.6), we obtain:

$$\lim_{k_m\to\infty} \frac{d_{m,m+1}}{d_{m+1,m+1}} = 0. \tag{5.9}$$

Then, combining (5.6) with (5.8) and (5.9), we are lead to:

$$\lim_{k_{m-1},k_m\to\infty} d_{mm}\mathbf{Q}_m^{(2)} = \mathbf{R}_m = \Delta\mathbf{I}_m - \Delta\mathbf{I}_{m-1}. \tag{5.10}$$

We are now ready to precisely quantify the asymptotic behaviour of the two terms in the right-hand side of (5.5) as both k_{m-1} and k_m tend to infinity. For the first term, (5.10) along with the defining relation (4.1) of the convexity indicator $\mathbf{P}_m$, gives:

$$\lim_{k_{m-1},k_m\to\infty} \Delta\mathbf{I}_m \times (d_{mm}\mathbf{Q}_m^{(2)}) = \frac{\mathbf{P}_m}{h_{m-1}h_m}.$$

For the second term, noting that $H_m^{(1)}(0; c_{m+1}) = d_{m,m+1}$ (see Eqs. (5.2) and the fourth of (2.10b) and (2.10c)) we can write:

$$\begin{aligned}
&\lim_{k_{m-1},k_m\to\infty} h_m H_m^{(1)}(0; c_{m+1})\mathbf{Q}_{m+1}^{(2)} \times (d_{mm}\mathbf{Q}_m^{(2)}) \\
&= \lim_{k_{m-1},k_m\to\infty} -\frac{d_{m,m+1}}{d_{m+1,m+1}}(d_{m+1,m+1}\mathbf{Q}_{m+1}^{(2)}) \times (d_{mm}\mathbf{Q}_m^{(2)}) = 0,
\end{aligned}$$

as a result of (5.9) and Lemma 3.1. This completes the proof of the Lemma. □

On the basis of the previous lemma we can state:

Corollary 5.1. (i) If $\mathbf{P}_m \cdot \mathbf{P}_{m+1} > 0$, then:

$$\begin{aligned}
\mathbf{P}_n \cdot \mathbf{w}_m > 0, \quad n = m, m+1, \quad \text{as } k_{m-1}, k_m \to \infty, \\
\mathbf{P}_n \cdot \mathbf{w}_{m+1} > 0, \quad n = m, m+1, \quad \text{as } k_m, k_{m+1} \to \infty.
\end{aligned} \tag{5.11}$$

(ii) If $\mathbf{P}_m \cdot \mathbf{P}_{m+1} < 0$, then:

$$\begin{aligned}
(-1)^{n-m}\mathbf{P}_n \cdot \mathbf{w}_m > 0, \quad n = m, m+1, \quad \text{as } k_{m-1}, k_m \to \infty, \\
(-1)^{m+1-n}\mathbf{P}_n \cdot \mathbf{w}_{m+1} > 0, \quad n = m, m+1, \quad \text{as } k_m, k_{m+1} \to \infty.
\end{aligned} \tag{5.12}$$

In other words, Corollary 5.1 guarantees that, if the pairs k_{m-1}, k_m and k_m, k_{m+1} are sufficiently large, then the convexity criterion will be satisfied at least at the nodes $u = u_m$ and $u = u_{m+1}$. The rest of the section is devoted to showing that, as k_{m-1}, k_m, k_{m+1} tend appropriately to infinity, the convexity criterion is satisfied in the open parametric interval (u_m, u_{m+1}) as well. To start with, inequality (4.2) of Part (i.1) of the convexity criterion can equivalently be written as follows:

$$\begin{aligned}
&\mathbf{P}_n \cdot \mathbf{w}_m H_m^{(2)}(1-t; c_m) + \mathbf{P}_n \cdot \mathbf{w}_{m+1} H_m^{(2)}(t; c_{m+1}) \\
&> h_m \mathbf{P}_n \cdot \left(\mathbf{Q}_m^{(2)} \times \mathbf{Q}_{m+1}^{(2)}\right)\psi(t), \quad t \in (0,1), \quad n = m, m+1,
\end{aligned} \tag{5.13}$$

as it is readily inferred from the representation (5.3) of $\mathbf{w}(u)$. Let us henceforth assume that k_{m-1}, k_m, and k_{m+1} are sufficiently large so that inequalities (5.11) of Corollary 5.1 are satisfied. Then, setting

$$s = \min(\mathbf{P}_n \cdot \mathbf{w}_m, \mathbf{P}_n \cdot \mathbf{w}_{m+1}), \tag{5.14}$$

the ensuing inequality is a *sufficient condition* for (5.13) to hold true:

$$s\left(H_m^{(2)}(1-t;c_m)+H_m^{(2)}(t;c_{m+1})\right) > h_m\mathbf{P}_n\cdot\left(\mathbf{Q}_m^{(2)}\times\mathbf{Q}_{m+1}^{(2)}\right)\psi(t). \tag{5.15}$$

Taking into account the inequalities (2.9), (5.2) implies that:

$$H_m^{(2)}(t;c) = \Theta_m^{(2)}(t) + h_m^2 c\Phi_m^{(2)}(t) > 0 \tag{5.16}$$

for $t \in (0,1)$, since c is negative. Thus

$$H_m^{(2)}(1-t;c_m)+H_m^{(2)}(t;c_{m+1}) > 0, \tag{5.17}$$

which enables us to rewrite (5.15) as follows:

$$s > h_m\mathbf{P}_n\cdot\left(\mathbf{Q}_m^{(2)}\times\mathbf{Q}_{m+1}^{(2)}\right)\xi(t), \quad \xi(t) = \frac{\psi(t)}{H_m^{(2)}(1-t;c_m)+H_m^{(2)}(t;c_{m+1})}. \tag{5.18}$$

By virtue of (5.4), the rational function $\xi(t)$ can be written as:

$$\begin{aligned}\xi(t) &= \omega(t)\left[H_m^{(1)}(1-t;c_m)-H_m^{(1)}(0;c_m)\right]\\ &\quad+(1-\omega(t))\left[H_m^{(1)}(t;c_{m+1})-H_m^{(1)}(0;c_{m+1})\right],\end{aligned} \tag{5.19}$$

with

$$\omega(t) = \frac{H_m^{(2)}(t;c_{m+1})}{H_m^{(2)}(1-t;c_m)+H_m^{(2)}(t;c_{m+1})}. \tag{5.20}$$

Since $\xi(0)=\xi(1)=0$, as a result of (5.18) and the fact that $\psi(0)=\psi(1)=0$, Rolle's Theorem readily implies that $\xi'(t)$ has at least one root, say t_0, on (0,1). Next, we turn to investigate the uniqueness question of the root t_0. For this purpose, we differentiate (5.19) and after some straightforward calculus we arrive at the following expression for the derivative:

$$\xi'(t) = \omega'(t)\rho(t), \tag{5.21}$$

where

$$\rho(t) = \left(H_m^{(1)}(1-t;c_m)-H_m^{(1)}(0;c_m)\right) - \left(H_m^{(1)}(t;c_{m+1})-H_m^{(1)}(0;c_{m+1})\right) \tag{5.22}$$

and

$$\omega'(t) = \frac{det(\Omega(t))}{\left(H_m^{(2)}(1-t;c_m) + H_m^{(2)}(t;c_{m+1})\right)^2},$$

$\Omega(t)$ being a t-depending 2×2 matrix defined as:

$$\Omega(t) = \begin{bmatrix} H_m^{(2)}(1-t;c_m) & -H_m^{(3)}(1-t;c_m) \\ H_m^{(2)}(t;c_{m+1}) & H_m^{(3)}(t;c_{m+1}) \end{bmatrix}. \tag{5.23}$$

Appealing to Lemma A.1, proved in the Appendix, we have

$$\omega'(t) > 0, \quad t \in (0,1),$$

which, in view of (5.21), implies that $\xi'(t)$ and $\rho(t)$ share the same roots. Now, since

$$\rho'(t) = -H_m^{(2)}(1-t;c_m) - H_m^{(2)}(t;c_{m+1}) < 0 \tag{5.24}$$

(see Eq. (5.17)), t_0 is unique. Thus, $\xi'(t)$ has a unique root on (0,1), where $\xi(t)$ achieves its global maximum, for

$$\xi''(t_0) = \omega''(t_0)\rho(t_0) + \omega'(t_0)\rho'(t_0) = \omega'(t_0)\rho'(t_0) < 0.$$

In the sequel, we shall investigate the asymptotic behaviour of $\xi(t_0)$. To start with, since t_0 is a zero of $\rho(t)$, (5.22) gives:

$$H_m^{(1)}(1-t;c_m) - H_m^{(1)}(0;c_m) = H_m^{(1)}(t;c_{m+1}) - H_m^{(1)}(0;c_{m+1}), \tag{5.25}$$

in view of which, (5.19) degenerates for $t = t_0$ as follows:

$$\xi(t_0) = H_m^{(1)}(1-t_0;c_m) - H_m^{(1)}(0;c_m). \tag{5.26}$$

Appealing to (5.2) and (2.4a)–(2.4b), the right-hand side of (5.26) takes the form

$$\xi(t_0) = \left(\frac{a_\Theta + c_m h_m^2 a_\Phi}{k_m + \ell - 1}(1-t_0)^\ell + \frac{b_\Theta + c_m h_m^2 b_\Phi}{k_m - 1}\right)(1-t_0)^{k_m - 1}. \tag{5.27}$$

Let us now derive an asymptotic estimate for the coefficient c_m, appearing in the right-hand side of (5.27).

Lemma 5.2. If k_{m-1} and k_m tend to infinity with $k_{m-1} \approx k_m$, then $c_m = O(k_m^2)$.

Proof: Since

$$\Theta_m^{(3)}(1) = (k_m - 2)(k_m + \ell - 2)\Phi_m^{(3)}(1),$$

the defining relation (2.13b) of c_m can be written as:

$$c_m = -\frac{h_{m-1}^{-1}(k_{m-1}-2)(k_{m-1}+\ell-2)\Phi_{m-1}^{(3)}(1) + h_m^{-1}(k_m-2)(k_m+\ell-2)\Phi_m^{(3)}(1)}{h_{m-1}^{-1}\Phi_{m-1}^{(3)}(1) + h_m^{-1}\Phi_m^{(3)}(1)}. \tag{5.28}$$

Given that $\Phi_{m-1}^{(3)}(1)$ and $\Phi_m^{(3)}(1)$ are positive, applying the triangle inequality on the right-hand side of (5.28), we are led to

$$|c_m| \leq h_{m-1}^{-2}(k_{m-1}-2)(k_{m-1}+\ell-2) + h_m^{-1}(k_m-2)(k_m+\ell-2),$$

which, by virtue of the hypothesis $k_{m-1} \approx k_m$, ensures the validity of the Lemma. □

Combining the previous lemma with the asymptotic estimates (see Eqs. (2.5)):

$$a_\Theta \approx b_\Theta \approx k_m, \quad a_\Phi \approx b_\Phi \approx k_m^{-1},$$

we arrive at

$$\left(\frac{a_\Theta + c_m h_m^2 a_\Phi}{k_m+\ell-1}(1-t_0)^\ell + \frac{b_\Theta + c_m h_m^2 b_\Phi}{k_m-1}\right) = O(1).$$

If t_0 stays away from 0 and 1, the above estimate would imply that $\xi(t_0)$ tends to zero with exponential rate, as $k_{m-1}, k_m \to \infty$ with $k_{m-1} \approx k_m$. In view of this remark and in order to focus on the asymptotic behaviour of the root t_0, we rewrite (5.25) with the aid of (2.5a) and (2.5b) as below:

$$\left(\frac{1-t_0}{t_0}\right)^{k_m-1} = r(t_0), \tag{5.29}$$

where

$$r(t_0) = \frac{(k_m-1)\left[-(k_m-2)(k_m-3)+c_{m+1}h_m^2\right]t_0^\ell + (k_m+\ell-1)\left[(k_m+\ell-2)(k_m+\ell-3)-c_{m+1}h_m^2\right]}{(k_m-1)[-(k_m-2)(k_m-3)+c_m h_m^2](l-t_0)^\ell + (k_m+\ell-1)[(k_m+\ell-2)(k_m+\ell-3)-c_m h_m^2]}.$$

Noting that the numerator (denominator) of the rational function $r(t_0)$ is positive and strictly increasing (decreasing) in (0,1), we get the bilateral bound:

$$r(1) \leq \left(\frac{1-t_0}{t_0}\right)^{k_m-1} \leq r(0). \tag{5.30}$$

Setting $t_0 = 0$ in the defining relation of $r(t_0)$, we get the following expression for

$$r(0) = (k_m + \ell - 1)\frac{(k_m + \ell - 2)(k_m + \ell - 3) - c_{m+1}h_m^2}{p(k_m) - \ell c_m h_m^2}, \tag{5.31}$$

where

$$p(k_m) = (k_m + \ell - 1)(k_m + \ell - 2)(k_m + \ell - 3) - (k_m - 1)(k_m - 2)(k_m - 3).$$

Since c_m is negative, (5.31) enables us to bound $r(0)$ from above as below:

$$r(0) < (k_m + \ell - 1)r_0(k_m, k_{m+1}),$$

$$r_0(k_m, k_{m+1}) = \frac{(k_m + \ell - 2)(k_m + \ell - 3) - c_{m+1}h_m^2}{p(k_m)}. \tag{5.32}$$

Now, due to the fact that the right-hand side of the above inequality depends on k_{m+1} as well, it is necessary to strengthen the adopted increase pattern by assuming that, along with k_{m-1} and k_m, k_{m+1} increases as well with $k_{m-1} \approx k_m \approx k_{m+1}$. Combining this hypothesis with Lemma 5.2 and the readily seen facts:

$$p(k_m) > 0, \quad p(k_m) \approx k_m^2,$$

it is straightforward to show that

$$r_0(k_m, k_{m+1}) > 0, \quad r_0(k_m, k_{m+1}) = O(1). \tag{5.33}$$

Working similarly for the other boundary value, $r(1)$, we get

$$r(1) > \frac{1}{k_m + \ell - 1} r_1(k_{m-1}, k_m),$$

$$r_1(k_{m-1}, k_m) = \frac{p(k_m)}{(k_m + \ell - 2)(k_m + \ell - 3) - c_m h_m^2}, \tag{5.34}$$

where

$$r_1(k_{m-1}, k_m) > 0, \quad r_1(k_{m-1}, k_m) = O(1). \tag{5.35}$$

In view of (5.32) and (5.34), (5.30) can be weakened as below:

$$\frac{1}{k_m + \ell - 1} r_1(k_{m-1}, k_m) < \left(\frac{1 - t_0}{t_0}\right)^{k_{m-1}} < (k_m + \ell - 1)r_0(k_m, k_{m+1}).$$

Taking now into account the asymptotic estimates appearing in (5.33) and (5.35), it is straightforward to conclude that

$$\lim_{\substack{k_{m-1},k_m,k_{m+1}\to\infty \\ k_{m-1}\approx k_m\approx k_{m+1}}} \frac{1-t_0}{t_0} = 1,$$

i.e., the root t_0 of $\xi(t) = 0$ tends to $\frac{1}{2}$ as k_{m-1}, k_m and k_{m+1} increase so that $k_{m-1} \approx k_m \approx k_{m+1}$. Grounded on this outcome, and recalling (5.27) and Lemma 5.2, we can state the following:

Lemma 5.3. Let $k_{m-1}, k_m, k_{m+1} \to \infty$ with $k_{m-1} \approx k_m \approx k_{m+1}$. Then $\xi(t_0) = O(2^{-k_m})$.

Let us now return to inequality (5.18), which is a sufficient condition for Part (i.1) of the convexity criterion to hold true. Multiplying both sides of (5.18) with the positive factor $d_{mm}d_{m+1,m+1}$, the latter can be written as:

$$d_{mm}d_{m+1,m+1}s > h_m\mathbf{P}_n \cdot \left(d_{mm}\mathbf{Q}_m^{(2)} \times d_{m+1,m+1}\mathbf{Q}_{m+1}^{(2)}\right)\xi(t). \tag{5.36}$$

Combining Lemma 5.1 with the defining relation (5.14) of s, we obtain

$$\begin{aligned} &\lim_{\substack{k_{m-1},k_m,k_{m+1}\to\infty \\ k_{m-1}\approx k_m\approx k_{m+1}}} d_{mm}d_{m+1,m+1}s \\ &\quad = \lim_{\substack{k_{m-1}k_m,k_{m+1}\to\infty \\ k_{m-1}\approx k_m\approx k_{m+1}}} \min\left(\frac{\mathbf{P}_n\mathbf{P}_m}{h_{m-1}h_m}d_{m+1,m+1}, \frac{\mathbf{P}_n\mathbf{P}_{m+1}}{h_m h_{m+1}}d_{mm}\right). \end{aligned}$$

Then appealing to Lemma 3.2(ii), we readily see that there exists a positive constant C_s such that the left-hand side of inequality (5.36) is in the limit, bounded from below as:

$$\lim_{\substack{k_{m-1},k_m,k_{m+1}\to\infty \\ k_{m-1}\approx k_m\approx k_{m+1}}} d_{mm}d_{m+1,m+1}s > \frac{C_s}{k_m}. \tag{5.37}$$

Regarding now the asymptotic behaviour of the right-hand side of (5.36), limiting relation (5.10) and Lemma 5.3 imply:

$$h_m\mathbf{P}_n \cdot \left(d_{mm}\mathbf{Q}_m^{(2)} \times d_{m+1,m+1}\mathbf{Q}_{m+1}^{(2)}\right)\xi(t) = O(2^{-k_m}). \tag{5.38}$$

Obviously, (5.37) and (5.38) secure that, if k_{m-1}, k_m and k_{m+1} increase in conformity with Lemma 5.3, the sought for inequality (5.15) will be eventually satisfied in (0, 1), equivalently Part (i.1) of the convexity criterion will be eventually fulfilled in (u_m, u_{m+1}). Combining this result with Part (i) of the Corollary 5.1 we can state:

Theorem 5.1. Let $\mathbf{P}_m \cdot \mathbf{P}_{m+1} > 0$. If $k_{m-1}, k_m, k_{m+1} \to \infty$ so that $k_{m-1} \approx k_m \approx k_{m+1}$, then Part (i.1) of the convexity criterion of Definition 4.1 will be eventually fulfilled.

We conclude this section by investigating the proper increase pattern that ensures the fulfillment of the second part, Part (i.2), of the convexity criterion of Definition 4.1. One should recall at this point that, due to Corollary 5.1, Part (i.2) is indeed fulfilled at the nodes $u = u_m$ and $u = u_{m+1}$; see relative comments just after Corollary 5.1. To proceed, we introduce the function:

$$\lambda(u) = \frac{\mathbf{P}_n \cdot \mathbf{w}(u)}{\psi(t)} = \mathbf{P}_n \cdot \mathbf{w}_m \frac{H_m^{(2)}(1-t;c_m)}{\psi(t)} + \mathbf{P}_n \cdot \mathbf{w}_{m+1} \frac{H_m^{(2)}(t;c_{m+1})}{\psi(t)} + h_m \mathbf{P}_n \cdot \left(\mathbf{Q}_{m+1}^{(2)} \times \mathbf{Q}_m^{(2)}\right), \tag{5.39}$$

where $\psi(t)$ is positive in (u_m, u_{m+1}), as it is readily seen from its defining relation (5.4), the positivity of $H_m^{(2)}(t;c)$ in (0, 1) (see inequality (5.16)) and the fact that:

$$H_m^{(1)}(t;c) - H_m^{(1)}(0;c) = \int_0^t H_m^{(2)}(\tau;c)d\tau \geq 0. \tag{5.40}$$

Differentiating the right-hand side of (5.39) and performing some straight-forward calculus, we end up with the following expression for

$$\lambda'(u) = \frac{det(\Omega(t))}{\psi^2(t)}[-\mathbf{P}_n \cdot \mathbf{w}_m(H_m^{(1)}(1-t;c_m) - H_m^{(1)}(0;c_m)) + \mathbf{P}_n \cdot \mathbf{w}_{m+1}(H_m^{(1)}(t;c_{m+1}) - H_m^{(1)}(0;c_{m+1}))],$$

where $\Omega(t)$ is the matrix already defined in (5.23). Then, combining the positivity of $det(\Omega(t))$ (see Lemma A.1 in the Appendix) with inequality (5.40), we conclude that $\lambda'(u)$ is of constant sign in (u_m, u_{m+1}) if and only if the quantities $-\mathbf{P}_n \cdot \mathbf{w}_m$ and $\mathbf{P}_n \cdot \mathbf{w}_{m+1}$ share the same sign. Corollary 5.1 (ii) guarantees that this condition will be satisfied for sufficiently large degrees k_{m-1}, k_m, k_{m+1}, securing the monotonicity of $\lambda(u)$ in (u_m, u_{m+1}). On the other hand, we can prove the following limiting relations:

$$\lim_{t\to 1} \frac{H_m^{(2)}(t;c)}{\psi(t)} = \infty, \quad \lim_{t\to 0}\left|\frac{H_m^{(2)}(t;c)}{\psi(t)}\right| < \infty$$

which, in conjunction with (5.39), lead to

$$\lim_{u\to u_m^+} \lambda(u) = sign(\mathbf{P}_n \cdot \mathbf{w}_m)\infty, \qquad \lim_{u\to u_{m+1}^-} \lambda(u) = sign(\mathbf{P}_n \cdot \mathbf{w}_{m+1})\infty. \tag{5.41}$$

Recalling once more Corollary 5.1 (ii), we can say that the above limiting relations imply that, if k_{m-1}, k_m, k_{m+1} are sufficiently large, the unbounded limits in (5.41) will be of opposite sign and, thus, by virtue of the monotonicity of $\lambda(u)$, the latter will exhibit only one root in $[u_m, u_{m+1}]$. Since $\psi(t)$ is non-negative on $[u_m, u_{m+1}]$, the previous outcome holds true for $\mathbf{P}_n \cdot \mathbf{w}(u)$ as well. Accordingly, we can state:

Theorem 5.2. Let $\mathbf{P}_m \cdot \mathbf{P}_{m+1} < 0$. If $k_{m-1}, k_m, k_{m+1} \to \infty$, then Part (i.2) of the convexity criterion of Definition 4.1 will be eventually satisfied.

6. Asymptotic Validity of the Torsion Criterion

Appealing to the alternative representation (5.1) of an element $\mathbf{Q}(u) \in \Gamma_4(\mathscr{K})$, the numerator $\sigma(u)$ (see Eq. (4.6)) of the torsion of $\mathbf{Q}(u)$, can be expressed as

$$\sigma(u) = h_m^{-1} det(\Omega(u)) det(\mathbb{T}_1), \tag{6.1}$$

where

$$\mathbb{T}_1 = \left[\frac{\mathbf{I}_{m+1} - \mathbf{I}_m}{h_m} \quad \mathbf{Q}_m^{(2)} \quad \mathbf{Q}_{m+1}^{(2)} \right]. \tag{6.2}$$

Since $det\ (\Omega(t))$ is positive for $t \in (0, 1)$ (see Lemma A.1 in the Appendix), while it vanishes for $t = 0, 1$, (6.1) implies the following:

(i) Part (ii.1) of the torsion criterion of Definition 4.1 will be satisfied, provided that the following discrete condition is fulfilled:

$$det(\mathbb{T}_1)\Delta_m > 0. \tag{6.3}$$

(ii) Part (ii.2) can never be fulfilled, the torsion numerator being always equal to zero at the nodes of $\mathscr{U}$.

Returning to Part (ii.1) of the torsion criterion, we scale $\mathbb{T}_1$ by the 3×3 diagonal matrix $\mathbb{F} = diag\{1, d_{mm}, d_{m+1,m+1}\}$, whose determinant is obviously positive. Then condition (6.3) is equivalent to

$$det(\mathbb{T}_1\mathbb{F})\Delta_m > 0. \tag{6.4}$$

Recalling now the limiting relation (5.10), we have that

$$\lim_{k_{m-1}, k_m, k_{m+1} \to \infty} det(\mathbb{T}_1\mathbb{F}) = \frac{\Delta_m}{h_{m-1} h_m h_{m+1}}, \tag{6.5}$$

which, in view of (6.4), leads to:

Theorem 6.1. Let $\Delta_m \neq 0$. If $k_{m-1}, k_m, k_{m+1} \to \infty$, then Part (ii.1) of the torsion criterion of Definition 4.1 will be eventually fulfilled.

7. An Algorithm for C^4 Shape-Preserving Interpolation

Exploiting the results derived in Sections 5 and 6, we proceed to formulate the ensuing algorithm that is able to yield, after a finite number of iterations,

C^4-continuous interpolants in $\Gamma_4(\mathscr{K})$, that conform with the convexity criterion and Part (ii.1) of the torsion criterion of Definition 4.1.

Step 0 Read the interpolation point-set $\mathscr{D}$, the parametrization $\mathscr{U}$ and the boundary conditions (approved types of boundary conditions: Type-I, Periodic; see §2.2).

Fix the parameter $\ell(\geq 1)$ and set initial values $k_m^{(0)}(\geq 5)$ for the variable part of the segment degrees $\mathscr{K} = \{k_m + \ell, m = 1, \ldots, N-1\}$.

Specify a constant $\mathbf{C} >> 1$.

Step 1 Compute the convexity indicators $\mathbf{P}_m, m = 1, \ldots, N$ (Eq. (4.1)) and the torsion indicators $\Delta_m, m = 1, \ldots, N-1$ (Eq. (4.4)).

Define the arrays: $\mathscr{I}_{tors} = \{m : \Delta_m \neq 0\}, \mathscr{I}_{conv} = \{m : \mathbf{P}_m \cdot \mathbf{P}_{m+1} > 0\}$, and $\mathscr{I}_{nonConv} = \{m : \mathbf{P}_m \cdot \mathbf{P}_{m+1} < 0\}$.

Define the linked lists: $\mathscr{J}_{nodalConv}, \mathscr{J}_{interConv}, \mathscr{J}_{tors}$.

Step 2 Compute the elements $d_{ij}, i, j = 1, \ldots N$, of matrix $\mathbb{D}$ and the vectors $\mathbf{R}_i$, $i = 1, \ldots, N$, of the right-hand side matrix $\mathbb{R}$ of the system (2.15) (Eqs. (2.10), (2.13)).

Solve the system (2.15).

Step 3a $\forall m \in \mathscr{I}_{conv}$:

If $(\mathbf{P}_{n_1} \cdot \mathbf{w}_{n_2} < 0, n_1, n_2 = m, m+1)$ then

append n_2 to $\mathscr{J}_{nodalConv}$

else

find the unique root t_0 of $\rho(t) = 0$ (Eq. (5.22)).

If (inequality (5.18) for $t = t_0$ is not fulfilled) then

append m to $\mathscr{J}_{interConv}$.

Step 3b $\forall m \in \mathscr{I}_{nonConv}$:

If $(\mathbf{P}_n \cdot \mathbf{w}_n < 0,\ n = m,\ m+1$ or

$\mathbf{P}_{n_1} \cdot \mathbf{w}_{n_2} > 0,\ n_1 \neq n_2,\ n_1, n_2 = m, m+1)$ then

append n or n_2 to $\mathscr{J}_{nodalConv}$, respectively.

Step 3c $\forall m \in \mathscr{I}_{tors}$:

If (inequality (6.4) is not fulfilled) then

append m to $\mathscr{J}_{tors}$.

Step 4 If $(\mathscr{I}_{nodalConv} = \emptyset \wedge \mathscr{I}_{interConv} = \emptyset \wedge \mathscr{I}_{tors} = \emptyset)$ then

STOP

else

$\forall m \in \mathscr{I}_{nodalConv}$ set $k_n^{(j+1)} = k_n^{(j)} + 1, \quad n = m-1, m,$

$\forall m \in \mathscr{I}_{interConv}$ set $k_n^{(j+1)} = k_n^{(j)} + 1, \quad n = m-1, m, m+1,$

$\forall m \in \mathscr{I}_{tors}$ set $k_n^{(j+1)} = k_n^{(j)} + 1, \quad n = m-1, m, m+1.$

Define a partition $\{\mathscr{P}_i\}_{i=1}^d$ of $\mathscr{I}_{interConv}$, with the property:

$\forall m_1, m_2 \in \mathscr{I}_{interConv}$ with $m_1 < m_2$,

$m_1, m_2 \in \mathscr{P}_i$, if and only if

$\forall \mu_\nu \in \mathscr{I}_{interConv}$ with $m_1 \le \mu_\nu \le m_2$, then $\mu_\nu \in \mathscr{P}_i$ and $\mu_{\nu+1} - \mu_\nu < 2$.

For $i = 1, \ldots, d$:

find the index $r \in \mathscr{P}_i$, such that: $k_r^{(j+1)} \ge k_m^{(j+1)}, \ m \in \mathscr{P}_i$.

$\forall m \in \mathscr{P}_i$:

If $\left(\dfrac{k_m^{(j+1)}}{k_r^{(j+1)}} < \dfrac{1}{C} \right)$ then

$$\text{set } k_m^{(j+1)} = \left[\frac{1}{C} k_r^{(j+1)} \right] + 1.$$

Empty the lists $\mathscr{I}_{nodalConv}$, $\mathscr{I}_{interConv}$, $\mathscr{I}_{tors}$ and $\{\mathscr{P}_i\}_{i=1}^d$.

Increase the iteration index j by one and go to Step 2.

If, after a number of iterations, $\mathscr{I}_{nodalConv} = \emptyset$, $\mathscr{I}_{interConv} = \emptyset$, and $\mathscr{I}_{tors} = \emptyset$, Lemma 5.1, Theorems 5.1, 5.2 and Theorem 6.1 guarantee that the corresponding outcome spline $\mathbf{Q}(u) \in \Gamma_4(\mathscr{K})$, provided by the above algorithm, will satisfy the convexity criterion and Part (ii.1) of the torsion criterion of Definition 4.1. The assertion that this will be indeed the case after a finite number of iterations, is grounded on the remark that the increase patterns, adopted in Step 4 of the algorithm, are in full conformity with those supposed in the lemma and theorems referred above.

8. Numerical Results

In this section we present and discuss the performance of the shape-preserving interpolation algorithm of §7 for a pair of benchmark data. More accurately, the C^4 outcome of the afore-mentioned algorithm is compared against the standard C^4 quintic interpolant as well as the C^2 shape-preserving interpolant provided by the algorithm presented in Kaklis and Karavelas [11].

The first example deals with the two-dimensional functional data taken from Späth [16]. The data set $\mathscr{D}$ consists in this case of ten points, whose x and y co-ordinates are given in Table 8.1. The imposed boundary conditions are of type I, with tangent vectors $\mathbf{v}_1 = (1,-1)^T, \mathbf{v}_N = (1,0.5)^T$, while the adopted parametrization is, naturally, the *x-parametrization*. The final degree distributions $\mathscr{K}_2$ and $\mathscr{K}_4$ of the shape preserving splines in $\Gamma_2(\mathscr{K}_2)$ and $\Gamma_4(\mathscr{K}_4; \ell = 1)$ are given in the third and the fourth column of Tabel 8.1, respectively. Coming now to the graphical output, Figure 8.1 depicts the interpolation points (rhombuses) along with the C^4 shape-preserving interpolating spline in $\Gamma_4(\mathscr{K}_4; \ell = 1)$ (solid line), the C^2 shape-preserving interpolating spline in $\Gamma_2(\mathscr{K}_2)$ (dashed line) as well as the C^4

Table 8.1. The x- and y-coordinates of the interpolation points along with the degree distributions $\mathscr{K}_2$ and $\mathscr{K}_4$ for the shape-preserving interpolation in $\Gamma_2(\mathscr{K}_2)$ and $\Gamma_4(\mathscr{K}_4; \ell = 1)$, respectively

x	y	$\mathscr{K}_2$	$\mathscr{K}_4$
0.0	10.0	5	7
1.0	8.0	7	10
1.5	5.0	10	10
2.5	4.0	10	10
4.0	3.5	10	10
4.5	3.4	10	5
5.5	6.0	7	13
6.0	7.1	7	13
8.0	8.0	7	13
10.0	8.5		

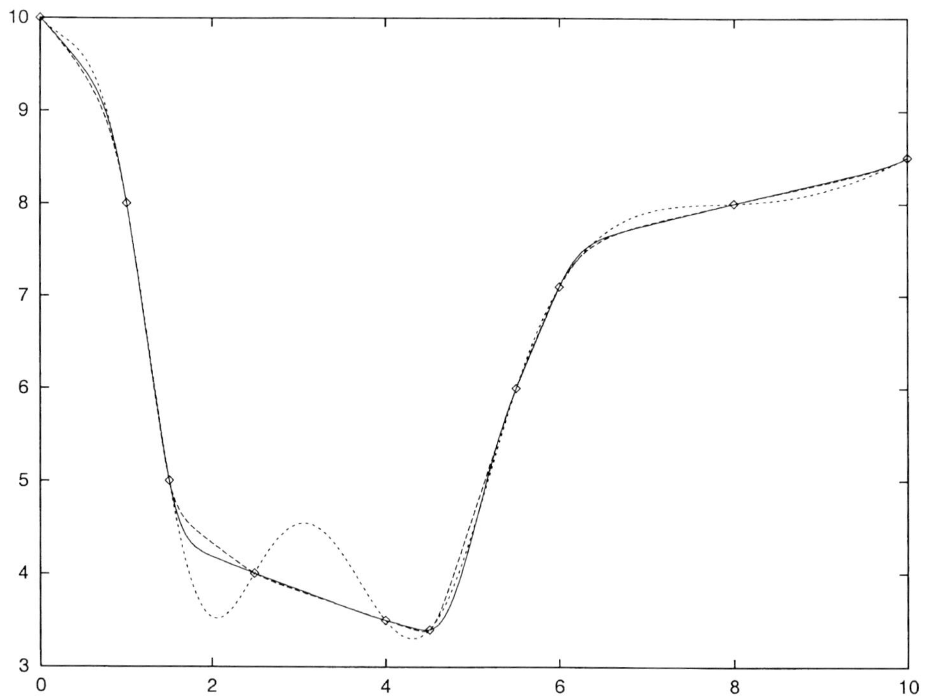

Figure 8.1. Interpolation points $\diamond$; the C^4 shape-preserving interpolant in $\Gamma_4(\mathscr{K}_4; \ell = 1)$ (–); the C^2 shape-preserving interpolant in $\Gamma_2(\mathscr{K}_2)$(- - -); the C^4 quintic interpolating spline ($\cdots$)

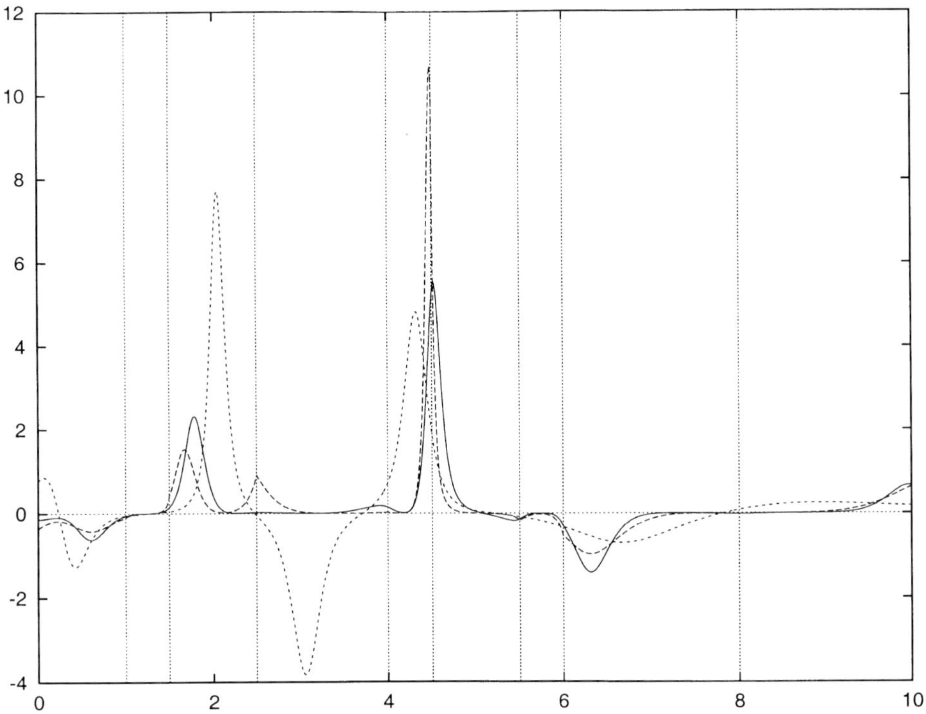

Figure 8.2. Curvature distribution of the curves in Fig. 8.1

quintic interpolating spline (dotted line). Figures 8.2 and 8.3 depict the curvature distribution and its arc-length derivatives, respectively, for each one of the curves in Fig. 8.1. The horizontal axis in Figs. 8.2 and 8.3 represents the u-parameter, while the dotted vertical lines indicate the nodes $u = u_m, m = 1, \ldots, N$, of the parametrization $\mathcal{U}$.

The second benchmark data set is a three dimensional point-set $\mathcal{D}$, consisting of eight ($N = 9$) points; see the rhombuses in Fig. 8.4. The x-, y- and z-coordinates of these points are given in the first three columns of Table 8.2. Due to the periodicity of the input data ($\mathbf{I}_1 = \mathbf{I}_9$), the imposed boundary conditions are periodic, while $\mathcal{U}$ is choosen to be the *chord-length parametrization*. The major part of the output of this numerical experiment is organized in direct analogy with that of the first one; see the last two columns of Table 8.2 and Figs. 8.4–8.6. Additionally, Figs. 8.7 and 8.8 provide the torsion distribution and its arc-length derivative, respectively, for each one of the curves in Fig. 8.4.

9. Remarks and Conclusions

On the basis of an extensive series of numerical experiments with the algorithm of §7, two of which have been presented in the previous section, we proceed to

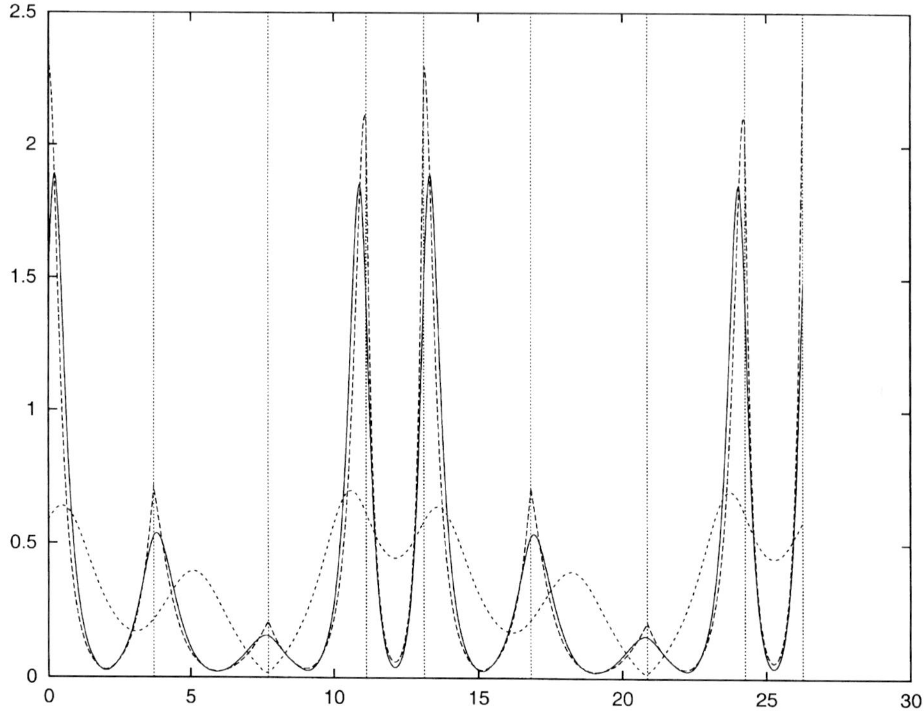

Figure 8.3. Arc-length derivative of the curvature distribution of the curves in Fig. 8.1

Table 8.2. The x-, y- and z-coordinates of the interpolation points along with the degree distributions $\mathscr{K}_2$ and $\mathscr{K}_4$ for shape-preserving interpolation in $\Gamma_2(\mathscr{K}_2)$ and $\Gamma_4(\mathscr{K}_4; \ell = 1)$, respectively

x	y	z	$\mathscr{K}_2$	$\mathscr{K}_4$
5.0	1.0	2.5	7	9
2.0	1.5	0.4	7	9
−2.0	1.5	1.0	6	8
−5.0	1.0	2.5	7	9
−5.0	−1.0	2.5	7	9
−2.0	−1.5	0.4	7	9
2.0	−1.5	1.0	6	8
5.0	−1.0	2.5	7	9
5.0	1.0	2.5		

provide a series of general remarks on the performance of the shape-preserving interpolation technique developed herein.

- For usual data sizes ($N \leq 100$) the run time of the algorithm on a Pentium processor, is very small (≤ 1 sec).
- In comparison with the algorithm for C^2 shape preserving interpolation, presented in Kaklis and Karavelas [11], the present algorithm exhibits the following features:

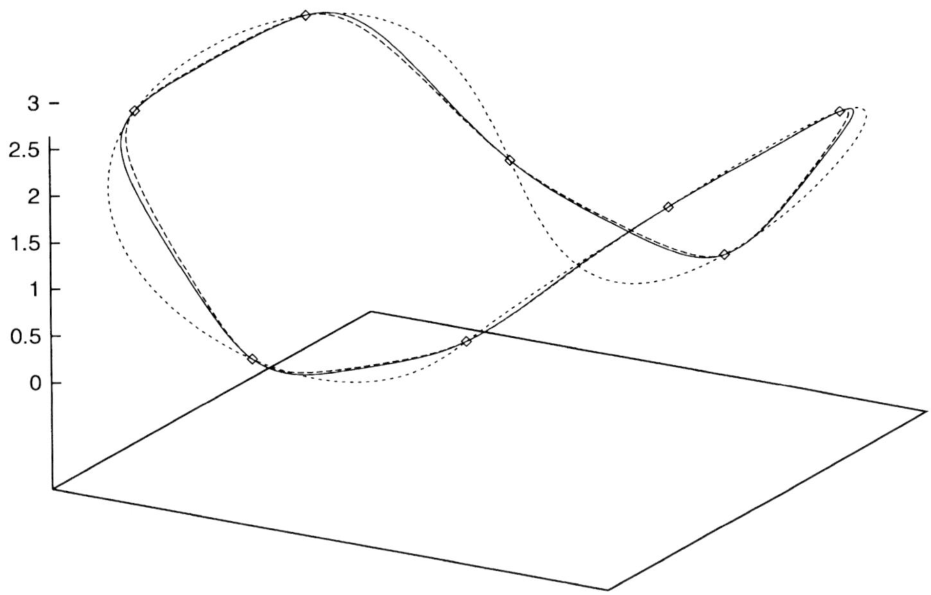

Figure 8.4. Interpolation points $\diamondsuit$; the C^4 shape-preserving interpolant in $\Gamma_4(\mathscr{K}_4; \ell = 1)$ (–); the C^2

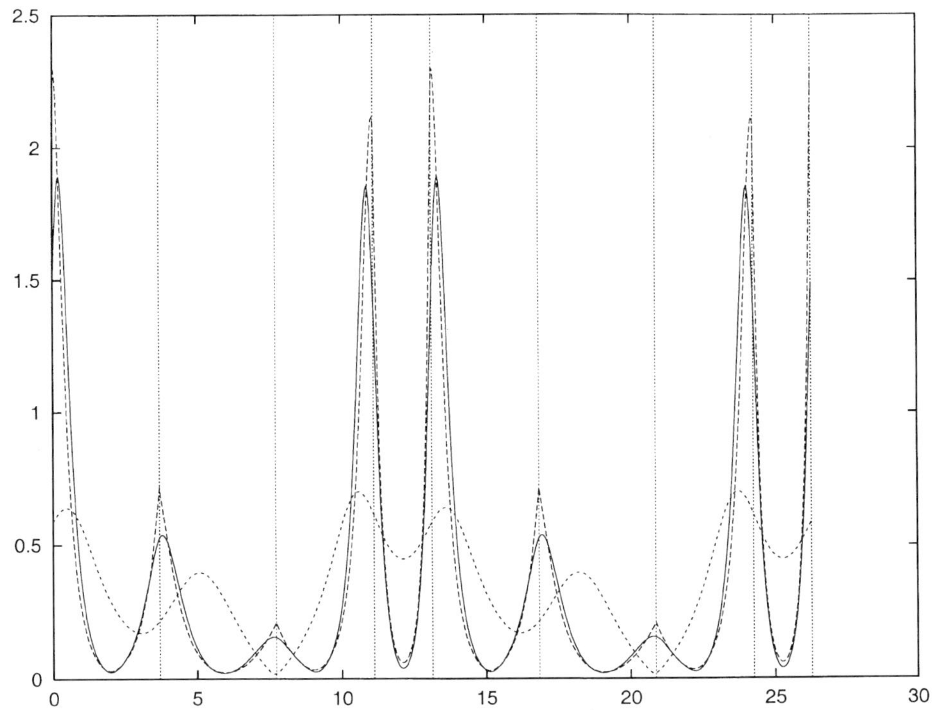

Figure 8.5. Curvature distribution of the curves in Fig. 8.4

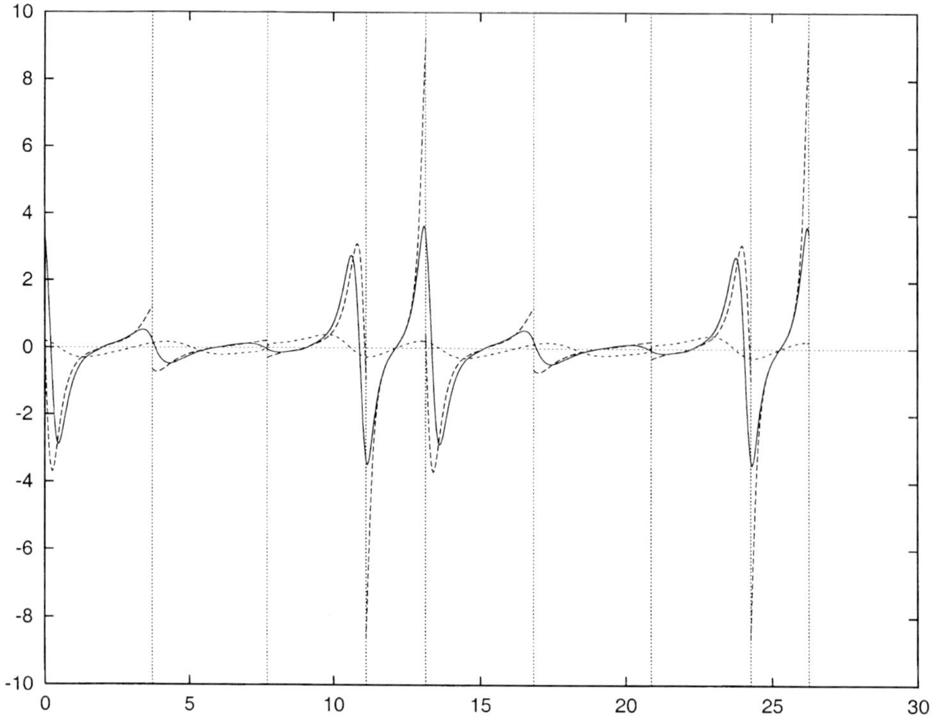

Figure 8.6. Arc-length derivative of the curvature distribution of the curves in Fig. 8.4

- It has the same memory requirements, of order $O(N)$.
- It requires, more or less, the same number of iterations for fulfilling the criteria of Definition 4.1 it can cope with.
- In addition, it requires the solution of a non-linear equation, namely $\rho(t) = 0,\ t \in (0, 1)$. Nevertheless, since this equation possesses a unique solution on (0,1), its root is determined efficiently and robustly via a Newton-Raphson method.
- It provides smoother curvature distributions, as implied by the given curvature plots (see Figs. 8.2, 8.5) and their arc-length derivative plots (see Figs. 8.3, 8.6). It is noticeable that, whenever a local maximum occurs in the vicinity of a knot, which is usually the case due to the linear like behaviour of these splines for large segment degrees, the present algorithm decreases it considerably.
- It leads to larger torsion values in the interior of the parametric intervals as a result of the fact that, due to the C^4 continuity, not only the torsion but its arc-length derivaive as well vanish at the parametric nodes.

In view of the above remarks it is legitimate to expect that, increasing further the continuity of the family of variable degree polynomial splines, by constructing the

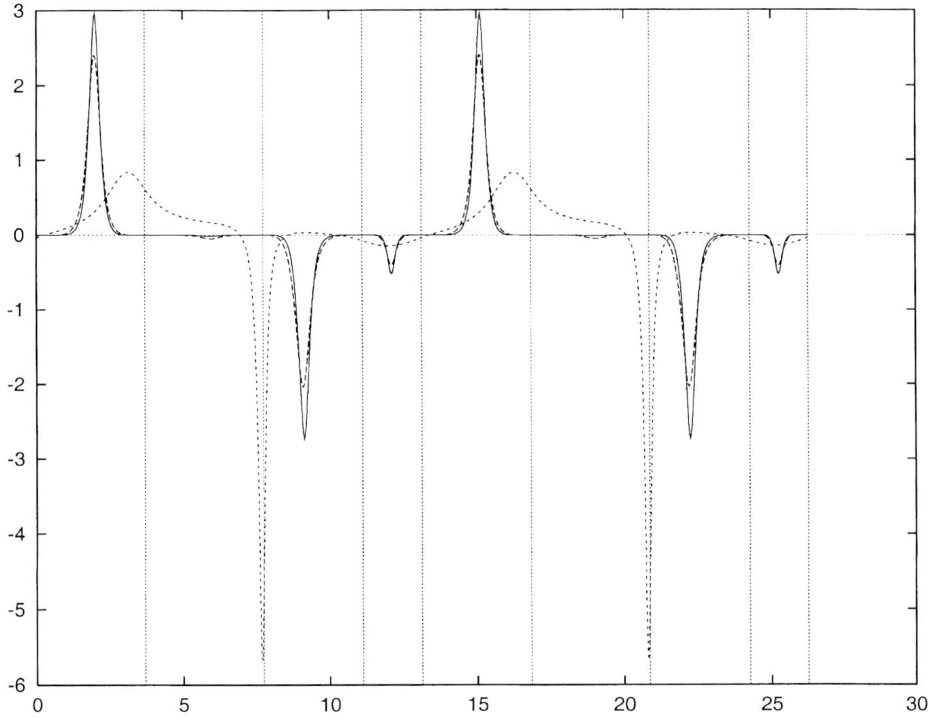

Figure 8.7. Torsion distribution of the curves in Fig. 8.4

spline family $\Gamma_{2n}(\mathscr{K})$, $n > 2$, would result in shape-preserving interpolants with smoother curvature plots and low curvature maxima. On the other hand, however, nodal torsion and its derivatives would vanish while torsion maxima would increase, which seems to limit the merit of generalization from $\Gamma_2(\mathscr{K})$ to $\Gamma_{2n}(\mathscr{K})$ $n = 2, 3, \ldots$ to planar shape-preserving interpolation.

Appendix

In this appendix we state and prove a lemma, that is necessary for establishing that the proposed family $\Gamma_4(\mathscr{K})$ of C^4 polynomial splines of non-uniform degree is able to conform with both parts of the convexity criterion of Definition 4.1 (see Ths 5.1, 5.2) and the first part (Part (ii.1)) of the corresponding torsion criterion; see Th. 6.1.

Lemma A.1. The determinant of

$$\Omega(t) = \begin{bmatrix} H_m^{(2)}(1-t; c_m) & -H_m^{(3)}(1-t; c_m) \\ H_m^{(2)}(t; c_{m+1}) & H_m^{(3)}(t; c_{m+1}) \end{bmatrix}$$

is positive on (0, 1), while it vanishes at $t = 0, 1$.

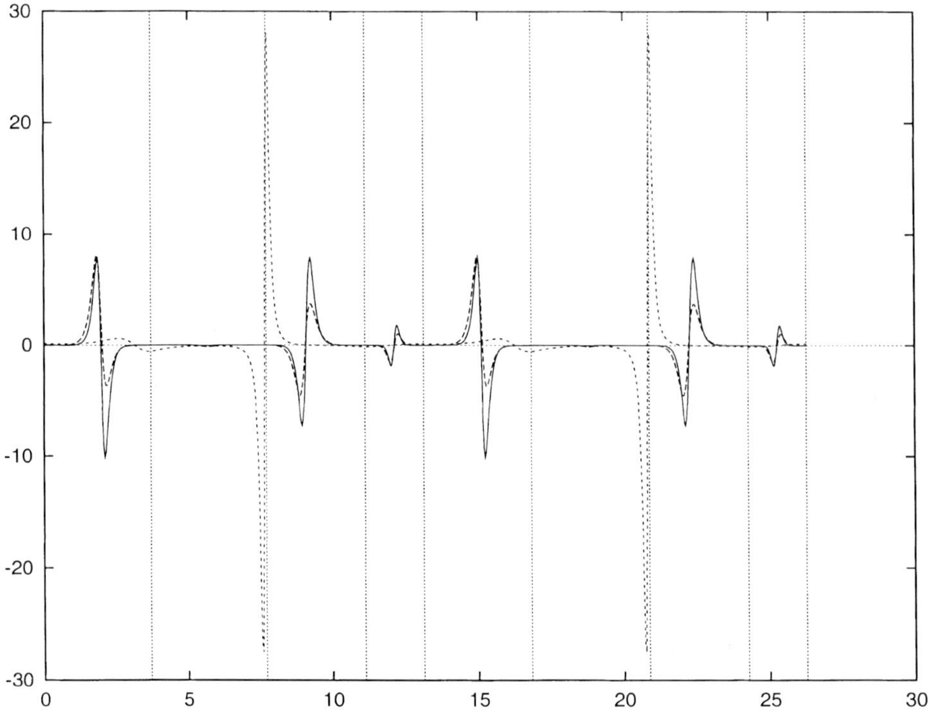

Figure 8.8. Arc-length derivative of the torsion distribution of the curves in Fig. 8.4

Proof: Appealing to the defining relation (5.2) of $H_m(t;c)$, the determinant $det(\Omega(t))$ of $\Omega(t)$ can be expressed as below:

$$det(\Omega(t)) = b_1(t) + h_m^2 c_{m+1} b_2(t) + h_m^2 c_m b_2(1-t) + h_m^4 c_m c_{m+1} b_3(t), \qquad \text{(A.1)}$$

where:

$$b_1(t) = \Theta_m^{(2)}(1-t)\Theta_m^{(3)}(t) + \Theta_m^{(2)}(t)\Theta_m^{(3)}(1-t),$$
$$b_2(t) = \Theta_m^{(2)}(1-t)\Phi_m^{(3)}(t) + \Phi_m^{(3)}(t)\Theta_m^{(2)}(1-t),$$
$$b_3(t) = \Phi_m^{(2)}(1-t)\Phi_m^{(3)}(t) + \Phi_m^{(2)}(t)\Phi_m^{(3)}(1-t).$$

As pointed out in (2.9), the second- and third-order derivatives of the auxiliary polynomial $\Theta_m(t)$ are both positive in (0,1), thus $b_1(t)$ is positive too. The analogous proof for $h_m^2 c_{m+1} b_2(t)$ and $h_m^4 c_m c_{m+1} b_3(t)$ is not so straightforward, for $\Phi_m^{(3)}(t)$ does not exhibit constant sign on (0,1). To reach this conclusion for, e.g., $h_m^2 c_{m+1} b_2(t)$, we rewrite $b_2(t)$ in the following form:

$$b_2(t) = (k_m - 2)(k_m + \ell - 2)(k_m - 3)\left(\frac{1}{\ell^2 + \ell(2k_m - 5)}\right)^2 t^{k_m-3}(1-t)^{k_m-3} p(t),$$

where

$$p(t) = (1-t)\left[\frac{l_1}{l_2} - (1-t)^\ell\right][t^\ell - l_2] + t[l_1 - (1-t)^\ell][t^\ell - 1],$$

and $l_1 = \frac{k_m+\ell-3}{k_m-3}$ and $l_2 = \frac{k_m-2}{k_m+\ell-2}$. Since $t^{k_m-3}(1-t)^{k_m-3}$ is positive, we have only to investigate the sign of $p(t)$. Noting that

$$p(0) = \left(\frac{l_1}{l_2} - 1\right)(-l_2) < 0, \quad p(1) = 0,$$

it suffices to prove that $p(t)$ is concave upwards, in order to deduce, in conjunction with the fact that c_{m+1} is negative, the positivity of $h_m^2 c_{m+1} b_2(t)$. Indeed, rewriting $p(t)$ as:

$$p(t) = 2\left[\frac{1-t}{2}\left[\frac{l_1}{l_2}t^\ell + l_2(1-t)^\ell\right] + \frac{t}{2}\left[l_1 t^\ell + (1-t)^\ell\right] + \frac{1}{2}\left[-l_1 - t^\ell(1-t)^\ell\right]\right],$$

it is readily seen that $p(t)/2$ is a convex combination of the concave upwards graphs:

$$\left[\frac{l_1}{l_2}t^\ell + l_2(1-t)^\ell\right], \quad \left[l_1 t^\ell + (1-t)^\ell\right], \quad \left[-l_1 - t^\ell(1-t)^\ell\right],$$

and, thus, $p(t)$ is concave upwards.

The positivity of the fourth term $h_m^4 c_m c_{m+1} b_3(t)$ in the right-hand side of (A.1), can be derived in a manner directly analogous to that $h_m^2 c_{m+1} b_2(t)$. Collecting the above results, we can state that $det(\Omega(t))$ is indeed positive in (0,1). As for the behaviour of $det(\Omega(t))$ at the boundary points $t = 0, 1$ it stems from the fact that $H_m^{(2)}(0; c) = H_m^{(3)}(0; c) = 0$, as it can be readily seen from (5.2) and Table 2.1. □

Acknowledgements

Thanks are due to both referees for their remarks. Especially, the authors are indebted to the anonymous referee for her/his suggestions that resulted in improving the preliminary version of this paper considerably.

References

[1] Asaturyan, S., Costantini, P., Manni, C.: Shape-preserving interpolating curves in $\mathbb{R}^3$: A local approach. In: Creating fair and shape-preserving curves and surfaces (Nowacki, H., Kaklis, P. D., eds.), pp. 99–108. Stuttgart: B.G. Teubner, 1998.

[2] Costantini, P.: Shape-preserving interpolation with variable degree polynomial splines. In: Advanced course on FAIRSHAPE (Hoschek, J., Kaklis, P. D., eds.), pp. 87–114. Stuttgart: B.G. Teubner, 1996.

[3] Costantini, P.: Variable degree polynomial splines. In: Curves and surfaces with applications in CAGD (Le Méhauté, A., Rabut, C., Schumaker, L. L., eds.), pp. 85–94. Nashville: Vanderbilt University Press, 1997.

[4] Costantini, P.: Curve and surface construction using variable degree polynomial splines. CAGD *17*, 419–446 (2000).

[5] Eckhaus, W.: Asymptotic analysis of singular perturbations. Amsterdam: North-Holland, 1979.
[6] Ginnis, A. I., Kaklis, P. D., Gabrielides, N. C.: Sectional-curvature preserving skinning surfaces with a 3D spine curve. In: Advanced topics in multivariate approximation (Fontanella, F., Jetter, K., Laurent, P.-J., eds.), pp. 113–123. Singapore: World Scientific, 1996.
[7] Goodman, T. N. T., Ong, B. H.: Shape preserving interpolation by G^2 curves in three dimensions. In: Curves and surfaces with applications in CAGD (Le Méhauté, A., Rabut, C., Schumaker, L. L., eds.), pp. 151–158. Nashville: Vanderbilt University Press, 1997.
[8] Goodman, T. N. T., Ong, B. H.: Shape preserving interpolation by space curves. CAGD *15*, 1–17 (1997).
[9] Hoschek, J., Lasser, D.: Fundamentals of computer aided geometric design. Wellesley: AK Peters, 1993.
[10] Kaklis, P. D., Ginnis, A. I.: Sectional-curvature preserving skinning surfaces. CAGD *13*, 583–671 (1996).
[11] Kaklis, P. D., Karavelas, M. I.: Shape-preserving interpolation in $\mathbb{R}^3$. IMA J. Numer. Anal. *17*, 373–419 (1997).
[12] Kaklis, P. D., Pandelis, D. G.: Convexity-preserving polynomial splines of non-uniform degree. IMA J. Numer. Anal. *10*, 223–234 (1990).
[13] Kaklis, P. D., Sapidis, N. S.: Convexity-preserving interpolatory parametric splines of non-uniform polynomial degree. CAGD *12*, 1–26 (1995).
[14] Messac, A., Sivanandan, A.: A new family of convex splines for data interpolation. CAGD *15*, 39–59 (1997).
[15] Sapidis, N. S., Kaklis, P. D.: A hybrid method for shape-preserving interpolation with curvature-continuous quintic splines. Computing [Suppl.] *10*, 285–301 (1995).
[16] Späth, H.: Exponential spline interpolation. Computing *4*, 225–233 (1969).

N. C. Gabrielides
P. D. Kaklis
Ship Design Laboratory
Department of Naval Architecture and Marine Engineering
National Technical University of Athens
9 Heroon Polytechneiou
GR-157 73 Zografou
Athens, Greece
e-mail: kaklis@deslab.ntua.gr

Computing [Suppl] 14, 155–184 (2001)

Computing

Blossoming and Divided Difference

R. Goldman, Houston, TX

Abstract

Blossoming and divided difference are shown to be characterized by a similar set of axioms. But the divided difference obeys a cancellation postulate which is not included in the standard blossoming axioms. Here the blossom is extended to incorporate a new set of parameters along with a cancellation axiom. Both the standard blossom and the divided difference operator are special cases of this new extended blossom. It follows that these dual functionals all satisfy a similar collection of formulas and identities, including a Marsden identity, a recurrence relation, a degree elevation formula, a multirational property, a differentiation identity, and expressions for partial derivatives with respect to their parameters. In addition, formulas are presented that express the divided differences of polynomials in terms of the blossom. Canonical examples are provided for the blossom, the divided difference, and the extended blossom, and general proof procedures are developed based on these characteristic functions.

AMS Subject Classifications: 65 D17, 41 A10.

Key Words: Blossom, divided difference, dual functionals, Marsden identity.

1. Introduction – Dual Functionals

Dual functionals are maps that compute the coefficients of arbitrary functions with respect to a fixed basis. For example, function evaluation furnishes the dual functionals for polynomials with respect to the Lagrange basis and differentiation provides the dual functionals for analytic functions relative to their Taylor expansion. In Approximation Theory and Computer Aided Geometric Design two of the most important examples of dual functionals are the divided difference operator, which provides the dual functionals for the Newton basis, and the blossom, which furnishes the dual functionals for the Bernstein and B-spline bases.

Often dual bases satisfy some simple properties that make them easier to manipulate than the primal bases. Thus in addition to providing the coefficients of functions relative to some primal basis, dual functionals are important tools because they can be used to develop algorithms for functions expressed relative to the primal basis. For example, subdivision algorithms for Bezier and knot insertion procedures for B-spline curves and surfaces can be developed quite easily using blossoming.

Superficially, blossoming and divided difference seem to be very different operators. The thesis of this paper is that there is a very deep connection between the blossom and the divided difference because these two dual functionals can be characterized by a very similar set of axioms. Indeed the divided difference turns out to be a special case of an extended version of the blossom and this extended blossom can be constructed explicitly in terms of divided differences. Some of these ideas were initially discussed in [11], [13]; this paper is a companion to [12], but with greater emphasis on the divided difference.

Since blossoming and divided difference share a similar set of axioms, these dual functionals also satisfy a very similar collection of formulas and identities, including a Marsden identity, a recurrence relation, a degree elevation formula, a differentiation identity, and expressions for partial differentiation with respect to their parameters. In addition, we shall obtain formulas that express the divided differences of polynomials in terms of the blossom. One of the leitmotifs of this paper is that there are many ways to derive such identities: (i) by appealing directly to the axioms, (ii) by checking that the axioms are satisfied and then invoking uniqueness, (iii) by verifying these identities on certain canonical examples and then extending to the entire space of applicable functions, or (iv) by employing explicit formulas for the blossom or the divided difference. We shall demonstrate all four of these proof techniques with examples.

We begin in Section 2 by reviewing the blossoming axioms and recalling a similar set of axioms that completely characterize the divided difference. The axioms for the divided difference contain a new rule, the cancellation axiom, which does not appear among the standard axioms of the blossom. To incorporate the divided difference into the blossoming paradigm, we extend the blossoming axioms to include a new set of parameters along with a cancellation axiom. We then show that both the standard blossom and the divided difference operator are special cases of this new extended form of the blossom.

The axiomatic approach to blossoming and divided difference is rather abstract, so in Section 3 we compute the blossom, the divided difference, and the extended blossom on an explicit set of canonical examples. We then apply these examples to derive a Marsden identity for each of these operators. Section 4 is devoted to deriving additional formulas and identities for the blossom and the divided difference, confirming our thesis that formulas and identities for one theory generally carry over in a straightforward manner to the other theory. We also exhibit a variety of proof techniques that can be adopted to derive such formulas and identities. We close in Section 5 with a brief summary of our work and a few open questions for future research.

2. Axioms for Blossoming and Divided Difference

2.1. The Blossoming Axioms

The *blossom* of a polynomial $P(x)$ of degree less than or equal to m is the unique, symmetric, multiaffine polynomial $p(u_1, \ldots, u_m)$ that reduces to $P(x)$ along the diagonal. Thus the multiaffine blossom satisfies the following axioms:

Standard Blossoming Axioms (Polynomials)

Symmetry

$$p(u_1, \ldots, u_m) = p(u_{\sigma(1)}, \ldots, u_{\sigma(m)})$$

Multiaffine

$$p(u_1, \ldots, (1-\alpha)u + \alpha w, \ldots, u_m) = (1-\alpha)p(u_1, \ldots, u, \ldots, u_m) + \alpha p(u_1, \ldots, w, \ldots, u_m)$$

Diagonal

$$p(\underbrace{x, \ldots, x}_{m}) = P(x)$$

This blossom is well known in mathematics: it is the classical polar form [25], [29]. Remarkably, the polar form provides the dual functionals for the Bernstein and B-spline bases. In particular, the Bezier coefficients of a polynomial curve are given by its blossom evaluated at zeros and ones. More generally, the B-spline coefficients of a piecewise polynomial curve are given by its local blossom evaluated at consecutive knots. Blossoming revolutionized the theory of polynomial and piecewise polynomial curves and surfaces by emphasizing the characteristic properties of the dual functionals – symmetric, multiaffine, diagonal – rather than explicit formulas, as tools for analyzing Bezier and B-spline curves and surfaces [3], [7], [8], [10], [16], [24], [27], [28]. Algorithms for subdivision and knot insertion for the Bezier and B-spline representations are readily derived from blossoming.

In addition to the axioms, the main facts about the blossom are existence, uniqueness, and the dual functional property. We provide a constructive proof for existence below, and we shall derive the dual functional property in Section 3. Additional formulas and identities will be provided in Section 4. For an alternative approach to these properties as well as a proof of uniqueness, see [23]–[25].

Ramshaw furnishes many explicit expressions for the blossom [25]. Perhaps the best known is the following formula of de Boor-Fix [1], [6].

Theorem 2.1. (*Existence*)

Let $P(x)$ be a polynomial of degree less than or equal to m. Then for all τ

$$p(u_1, \ldots, u_m) = \sum_j \frac{(-1)^{m-j}}{m!} \psi^{(j)}(\tau) P^{(m-j)}(\tau) \qquad (2.1)$$
$$\psi(x) = (x - u_1) \cdots (x - u_m).$$

Proof: It is easy to see that the right hand side of Eq. (2.1) for $p(u_1, \ldots, u_m)$ is symmetric and multiaffine in the u parameters, since $\psi(x)$ is symmetric and multiaffine in $u_1, \ldots, u_m$. The diagonal property follows by observing that when $u_1 = \cdots = u_m = t$, the right hand side reduces to the Taylor expansion of $P(t)$ at $t = \tau$. □

It follows from Eq. (2.1) that blossoming is a linear operator. This result is also a consequence of the uniqueness of the blossom.

2.2. *Axioms for the Divided Difference*

Just like the blossom $p(u_1, \ldots, u_m)$ of a polynomial $P(x)$, the divided difference $F[v_0, \ldots, v_n]$ of a differentiable function $F(x)$ can be completely characterized by a simple set of axioms.

Axioms for the Divided Difference (Differentiable Functions)

Symmetry

$$F[v_0, \ldots, v_n] = F[v_{\sigma(0)}, \ldots, v_{\sigma(n)}]$$

Affinity

If $u = (1-\alpha)u_1 + \alpha u_2$, then

$$\{(x-u)F(x)\}[v_0, \ldots, v_n] = (1-\alpha)\{(x-u_1)F(x)\}[v_0, \ldots, v_n] + \alpha\{(x-u_2)F(x)\}[v_0, \ldots, v_n]$$

Cancellation

$$\{(x-t)F(x)\}[v_0, \ldots, v_n, t] = F[v_0, \ldots, v_n]$$

Differentiation

$$F[\underbrace{x, \ldots, x}_{n+1}] = \frac{F^{(n)}(x)}{n!}$$

The divided difference is the unique operator satisfying these four properties [15]. Alternative axioms for the divided difference are also provided in [15]. Notice, in particular, that the affinity axiom is a simple consequence of the linearity of the divided difference operator, but we have chosen this axiom in place of linearity to emphasize the similarity between the divided difference axioms and the blossoming axioms. Indeed, what is remarkable here is that in the presence of the other three divided difference axioms this weak form of linearity is actually equivalent to linearity.

The divided difference axioms of symmetry, affinity, and differentiation closely resemble the blossoming axioms of symmetry, multiaffinity, and evaluation along the diagonal. But the divided difference has one additional axiom not incorporated in blossoming: the cancellation axiom. In Section 2.3 we shall show how to extend the blossom to accommodate an additional set of parameters along with a cancellation axiom, thus unifying within a single framework both blossoming and divided difference.

The divided difference is ubiquitous in numerical analysis and approximation theory, and is related both to Newton interpolation and to B-spline approximation [26]. Indeed the divided difference provides the dual functionals for the Newton basis, and classically the B-splines are defined specifically in terms of divided differences [4]. For analytic functions, the divided difference can be constructed explicitly using complex contour integration [9]. This explicit integration formula establishes the existence of the divided difference of an analytic function, and since this formula and two other related integration formulas from complex analysis will play an important role later in this paper we shall now recall these three identities.

Cauchy's Integral Formula

$$F(t) = \frac{1}{2\pi i}\oint_C \frac{F(z)}{(z-t)}dz \tag{2.2}$$

Cauchy's Integral Formula for Derivatives

$$\frac{F^{(n)}(t)}{n!} = \frac{1}{2\pi i}\oint_C \frac{F(z)}{(z-t)^{n+1}}dz \tag{2.3}$$

Complex Contour Integration Formula for Divided Difference

$$F[v_0, \ldots, v_n] = \frac{1}{2\pi i}\oint_C \frac{F(z)dz}{(z-v_0)\cdots(z-v_n)} \tag{2.4}$$

Cauchy's two integral formulas are fundamental tools in complex analysis [19]. In Cauchy's two formulas C is any simple closed contour containing the parameter t, and in the divided difference formula C is any simple closed contour containing the parameters $v_0, \ldots, v_n$. In all three identities $F(z)$ is a function that is analytic in an open disk containing C. The complex integration formula for the divided difference follows from the divided difference axioms and Cauchy's integral formula for the derivative. Indeed to establish this result, all we need to do is to show that the right hand side of Eq. (2.4) satisfies the four divided difference axioms. But symmetry, affinity, and cancellation are easy to verify. Moreover, by Cauchy's integral formula for the derivative, when $v_0 = v_1 = \cdots = v_n = t$,

$$\frac{1}{2\pi i}\oint_C \frac{F(z)dz}{(z-v_0)\cdots(z-v_n)} = \frac{1}{2\pi i}\oint_C \frac{F(z)dz}{(z-t)^{n+1}} = \frac{F^{(n)}(t)}{n!}.$$

Thus the right hand side of Equation (2.4) satisfies the four divided difference axioms, so by uniqueness the right hand side must be equal to the divided difference. We shall provide an alternative derivation of this identity in Section 3.2.

2.3. Extending the Blossoming Axioms

Both the blossom and the divided difference can be extended to incorporate additional parameters. The link between these two dual functionals is most clearly seen through these extensions, which we shall now introduce.

The *extended blossom of order* $k \in \mathbf{Z}$ of a function $F(x)$ is a function $f(u, v)$ – specified on all pairs $u = (u_1, \ldots u_m)$ and $v = (v_1, \ldots, v_n)$ with $m - n = k$ – that satisfies the following properties: $f(u, v) = f(u_1, \ldots, u_m/v_1, \ldots, v_n)$ is bisymmetric in the u and v parameters, multiaffine in the u parameters, satisfies a cancellation property, and reduces to $F(x)$ along the diagonal. Thus the extended blossom satisfies the following axioms:

Extended Blossoming Axioms

Bisymmetry

$$f(u_1, \ldots, u_m/v_1, \ldots, v_n) = f(u_{\sigma(1)}, \ldots, u_{\sigma(m)}/v_{\tau(1)}, \ldots, v_{\tau(n)})$$

Multiaffine in u

$$\begin{aligned} f(u_1, \ldots, (1-\alpha)u + \alpha w, \ldots, u_m/v_1, \ldots, v_n) \\ = (1-\alpha) f(u_1, \ldots, u, \ldots, u_m/v_1, \ldots, v_n) \\ + \alpha f(u_1, \ldots, w, \ldots, u_m/v_1, \ldots, v_n) \end{aligned}$$

Cancellation

$$f(u_1, \ldots, u_m, w/v_1, \ldots, v_n, w) = f(u_1, \ldots, u_m/v_1, \ldots, v_n)$$

Diagonal

$$f(\underbrace{x, \ldots, x}_{m} / \underbrace{x, \ldots, x}_{n}) = F(x)$$

When $k = m - n \geq 0$, it follows easily from these axioms that $F(x)$ must be a polynomial in x of degree less than or equal to k. In this case we shall also insist that the blossom $f(u_1, \ldots, u_m/v_1, \ldots, v_n)$ must be a polynomial of *degree*(F) in the u and v parameters. Thus when $k \geq 0$, blossoming is strictly a polynomial theory. Moreover, notice than when $n = 0$, the polynomial $p(u_1, \ldots, u_k/)$ is symmetric, multiaffine, and reduces to $P(x)$ along the diagonal. Thus $p(u_1, \ldots, u_k/)$ is the standard blossom of $P(x)$. Hence the extended blossom of positive order contains within it the standard blossom. Notice too that $p(u_1, \ldots, u_m/v_1, \ldots, v_n)$ is defined only if $k = m - n \geq$ *degree*(P), for otherwise $p(u_1, \ldots, u_k/)$ cannot be the standard blossom of $P(x)$. Finally observe that when $k = 0$, $P(x)$ is a constant, so $p(u_1, \ldots, u_m/v_1, \ldots, v_m) = P(x)$ for all values of the parameters $u_1, \ldots, u_m$ and $v_1, \ldots, v_m$.

We shall now establish that for any fixed value of $k \geq$ *degree*(P), the extended blossom of $P(x)$ exists for all values of $n \geq 0$. The extended blossom is also unique for $k \geq 0$; for a proof see [12].

Theorem 2.2. (*Existence*)

Let $P(x)$ be a polynomial of degree less than or equal to k, and let $p^*(u_1, \ldots, u_k)$ denote the standard blossom of $P(x)$. Then the extended blossom of $P(x)$ of order k is given by

$$p(u_1, \ldots, u_m/v_1, \ldots, v_n) = \sum (-1)^{\beta} p^*(u_{i_1}, \ldots, u_{i_\alpha}, v_{j_1}, \ldots, v_{j_\beta}) \tag{2.5}$$

where the sum is taken over all collections of indices $\{i_1, \ldots, i_\alpha\}$ *and* $\{j_1, \ldots, j_\beta\}$ *such that*

i. $i_1, \ldots, i_\alpha$ *are distinct,*

ii. $j_1, \ldots, j_\beta$ *need not be distinct,*

iii. $\alpha + \beta = k = m - n$.

Proof: Let $\tilde{p}(u_1, \ldots, u_m/v_1, \ldots, v_n)$ denote the right hand side of Eq. (2.5). We must check that $\tilde{p}$ satisfies the axioms of the extended blossom of order k. Clearly, by construction, $\tilde{p}(u_1, \ldots, u_m/v_1, \ldots, v_n)$ is a bisymmetric polynomial that is multiaffine in the u parameters. Moreover $\tilde{p}$ satisfies the cancellation property for the following reason. Suppose, without lost of generality, that $u_1 = v_1$. Then, by symmetry,

$$(-1)^\beta p^*(u_1, u_{i_2}, \ldots, u_{i_\alpha}, v_{j_1}, \ldots, v_{j_\beta}) + (-1)^{\beta+1} p^*(u_{i_2}, \ldots, u_{i_\alpha}, v_1, v_{j_1}, \ldots, v_{j_\beta}) = 0.$$

Hence all the terms containing u_1 or v_1 cancel. The remaining sum is exactly equal to $\tilde{p}(u_2, \ldots, u_{n+k}/v_2, \ldots, v_k)$, so $\tilde{p}$ satisfies the cancellation property. Finally, $\tilde{p}$ reduces to P along the diagonal because by the cancellation property,

$$\tilde{p}(\underbrace{x, \ldots, x}_{m} / \underbrace{x, \ldots, x}_{n}) = \tilde{p}(\underbrace{x, \ldots, x}_{k}/) = p^*(\underbrace{x, \ldots, x}_{k}) = P(x). \qquad \square$$

When $k = m - n < 0$, the function $F(x)$ need no longer be a polynomial in x, and the blossom $f(u_1, \ldots, u_m/v_1, \ldots, v_n)$ is not required to be a polynomial in the v parameters, although by the multiaffine property it must still be a polynomial in the u parameters. Blossoms of negative order play the same role for analytic functions and negative degree Bernstein and B-spline bases that the standard blossom plays for polynomials and positive degree Bernstein and B-spline bases [11, 13]. In particular, the coefficients of an arbitrary analytic function relative to the degree $-d < 0$ Bernstein basis functions are given by its order $k = -d$ blossom evaluated at zeros and ones [11]. More generally, the coefficients of an arbitrary piecewise analytic function relative to the B-splines of degree $-d$ are given by its order $k = -d$ blossom evaluated at consecutive knots [13]. Thus the blossoms of negative order provide the dual functionals for the Bezier and B-spline bases of negative degree. Algorithms for differentiation and other change of basis procedures can be derived from this blossom [11].

The axioms for the extended blossom resemble quite closely the axioms for the divided difference. Therefore it should come as no surprise that we can express the extended blossom of negative order in terms of the divided difference. Moreover we shall see shortly that the divided difference is actually just a special instance of the extended blossom.

For positive order we have seen that the extended blossom of a polynomial can be constructed from its standard blossom by introducing an additional set of v parameters. Similarly we shall now show that the blossom of negative order can be constructed from the divided difference by introducing an additional set of u parameters. This formula will establish the existence of the blossom of negative order for all differentiable functions. An alternative derivation of this identity for analytic functions is provided in Section 3.3. The extended blossom of a differentiable function is also unique for any order $k < 0$, provided we assume continuity in the v parameters; for a proof see [14].

Theorem 2.3. (*Existence*)

Let $F(x)$ be a differentiable function and let $F^{-(n-m-1)}(x)$ denote the $(n-m-1)^{st}$ antiderivative of $F(x)$. If $k = m - n < 0$, then

$$f(u_1, \ldots, u_m / v_1, \ldots, v_n) = \{(n-m-1)!(x-u_1)\cdots(x-u_m)F^{-(n-m-1)}(x)\}[v_1, \ldots, v_n] \quad (2.6)$$

Proof: To establish this result, all we need to do is to verify that the right hand side of Eq. (2.6) satisfies the four axioms of the extended blossom of negative order. But these four properties all follow immediately from the corresponding properties of the divided difference. □

Since the extended blossom is a polynomial in the u parameters, we can homogenize with respect to the u parameters. Homogenizing Eq. (2.6) yields

$$f((u_1, w_1), \ldots, (u_m, w_m) / v_1, \ldots, v_n) = \{(n-m-1)!(w_1x-u_1)\cdots(w_mx-u_m)F^{-(n-m-1)}(x)\}[v_1, \ldots, v_n]. \quad (2.7)$$

Now we can write the divided difference as a homogenized version of the extended blossom of order -1.

Theorem 2.4.

$$F[v_1, \ldots, v_{m+1}] = (-1)^m f(\underbrace{\delta, \ldots, \delta}_{m} / v_1, \ldots, v_{m+1}), \quad (2.8)$$

where $\delta = (1, 0)$. That is, up to sign, the divided difference operator is the homogenized extended blossom of degree -1 evaluated at $(u_i, w_i) = \delta = (1, 0)$, $i = 1, \ldots, m$.

Proof: This result follows immediately from Eq. (2.7) with $n = m + 1$.

This last result suggests that identities for the blossom and identities for divided difference must have much in common. We shall see shortly that this is indeed the case.

3. Canonical Examples, Marsden Identities, and Dual Functionals

The axiomatic approach to the blossom and the divided difference is rather abstract. To make these theories more concrete, we will now consider some specific examples. We shall see that these examples are canonical in the sense that once we know the blossom or the divided difference for these particular functions, we know it for all functions to which the theory applies.

3.1. The Standard Blossom and the Power Basis

Consider the polynomials $P(x) = (x - t)^m$, where t is a fixed but arbitrary constant. The blossom of these polynomials is obtained simply by replacing x by a different parameter u_k in each of the m factors of $(x - t)^m$. That is, we have:

$$\begin{aligned} P(x) &= (x - t)^m \\ p(u_1, \ldots, u_m) &= (u_1 - t) \cdots (u_m - t). \end{aligned} \tag{3.1}$$

We can easily check that $p(u_1, \ldots, u_m)$ has the three required properties. Indeed:

1. $p(u_1, \ldots, u_m)$ is symmetric because multiplication is commutative;
2. $p(u_1, \ldots, u_m)$ is multiaffine because:
   (i) $(1 - \alpha)u + \alpha w - t = (1 - \alpha)(u - t) + \alpha(w - t)$,
   (ii) multiplication distributes though addition;
3. $p(u_1, \ldots, u_m)$ satisfies the diagonal property by substitution.

Once we have the blossom for polynomials of the form $(x - t)^m$, it is an easy matter to construct the blossom for arbitrary polynomials of degree m. Select any $m + 1$ distinct parameters $t_0, \ldots, t_m$. Then the polynomials $(x - t_0)^m, \ldots, (x - t_m)^m$ form a basis for the polynomials of degree m, so we can write any polynomial $Q(x)$ of degree less than or equal to m as a linear combination of these basis functions. Since blossoming is a linear operator,

$$\begin{aligned} Q(x) &= \sum\nolimits_j c_j (x - t_j)^m \\ q(u_1, \ldots, u_m) &= \sum\nolimits_j c_j (u_1 - t_j) \cdots (u_m - t_j). \end{aligned}$$

These observations demonstrate once again the existence of the standard blossom.

We can also use the polynomials $P(x) = (x - t)^m$ to establish the dual functional property of the blossom – that is, that the blossom evaluated at the knots provides the dual functionals for the B-splines. Recall that given a knot vector $\{x_k\}$, the B-splines $\{N_{k,m}(x)\}$ of degree m can be defined recursively by:

$$\begin{aligned} N_{j,0}(x) &= 1 \qquad x_j \le x < x_{j+1} \\ N_{j,m}(x) &= \frac{x - x_j}{x_{j+m} - x_j} N_{j,m-1}(x) + \frac{x_{j+m+1} - x}{x_{j+m+1} - x_{j+1}} N_{j+1,m-1}(x). \end{aligned} \tag{3.2}$$

The dual functional property for the polynomials $(x - t)^m$ is the Marsden identity [20].

Theorem 3.1. (*Marsden Identity*)

$$(x-t)^m = \sum\nolimits_j (x_{j+1}-t)\cdots(x_{j+m}-t)N_{j,m}(x) \tag{3.3}$$

Proof: Although this result is well known, here we provide an inductive argument so that later on we can see the similarity between this proof and the proof in Section 3.2 of the Marsden identity for the divided difference and the Newton basis and the proof in Section 3.3 of the Marsden identity for the extended blossom and B-splines of negative degree. To simplify our notation, let

$$\psi_{j,m}(t) = (x_{j+1}-t)\cdots(x_{j+m}-t).$$

We must show that

$$(x-t)^m = \sum\nolimits_j \psi_{j,m}(t)N_{j,m}(x).$$

For $m=0$, this result is obvious. Now we proceed by induction on m. To begin, observe that

$$\begin{aligned}\psi_{j,m}(t) &= (x_{j+m}-t)\psi_{j,m-1}(t)\\ \psi_{j,m}(t) &= (x_{j+1}-t)\psi_{j+1,m-1}(t).\end{aligned}$$

Hence by the B-spline recurrence

$$\begin{aligned}\sum\nolimits_j \psi_{j,m}(t)N_{j,m}(x) &= \sum\nolimits_j \psi_{j,m}(t)\left\{\frac{x-x_j}{x_{j+m}-x_j}N_{j,m-1}(x)+\frac{x_{j+m+1}-x}{x_{j+m+1}-x_{j+1}}N_{j+1,m-1}(x)\right\}\\ &= \sum\nolimits_j \frac{x-x_j}{x_{j+m}-x_j}(x_{j+m}-t)\psi_{j,m-1}(t)N_{j,m-1}(x)\\ &\quad+\sum\nolimits_j \frac{x_{j+m+1}-x}{x_{j+m+1}-x_{j+1}}(x_{j+1}-t)\psi_{j+1,m-1}(t)N_{j+1,m-1}(x)\\ &= \sum\nolimits_j \left\{\frac{x-x_j}{x_{j+m}-x_j}(x_{j+m}-t)+\frac{x_{j+m}-x}{x_{j+m}-x_j}(x_j-t)\right\}\\ &\quad\times\psi_{j,m-1}(t)N_{j,m-1}(x)\end{aligned}$$

But

$$x-t = \frac{x-x_j}{x_{j+m}-x_j}(x_{j+m}-t)+\frac{x_{j+m}-x}{x_{j+m}-x_j}(x_j-t).$$

Therefore by the inductive hypothesis

$$\sum\nolimits_j \psi_{j,m}(t)N_{j,m}(x) = (x-t)\sum\nolimits_j \psi_{j,m-1}(t)N_{j,m-1}(x) = (x-t)^m. \qquad \square$$

Theorem 3.3. (*Marsden Identity – Newton Basis*)

$$(x-t)^{-1} = \sum_{n\geq 0} \frac{(-1)^n N_n(x)}{(v_1-t)\cdots(v_{n+1}-t)} \tag{3.9}$$

provided that the nodes $v_1, v_2, \ldots$ are chosen so that the right hand side converges.

Proof: We proceed much as in the proof of Theorem 3.1, but with a simpler recurrence for the basis functions (see below). To simplify our notation, let

$$\psi_n(t) = \frac{(-1)^n}{(v_1-t)\cdots(v_{n+1}-t)}.$$

Our goal is to prove that

$$(x-t)^{-1} = \sum_{n\geq 0} \psi_n(t) N_n(x)$$

or equivalently that

$$1 = (x-t)\sum_{n\geq 0} \psi_n(t) N_n(x).$$

Now observe that

$$\begin{aligned}
&N_n(x) = (x-v_n)N_{n-1}(x) \\
&-\psi_{n-1}(t) = (v_{n+1}-t)\psi_n(t) \\
&(x-t) = (x-v_{n+1}) + (v_{n+1}-t).
\end{aligned}$$

Therefore, since by assumption the right hand side of Eq. (3.9) converges,

$$\begin{aligned}
(x-t)\sum_{n\geq 0}\psi_n(t)N_n(x) &= \sum_{n\geq 0}\{(x-v_{n+1}) + (v_{n+1}-t)\}\psi_n(t)N_n(x) \\
&= \sum_{n\geq 1}(x-v_n)\psi_{n-1}(t)N_{n-1}(x) + \sum_{n\geq 0}(v_{n+1}-t)\psi_n(t)N_n(x) \\
&= 1 + \sum_{n\geq 1}\{\psi_{n-1}(t) - \psi_{n-1}(t)\}N_n(x) \\
&= 1.
\end{aligned}$$

Dividing both sides by $x-t$ yields the result. □

The right hand side of Eq. (3.9) will converge absolutely if

$$Lim_{n\to\infty}\left|\frac{\frac{N_n(x)}{(v_1-t)\cdots(v_{n+1}-t)}}{\frac{N_{n-1}(x)}{(v_1-t)\cdots(v_n-t)}}\right| = Lim_{n\to\infty}\left|\frac{(v_n-x)}{(v_{n+1}-t)}\right| < 1.$$

In particular, suppose that $t \neq v_j$ for all j. If $v_n \to v$ and $v > x > t$, then the right hand side of Eq. (3.9) will converge absolutely, so at least in this case the Marsden identity of Theorem 3.3 is guaranteed to hold.

Corollary 3.4. (*Dual Functionals*)

Suppose that the nodes $\{v_j\}$ are bounded and that the Marsden identity converges (e.g. see the preceding remark). Let $G(x)$ be an analytic function inside some open disk D containing the nodes $\{v_j\}$. Then

$$G(x) = \sum_{n \geq 0} G[v_1, \ldots, v_{n+1}] N_n(x). \tag{3.10}$$

Proof: Start by multiplying both sides of the Marsden identity for the Newton basis (Eq. (3.9)) by $G(t)$ to obtain

$$\frac{G(t)}{(x-t)} = \sum_{n \geq 0} \frac{(-1)^n G(t) N_n(x)}{(v_1 - t) \cdots (v_{n+1} - t)}. \tag{3.11}$$

Let $C \subset D$ be a simple closed contour containing the nodes $\{v_j\}$ and the parameter t. Integrating Eq. (3.11) around C yields

$$\frac{1}{2\pi i} \oint_C \frac{G(t)dt}{(t-x)} = \sum_{n \geq 0} \frac{1}{2\pi i} \oint_C \frac{G(t)dt}{(t - v_1) \cdots (t - v_{n+1})} N_n(x).$$

Applying Cauchy's integral formula (Eq. (2.2)) to the left hand side and the complex integration formula for the divided difference (Eq. (2.4)) to the right hand side, we arrive at

$$G(x) = \sum_{n \geq 0} G[v_1, \ldots, v_{n+1}] N_n(x). \quad \square$$

3.3. *The Extended Blossom of Negative Order and the Power Functions of Negative Degree*

For the extended blossom of order $k < 0$, let us again proceed in analogy with polynomials and take as our canonical functions $F(x) = (x - t)^k$, where t is a fixed but arbitrary, possibly complex, constant. When $m - n = k < 0$, there is a very simple formula for the blossom $f(u_1, \ldots, u_m / v_1, \ldots, v_n)$. Indeed, we have:

$$\begin{aligned} F(x) &= (x-t)^k \\ f(u_1, \ldots, u_m / v_1, \ldots, v_n) &= \frac{(u_1 - t) \cdots (u_m - t)}{(v_1 - t) \cdots (v_n - t)}. \end{aligned} \tag{3.12}$$

It is easy to verify that $f(u_1,\ldots,u_m/v_1,\ldots,v_n)$ has the four required properties.

1. $f(u_1,\ldots,u_m/v_1,\ldots,v_n)$ is bisymmetric because multiplication is commutative;
2. $f(u_1,\ldots,u_m/v_1,\ldots,v_n)$ is multiaffine in the u parameters because:
   (i) $(1-\alpha)u+\alpha w-t=(1-\alpha)(u-t)+\alpha(w-t)$,
   (ii) multiplication distributes though addition;
3. $f(u_1,\ldots,u_m/v_1,\ldots,v_n)$ satisfies the cancellation property by division of polynomials;
4. $f(u_1,\ldots,u_m/v_1,\ldots,v_n)$ satisfies the diagonal property by substitution and cancellation.

Notice, however, that if $F(x)=(x-t)^k$, $k=m-n\geq 0$, then

$$f(u_1,\ldots,u_m/v_1,\ldots,v_n)\neq\frac{(u_1-t)\cdots(u_m-t)}{(v_1-t)\cdots(v_n-t)}$$

even though the right hand side satisfies all four blossoming axioms because the right hand side is not a polynomial in the v parameters. Thus this polynomial assumption is required to ensure that the blossom is unique when $k\geq 0$.

As with divided difference, it follows by Cauchy's integral formula for derivatives (Eq. (2.3)), that once we know the extended blossom for these canonical functions, we can derive a formula for the extended blossom of arbitrary functions that are analytic in an open disk containing the v parameters. This we now proceed to do. Again this leads to a general proof technique, which we now exhibit by computing the extended blossom of an arbitrary function $G(x)$ that is analytic inside some open disk D containing the parameters $v_1,\ldots,v_n$.

To proceed, multiply Eq. (3.12) by $G^{(k+1)}(t)$ to obtain

$$\left\{\frac{G^{(k+1)}(t)}{(x-t)^{-k}}\right\}(u_1,\ldots,u_m/v_1,\ldots,v_n)=\frac{(u_1-t)\cdots(u_m-t)G^{(k+1)}(t)}{(v_1-t)\cdots(v_n-t)}. \tag{3.13}$$

Now let $C\subset D$ be a simple closed contour containing the parameters $v_1,\ldots,v_n,t$. Integrating Eq. (3.13) around C yields

$$\begin{aligned}&\frac{1}{2\pi i}\oint_C\frac{G^{(k+1)}(t)}{(x-t)^{-k}}(u_1,\ldots,u_m/v_1,\ldots,v_n)dt\\&=\frac{1}{2\pi i}\oint_C\frac{(u_1-t)\cdots(u_m-t)G^{(k+1)}(t)}{(v_1-t)\cdots(v_n-t)}dt.\end{aligned} \tag{3.14}$$

Since the extended blossom is with respect to x and the integral is with respect to t, blossoming and integration commute on the left hand side of Eq. (3.14). Therefore applying Cauchy's integral formula for the derivative (Eq. (2.3)) to the left hand side of Eq. (3.14), we get

$$g(u_1, \ldots, u_m/v_1, \ldots, v_n)$$
$$= \frac{1}{2\pi i} \oint_C \frac{(-k-1)!(t-u_1)\cdots(t-u_m)G^{(k+1)}(t)dt}{(t-v_1)\cdots(t-v_n)}. \tag{3.15}$$

Now recalling the complex integration formula for the divided difference (Eq. (2.4)) and substituting $k+1 = m-n+1$, we arrive at

$$g(u_1, \ldots, u_m/v_1, \ldots, v_n)$$
$$= \{(n-m-1)!(t-u_1)\cdots(t-u_m)G^{-(n-m-1)}(t)\}[v_1, \ldots, v_n]$$

which is exactly the result in Theorem 2.3.

The extended blossom of negative order provides the dual functionals for the B-splines of negative degree. Given knot sequences $\{u_i\}$ and $\{v_j\}$, these B-splines of degree $k < 0$ satisfy the recurrence [13]:

$$\begin{aligned} N_{m,0}(x) &= 1 \quad m = 0 \\ &= 0 \quad m \neq 0 \\ N_{m,k}(x) &= \frac{(x - v_{m-k})}{(u_m - v_{m-k})} N_{m-1,k-1}(x) + \frac{(u_{m+1} - x)}{(u_{m+1} - v_{m-k+1})} N_{m,k-1}(x). \end{aligned} \tag{3.16}$$

The dual functional property for the functions $(x-t)^k$, $k < 0$, is the analogue of the Marsden identity for the B-splines of negative degree.

Theorem 3.5. (*Marsden Identity for B-splines of Negative Degree*)

$$(x-t)^k = \sum_m \frac{(u_1 - t)\cdots(u_m - t)}{(v_1 - t)\cdots(v_{m-k} - t)} N_{m,k}(x) \quad k < 0 \tag{3.17}$$

provided that the knot sequences $\{u_i\}$, $\{v_j\}$ *are chosen so that the right hand side converges.*

Proof: We proceed as in the proof of Theorem 3.1 by induction on $|k|$, using here the recurrence (Eq. (3.16)) for the B-splines of negative degree. When $k = 0$, the result is obvious. To simplify our notation, for the remainder of this proof let

$$\psi_{m,k}(t) = \frac{(u_1 - t)\cdots(u_m - t)}{(v_1 - t)\cdots(v_{m-k} - t)}.$$

Thus when $|k| > 0$, our goal is to prove that

$$(x-t)^k = \sum_j \psi_{m,k}(t) N_{m,k}(x).$$

To proceed, observe that

$$\psi_{m,k}(t) = (u_m - t)\psi_{m-1,k-1}(t)$$
$$\psi_{m,k}(t) = (v_{m-k+1} - t)\psi_{m,k-1}(t).$$

Therefore by the inductive hypothesis and the recurrence (Eq. (3.16)) for B-splines of negative degree:

$$\begin{aligned}
(x-t)^k &= \sum_m \psi_{m,k}(t)N_{m,k}(x) \\
&= \sum_m (u_m - t)\psi_{m-1,k-1}(t)\left\{\frac{(x - v_{m-k})}{(u_m - v_{m-k})}N_{m-1,k-1}(x)\right\} \\
&\quad + \sum_m (v_{m-k+1} - t)\psi_{m,k-1}(t)\left\{\frac{(u_{m+1} - x)}{(u_{m+1} - v_{m-k+1})}N_{m,k-1}(x)\right\} \\
&= \sum_m (u_{m+1} - t)\psi_{m,k-1}(t)\left\{\frac{(x - v_{m-k+1})}{(u_{m+1} - v_{m-k+1})}N_{m,k-1}(x)\right\} \\
&\quad + \sum_m (v_{m-k+1} - t)\psi_{m,k-1}(t)\left\{\frac{(u_{m+1} - x)}{(u_{m+1} - v_{m-k+1})}N_{m,k-1}(x)\right\} \\
&= \sum_m \left\{(u_{m+1} - t)\frac{(x - v_{m-k+1})}{(u_{m+1} - v_{m-k+1})} + (v_{m-k+1} - t)\frac{(u_{m+1} - x)}{(u_{m+1} - v_{m-k+1})}\right\} \\
&\quad \times \psi_{m,k-1}(t)N_{m,k-1}(x).
\end{aligned}$$

But

$$x - t = \frac{(x - v_{m-k+1})}{(u_{m+1} - v_{m-k+1})}(u_{m+1} - t) + \frac{(u_{m+1} - x)}{(u_{m+1} - v_{m-k+1})}(v_{m-k+1} - t).$$

Hence

$$(x-t)^k = (x-t)\sum_m \psi_{m,k-1}(t)N_{m,k-1}(x).$$

Dividing both sides by $x - t$ completes the induction. □

Corollary 3.6. (*Dual Functionals*)

Suppose that the knots $\{v_j\}$ are bounded and that for these knots the Marsden identity for B-splines of negative degree converges. Let $G(x)$ be an analytic function inside some open disk D containing the knots $\{v_j\}$. Then

$$G(x) = \sum_m g(u_1, \ldots, u_m/v_1, \ldots, v_{m-k})N_{m,k}(x) \quad k < 0 \tag{3.18}$$

Proof: Here we mimic the proof of Corollary 3.4. Start by multiplying both sides of the Marsden identity for the negative degree B-splines by $G^{(k+1)}(t)$ to obtain

$$\frac{G^{(k+1)}(t)}{(x-t)^{-k}} = \sum_m \frac{(u_1-t)\cdots(u_m-t)G^{(k+1)}(t)}{(v_1-t)\cdots(v_{m-k}-t)} N_{m,k}(x) \tag{3.19}$$

Now let $C \subset D$ be a simple closed contour containing the knots $\{v_j\}$ and the parameter t. Integrating Eq. (3.19) around C yields

$$\frac{1}{2\pi i}\oint_C \frac{G^{(k+1)}(t)dt}{(x-t)^{-k}} = \sum_n \frac{1}{2\pi i}\oint_C \frac{(u_1-t)\cdots(u_m-t)G^{(k+1)}(t)dt}{(v_1-t)\cdots(v_{m-k}-t)} N_{m,k}(x).$$

Applying Cauchy's integral formula for derivatives – Eq. (2.3) – to the left hand side and the complex integration formula for the extended blossom – Eq. (3.15) – to the right hand side, we arrive at

$$G(x) = \sum_m g(u_1,\ldots,u_m/v_1,\ldots,v_{m-k})N_{m,k}(x). \quad \square$$

4. Additional Identities

Here we shall derive some additional common identities shared by blossoming and divided difference, including an analogue of the multiaffine property for the v parameters, a general recurrence relation, and formulas for degree elevation and differentiation. To get a better feel for each of these identities, we shall, when applicable, state the special cases for the standard blossom and for the divided difference alongside the general identity for the extended blossom.

One of the subsidiary goals of this section is to illustrate different proof techniques for deriving such identities. We will present four different methods:

i. appealing directly to the axioms;

ii. checking that the axioms are satisfied and then invoking uniqueness;

iii. verifying these identities on the canonical examples and then extending to the entire space of applicable functions using the methods introduced in Section 3;

iv. exploiting explicit formulas for the (extended) blossom or the divided difference.

We shall demonstrate each of these methods with at least one example. Note that often more than one proof technique may apply, though in each case we shall content ourselves with a single proof.

In the following results, $P(x)$ always represents a polynomial of degree d and $F(x)$ is always an arbitrary function that is analytic in some open disk D containing the v parameters.

The axioms for the extended blossom are not symmetric in the u and v parameters. For our first result, we derive an analogue of the multiaffine property for the v parameters. This *multirational property* can be used to replace the multiaffine axiom in the extended blossoming schemes. For a proof of this fact as well as additional alternative blossoming axioms, see [14].

Proposition 4.1. (*The Multirational Property*)

Let $v = (1-\alpha)v_i + \alpha v_j$, *then*

$$F[v_1, \ldots, v_n] = (1-\alpha)F[v_1, \ldots, v_{i-1}, v, v_{i+1}, \ldots, v_n] + \alpha F[v_1, \ldots, v_{j-1}, v, v_{j+1}, \ldots, v_n] \tag{4.1a}$$

$$\begin{aligned} &p(u_1, \ldots, u_m/v_1, \ldots, v_n) \\ &\quad = (1-\alpha)p(u_1, \ldots, u_m/v_1, \ldots, v_{i-1}, v, v_{i+1}, \ldots, v_n) \\ &\quad + \alpha p(u_1, \ldots, u_m/v_1, \ldots, v_{j-1}, v, v_{j+1}, \ldots, v_n) \quad m-n \geq d \end{aligned} \tag{4.1b}$$

$$\begin{aligned} &f(u_1, \ldots, u_m/v_1, \ldots, v_n) \\ &\quad = (1-\alpha)f(u_1, \ldots, u_m/v_1, \ldots, v_{i-1}, v, v_{i+1}, \ldots, v_n) \\ &\quad + \alpha f(u_1, \ldots, u_m/v_1, \ldots, v_{j-1}, v, v_{j+1}, \ldots, v_n). \quad m-n < 0 \end{aligned} \tag{4.1c}$$

Proof: The proofs of these three identities are much the same, so we shall prove only Eq. (4.1b). Here we invoke Method (i). Applying the cancellation, multiaffine, and symmetry properties:

$$\begin{aligned} p(u_1, \ldots, u_m/v_1, \ldots, v_n) &= p(u_1, \ldots, u_m, v/v_1, \ldots, v_n, v) \\ &= (1-\alpha)p(u_1, \ldots, u_m, v_i/v_1, \ldots, v_{i-1}, v_i, v_{i+1}, \ldots, v_n, v) \\ &\quad + \alpha p(u_1, \ldots, u_m, v_j/v_1, \ldots, v_{j-1}, v_j, v_{j+1}, \ldots, v_n, v) \\ &= (1-\alpha)p(u_1, \ldots, u_m/v_1, \ldots, v_{i-1}, v, v_{i+1}, \ldots, v_n) \\ &\quad + \alpha p(u_1, \ldots, u_m/v_1, \ldots, v_{j-1}, v, v_{j+1}, \ldots, v_n). \end{aligned}$$ □

Next we extend a well known degree elevation formula for the blossom, which turns out to be intimately related to a differentiation formula for the divided difference.

Proposition 4.2. (*Degree Elevation*)

$$p(u_1, \ldots, u_m) = \frac{\sum_{i=1}^{m} p(u_1, \ldots, u_{i-1}, u_{i+1}, \ldots, u_m)}{m} \quad m \geq d+1 \tag{4.2a}$$

$$F'[v_1, \ldots, v_n] = \sum_{j=1}^{n} F[v_1, \ldots, v_j, v_j, \ldots, v_n] \tag{4.2b}$$

$$p(u_1,\ldots,u_m/v_1,\ldots,v_n) = \frac{\sum_{i=1}^{m} p(u_1,\ldots,u_{i-1},u_{i+1},\ldots,u_m/v_1,\ldots,v_n)}{m-n} - \frac{\sum_{j=1}^{n} p(u_1,\ldots,u_m/v_1,\ldots,v_j,v_j,\ldots,v_n)}{m-n}, \quad k=m-n \geq d+1 \tag{4.2c}$$

$$f(u_1,\ldots,u_m/v_1,\ldots,v_n) = \frac{\sum_{i=1}^{m} f(u_1,\ldots,u_{i-1},u_{i+1},\ldots,u_m/v_1,\ldots,v_n)}{m-n} - \frac{\sum_{j=1}^{n} f(u_1,\ldots,u_m/v_1,\ldots,v_j,v_j,\ldots,v_n)}{m-n}, \quad k=m-n<0. \tag{4.2d}$$

Proof: Again the proofs of these four identities are all much the same, so we shall prove only Eq. (4.2d). Here we apply Method (ii). That is, we observe that since the extended blossom is unique, it is enough to show that the right hand side of Eq. (4.2d) satisfies the four axioms of the extended blossom for $k = m - n < 0$. But clearly the right hand side of Eq. (4.2d) is bisymmetric in the u and v parameters and multiaffine in the u parameters. To show that the cancellation axiom is also satisfied, suppose, without loss of generality, that $u_1 = v_1$. Then in all the terms on the right hand side that contain both u_1 and v_1 exactly once, these parameters cancel. What remains are just two terms in which u_1 or v_1 appear, and these terms sum to zero, since

$$f(u_2,\ldots,u_m/v_1,\ldots,v_n) = f(u_1,\ldots,u_m/v_1,v_1,\ldots,v_n).$$

Finally, along the diagonal the right hand side reduces to

$$\frac{m}{m-n}F(x) - \frac{n}{m-n}F(x) = F(x). \quad \square$$

Proposition 4.3. (*Partial Derivatives – u Parameters*)

Let μ_j denote the multiplicity of the parameter u_j in the sequence $u = (u_1,\ldots,u_m)$, and let p', f' denote the blossoms of P', F'. If $k = m - n \neq 0$, then

$$\frac{\partial p(u_1,\ldots,u_m)}{\partial u_j} = \frac{\mu_j}{m} p'(u_1,\ldots,\underbrace{u_j,\ldots,u_j}_{\mu_j-1},\ldots,u_m) \tag{4.3a}$$

$$\frac{\partial p(u_1,\ldots,u_m/v_1,\ldots,v_n)}{\partial u_j} = \frac{\mu_j}{m-n} p'(u_1,\ldots,\underbrace{u_j,\ldots,u_j}_{\mu_j-1},\ldots,u_m/v_1,\ldots,v_n) \tag{4.3b}$$

$$\frac{\partial f(u_1, \ldots, u_m/v_1, \ldots, v_n)}{\partial u_j} = \frac{\mu_j}{m-n} f'(u_1, \ldots, \underbrace{u_j, \ldots, u_j}_{\mu_j - 1}, \ldots, u_m/v_1, \ldots, v_n) \tag{4.3c}$$

Proof: Again the proofs of these three identities are much the same (see too the proof of Proposition 4.4), so we shall prove only Eq. (4.3a). Method (iii) is easiest to apply here. We begin then by verifying this identity on the canonical example

$$P(x) = (x-t)^m$$
$$p(u_1, \ldots, u_m) = (u_1 - t) \cdots (u_m - t).$$

Since μ_j is the multiplicity of the parameter u_j in the sequence $u = (u_1, \ldots, u_m)$,

$$p(u_1, \ldots, u_m) = (u_1 - t) \cdots (u_j - t)^{\mu_j} \cdots (u_m - t)$$
$$\frac{\partial p(u_1, \ldots, u_m)}{\partial u_j} = \mu_j (u_1 - t) \cdots (u_j - t)^{\mu_j - 1} \cdots (u_m - t).$$

On the other hand,

$$P'(x) = m(x-t)^{m-1}$$
$$p'(u_1, \ldots, \underbrace{u_j, \ldots, u_j}_{\mu_j - 1}, \ldots, u_m) = m(u_1 - t) \cdots (u_j - t)^{\mu_j - 1} \cdots (u_m - t).$$

Comparing these two formulas, we can see immediately that for the polynomials $P(x) = (x-t)^m$

$$\frac{\partial p(u_1, \ldots, u_m)}{\partial u_j} = \frac{\mu_j}{m} p'(u_1, \ldots, \underbrace{u_j, \ldots, u_j}_{\mu_j - 1}, \ldots, u_m).$$

For an arbitrary polynomial of degree at most m, we can reason as follows. Select any $m+1$ distinct parameters $t_0, \ldots, t_m$. Then the polynomials $(x - t_0)^m, \ldots, (x - t_m)^m$ form a basis for the polynomials of degree m. Hence we can write any polynomial $Q(x)$ of degree less than or equal to m as a linear combination of these basis functions. Since Eq. (4.3a) holds for this basis, it follows by the linearity of the blossom that it will also hold for $Q(x)$. □

Proposition 4.4. (*Partial Derivatives – v Parameters*)

Let μ_j denote the multiplicity of the parameter v_j in the sequence $v = (v_1, \ldots, v_n)$, and let p', f' denote the blossoms of P', F'. Then

$$\frac{\partial F[v_1, \ldots, v_n]}{\partial v_j} = \mu_j F[v_1, \ldots, \underbrace{v_j, \ldots, v_j}_{\mu_j + 1}, \ldots, v_n] \tag{4.4a}$$

$$\frac{\partial f(u_1,\ldots,u_m/v_1,\ldots,v_n)}{\partial v_j} = \frac{\mu_j}{n-m} f'(u_1,\ldots,u_m/v_1,\ldots,\underbrace{v_j,\ldots,v_j}_{\mu_j+1},\ldots,v_n) \tag{4.4b}$$

$$\frac{\partial p(u_1,\ldots,u_m/v_1,\ldots,v_n)}{\partial v_j} = \frac{\mu_j}{n-m} p'(u_1,\ldots,u_m/v_1,\ldots,\underbrace{v_j,\ldots,v_j}_{\mu_j+1},\ldots,v_n) \tag{4.4c}$$

Proof: The proofs of these three identities are much the same, so here we shall prove only Eq. (4.4a). Again Method (iii) is easiest to apply. Therefore, first let us verify this identity on the canonical examples

$$F(x) = (x-t)^{-1}$$

$$F[v_1,\ldots,v_n] = \frac{(-1)^{n-1}}{(v_1-t)\cdots(v_n-t)}.$$

Since μ_j is the multiplicity of the parameter v_j in the sequence $v = (v_1,\ldots,v_n)$,

$$F[v_1,\ldots,v_n] = \frac{(-1)^{n-1}}{(v_1-t)\cdots(v_j-t)^{\mu_j}\cdots(v_n-t)}$$

$$\frac{\partial F[v_1,\ldots,v_n]}{\partial v_j} = \frac{(-1)^n \mu_j}{(v_1-t)\cdots(v_j-t)^{\mu_j+1}\cdots(v_n-t)}.$$

On the other hand,

$$F(x) = (x-t)^{-1}$$

$$F[v_1,\ldots,\underbrace{v_j,\ldots,v_j}_{\mu_j+1},\ldots,v_n] = \frac{(-1)^n}{(v_1-t)\cdots(v_j-t)^{\mu_j+1}\cdots(v_n-t)}$$

Comparing these results, it follows immediately by inspection that

$$\frac{\partial F[v_1,\ldots,v_n]}{\partial v_j} = \mu_j F[v_1,\ldots,\underbrace{v_j,\ldots,v_j}_{\mu_j+1},\ldots,v_n].$$

For arbitrary analytic functions G, we can now reason as follows. Multiply both sides of Eq. (4.4a) for the function $(x-t)^{-1}$ by $G(t)$ to obtain

$$\frac{\partial}{\partial v_j}\left\{\frac{G(t)}{(x-t)}\right\}[v_1,\ldots,v_n] = \mu_j\left\{\frac{G(t)}{(x-t)}\right\}[v_1,\ldots,\underbrace{v_j,\ldots,v_j}_{\mu_j+1},\ldots,v_n].$$

Now let C be a simple closed contour containing the parameter x. Integrate this equation around C with respect to t. Then, since integration and divided difference commute because the divided difference is with respect to x and the integral is with respect to t:

$$\frac{\partial}{\partial v_j}\left\{\oint_C \frac{G(t)}{(x-t)}dt\right\}[v_1,\ldots,v_n] = \mu_j\left\{\oint_C \frac{G(t)}{(x-t)}dt\right\}[v_1,\ldots,\underbrace{v_j,\ldots,v_j}_{\mu_j+1},\ldots,v_n].$$

Applying Cauchy's integral formula to both sides of this equation, we arrive at

$$\frac{\partial G[v_1,\ldots,v_n]}{\partial v_j} = \mu_j G[v_1,\ldots,\underbrace{v_j,\ldots,v_j}_{\mu_j+1},\ldots,v_n]. \quad \square$$

Proposition 4.5. (*Differentiation*)

$$\binom{m}{j} p(\underbrace{\delta,\ldots,\delta}_{j},\underbrace{x,\ldots,x}_{m-j}) = \frac{P^{(j)}(x)}{j!} \tag{4.5a}$$

$$\binom{k}{j} p(\underbrace{\delta,\ldots,\delta}_{j},\underbrace{x,\ldots,x}_{m-j} / \underbrace{x,\ldots,x}_{n}) = \frac{P^{(j)}(x)}{j!} \qquad k = m-n \geq 0 \tag{4.5b}$$

$$\binom{k}{j} f(\underbrace{\delta,\ldots,\delta}_{j},\underbrace{x,\ldots,x}_{m-j} / \underbrace{x,\ldots,x}_{n}) = \frac{F^{(j)}(x)}{j!} \qquad k = m-n < 0 \tag{4.5c}$$

Proof: Again as the proofs of these three identities are much the same, we shall prove only Eq. (4.5c). Here we shall use mainly Method (iv). From the cancellation property and the explicit formula for the extended blossom:

$$\begin{aligned} f(\underbrace{\delta,\ldots,\delta}_{j},\underbrace{x,\ldots,x}_{m-j} / \underbrace{x,\ldots,x}_{n}) &= f(\underbrace{\delta,\ldots,\delta}_{j} / \underbrace{x,\ldots,x}_{n-m+j}) \\ &= (-1)^j (n-m-1)! F^{-(n-m-1)}[\underbrace{x,\ldots,x}_{n-m+j}] \\ &= \frac{(-1)^j (n-m-1)!}{(n-m+j-1)!} F^{(j)}(x). \end{aligned}$$

But since $k = m - n < 0$,

$$\binom{k}{j} = (-1)^j \binom{|k|+j-1}{j} = \frac{(-1)^j (n-m+j-1)!}{j!(n-m-1)!},$$

so

$$\binom{k}{j} f(\underbrace{\delta,\ldots,\delta}_{j}, \underbrace{x,\ldots,x}_{m-j} / \underbrace{x,\ldots,x}_{n}) = \frac{F^{(j)}(x)}{j!} \qquad k = m - n < 0. \qquad \square$$

Notice that the differentiation property for the divided difference

$$F[\underbrace{x,\ldots,x}_{m+1}] = \frac{F^{(m)}(x)}{m!}$$

is the special case of Eq. (4.5c) where $n = m + 1$ and $j = m$, since

$$F[\underbrace{x,\ldots,x}_{m+1}] = (-1)^m f(\underbrace{\delta,\ldots,\delta}_{m} / \underbrace{x,\ldots,x}_{m+1}).$$

Proposition 4.6. (*Recursion*)

Suppose that $v_n \neq v_1$. *Then*

$$F[v_1,\ldots,v_n] = \frac{F[v_2,\ldots,v_n] - F[v_1,\ldots,v_{n-1}]}{v_n - v_1} \tag{4.6a}$$

$$p(\underbrace{\delta,\ldots,\delta}_{j}, u_1,\ldots,u_{m-j}/v_1,\ldots,v_n) = \frac{p(\underbrace{\delta,\ldots,\delta}_{j-1}, u_1,\ldots,u_{m-j}/v_1,\ldots,v_{n-1}) - p(\underbrace{\delta,\ldots,\delta}_{j-1}, u_1,\ldots,u_{m-j}/v_2,\ldots,v_n)}{v_n - v_1} \tag{4.6b}$$

$$f(\underbrace{\delta,\ldots,\delta}_{j}, u_1,\ldots,u_{m-j}/v_1,\ldots,v_n) = \frac{f(\underbrace{\delta,\ldots,\delta}_{j-1}, u_1,\ldots,u_{m-j}/v_1,\ldots,v_{n-1}) - f(\underbrace{\delta,\ldots,\delta}_{j-1}, u_1,\ldots,u_{m-j}/v_2,\ldots,v_n)}{v_n - v_1} \tag{4.6c}$$

Proof: Again as the proofs of these three identities are much the same, we shall prove only Eq. (4.6b). Here we shall use Method (i). Applying the multilinear and cancellation properties of the homogenized blossom, we obtain

$$\begin{aligned}
(v_n - v_1)p(\underbrace{\delta, \ldots, \delta}_{j}, u_1, \ldots, u_{m-j}/v_1, \ldots, v_n) \\
&= p(\underbrace{\delta, \ldots, \delta}_{j-1}, v_n, u_1, \ldots, u_{m-j}/v_1, \ldots, v_n) \\
&\quad - p(\underbrace{\delta, \ldots, \delta}_{j-1}, v_1, u_1, \ldots, u_{m-j}/v_1, \ldots, v_n) \\
&= p(\underbrace{\delta, \ldots, \delta}_{j-1}, u_1, \ldots, u_{m-j}/v_1, \ldots, v_{n-1}) \\
&\quad - p(\underbrace{\delta, \ldots, \delta}_{j-1}, u_1, \ldots, u_{m-j}/v_2, \ldots, v_n).
\end{aligned}$$

Dividing both sides by $v_n - v_1$ yields the result. □

We close this section by deriving a formula that expresses the divided difference of a polynomial in terms of the blossom. We begin with an extension of the blossoming identity in Eq. (2.5).

Proposition 4.7. (*Blossoming Formulas for Polynomials*)

Let $p^*(u_1, \ldots, u_d)$ denote the standard blossom of $P(x)$. Then

$$p(u_1, \ldots, u_m/v_1, \ldots, v_n) = \frac{\sum(-1)^\beta p^*\left(u_{i_1}, \ldots, u_{i_\alpha}, v_{j_1}, \ldots, v_{j_\beta}\right)}{\binom{k}{d}}, \qquad k = m - n \geq d, \tag{4.7a}$$

$$p(u_1, \ldots, u_m/v_1, \ldots, v_n) = \frac{\sum(-1)^\beta p^*\left(u_{i_1}, \ldots, u_{i_\alpha}, v_{j_1}, \ldots, v_{j_\beta}\right)}{\binom{k}{d}}, \qquad k = m - n < 0, \tag{4.7b}$$

where the sums are taken over all collections of indices $\{i_1, \ldots, i_\alpha\}$ and $\{j_1, \ldots, j_\beta\}$ *such that*

i. $i_1, \ldots, i_\alpha$ *are distinct,*

ii. $j_1, \ldots, j_\beta$ *need not be distinct,*

iii. $\alpha + \beta = d$.

Proof: We have already proved Eq. (4.7a) for the case $k = d$ in Theorem 2.2, using Method (ii). That is, we observed that since the extended blossom is unique, it is enough to show that the right hand sides of these equations satisfy the four axioms of the extended blossom. The proof for $k \neq d$ is much the same, except that when we verify the diagonal property, we need to account for the constant coefficient $\binom{k}{d}^{-1}$. This can be achieved by straightforward counting arguments, so this analysis is left to the reader. For further details, see [12]. □

Corollary 4.8. *Let $P(x)$ be a polynomial of degree d, and let $p^{(n-1)}$ denote the multiaffine blossom of $P^{(n-1)}$. Then*

$$P[v_1, \ldots, v_n] = \left\{\frac{(d-n+1)!}{d!}\right\} \sum p^{(n-1)}(v_{j_1}, \ldots, v_{j_{d-n+1}}) \tag{4.8}$$

where the sum is taken over all indices $j_1, \ldots, j_{d-n+1}$ such that $1 \leq j_1 \leq \cdots \leq j_{d-n+1} \leq n$.

Proof: By Theorem (2.4)

$$P[v_1, \ldots, v_n] = (-1)^{n-1} p(\underbrace{\delta, \ldots, \delta}_{n-1} / v_1, \ldots, v_n),$$

and by Proposition (4.7)

$$p(\underbrace{\delta, \ldots, \delta}_{n-1} / v_1, \ldots, v_n) = \frac{\sum (-1)^{\beta} p^*(\underbrace{\delta, \ldots, \delta}_{\alpha}, v_{j_1}, \ldots, v_{j_\beta})}{(-1)^d},$$

where p^* is the standard blossom of P and $\alpha + \beta = d$. But since the right hand side is homogeneous in the u parameters (the $\delta' s$), all the terms $p^*(\underbrace{\delta, \ldots, \delta}_{\alpha}, v_{j_1}, \ldots, v_{j_\beta})$, with $\alpha < n-1$, vanish, since they contain a factor of zero. Thus

$$P[v_1, \ldots, v_n] = \sum p^*(\underbrace{\delta, \ldots, \delta}_{n-1}, v_{j_1}, \ldots, v_{j_{d-n+1}}). \tag{*}$$

Moreover by Proposition (4.5)

$$\frac{d!}{(d-n+1)!} p^*(\underbrace{\delta, \ldots, \delta}_{n-1}, \underbrace{x, \ldots, x}_{d-n+1}) = P^{(n-1)}(x).$$

Therefore

$$\frac{d!}{(d-n+1)!} p^*(\underbrace{\delta, \ldots, \delta}_{n-1}, u_1, \ldots, u_{d-n+1}) = p^{(n-1)}(u_1, \ldots, u_{d-n+1}),$$

since as a function of the u parameters, the left hand side is symmetric, multiaffine and reduces to $P^{(n-1)}(x)$ along the diagonal. Substituting this result into (*), we conclude that

$$P[v_1, \ldots, v_n] = \left\{\frac{(d-n+1)!}{d!}\right\} \sum p^{(n-1)}\left(v_{j1}, \ldots, v_{j_{d-n+1}}\right). \qquad \square$$

5. Summary, Conclusions, and Open Questions

Blossoming and divided difference share many properties because they satisfy a similar set of axioms. To unify these two theories, we have extended the standard blossoming axioms to incorporate two distinct sets of parameters – u parameters and v parameters – linked by a cancellation axiom. This extended blossom contains both the standard blossom and the divided difference as special cases: the u parameters are blossoming parameters, the v parameters are divided difference parameters.

There is, however, one very important difference between blossoming and divided difference. Blossoming, in its standard form, is essentially a polynomial theory. This is not the case for the divided difference operator which can be applied to a much wider range of functions. Remarkably, the extended blossom incorporates both theories. For positive order it is a polynomial theory; for negative order a theory of differentiable and analytic functions.

Blossoming provides the dual functionals for the standard Bernstein and B-spline bases, and the divided difference furnishes the dual functionals for the Newton bases. The extended blossom of negative order supplies the dual functionals for the Bernstein and B-spline bases of negative degree. In this context, the Newton bases emerge as homogenized B-splines of degree -1, and the divided difference as a homogeneous blossom of order -1.

The extended blossom goes a long way towards unifying and generalizing the theories of the blossom and the divided difference. These dual functionals all satisfy a Marsden identity, a multirational property, a recurrence relation, a degree elevation formula, a differentiation identity, and formulas for partial differentiation with respect to their parameters. Nevertheless some important issues remain unresolved.

In Section 2.2 we observed that the affinity axiom for the divided difference is a simple consequence of the linearity of the divided difference operator. Notice, however, that this implication fails to hold for blossoming. Why is it that linearity implies affinity for the divided difference, but not for the blossom? The answer seems to be as follows. Let us homogenize the blossom with respect to the u parameters. Then there are two kinds of linearity: linearity of the operator with respect to functions – i.e. the blossom of the sum of two functions is the sum of their blossoms – and linearity with respect to the homogeneous parameters. For the blossom these two kinds of linearity are distinct, but for divided difference these two forms of linearity actually coincide. In the homogeneous theory, the multiaffine axiom is replaced by the multilinear axiom, so in this sense multilinearity is equivalent to multiaffinity. Now by Theorem (2.4) divided difference is a homogeneous theory. So in divided difference, just as in blossoming, the affinity axiom can be replaced by linearity in the homogeneous parameters. But for divided difference linearity in the homogeneous parameters is equivalent to linearity with respect to functions. This equivalence explains why linearity of the operator with respect to the functions implies affinity for the divided difference, but not for the blossom.

All such anomalies are not so easily explained. One of the underlying themes of this paper is that identities for the blossom or for the divided difference typically generalize in a natural way to the extended blossom of both positive and negative order. There are, however, two well-known formulas that do not seem to generalize in this way: the product rule for the blossom and Leibniz's rule for the divided difference. Indeed, suppose that $P(x)$ and $Q(x)$ are polynomials with $\text{degree}(P) = m$ and $degree(Q) = n$, and let $R(x) = P(x)Q(x)$. In addition, let $F(x)$ and $G(x)$ be arbitrary differentiable functions. Then the following identities are known:

$$r(u_1, \ldots, u_{m+n}) = \frac{\sum_\sigma p\big(u_{\sigma(1)}, \ldots, u_{\sigma(m)}\big) q\big(u_{\sigma(m+1)}, \ldots, u_{\sigma(m+n)}\big)}{(m+n)!} \qquad \text{(Product Rule)}$$

$$(FG)[v_1, \ldots, v_n] = \sum\nolimits_j F\big[v_1, \ldots, v_j\big] G\big[v_j, \ldots, v_n\big] \quad \text{(Leibniz Rule)}$$

The first formula follows by the uniqueness of the blossom, since it is easy to check that the right hand side satisfies the three blossoming axioms; the second formula is well known and follows readily from the axioms and the interpolatory properties of the divided difference [17]. Nevertheless, there seems to be no simple generalization of these identities to the extended blossom. One reason for this difficulty could be that for negative order the explicit formula (Eq. (2.6)) for the extended blossom may involve a high order antiderivative and there is no simple expression for the high order antiderivative of the product of two functions. Another reason could be that the proof of Leibniz's rule is not straightforward, but appeals to the interpolatory properties of the divided difference. In any case, this failure is rather disappointing.

The theory of blossoming has recently been extended to trigonometric splines [18] and to Chebyshev splines [2], [21], [22]. These schemes also possess analogues of the divided difference [26]. Typically these blossoming theories retain the symmetry and diagonal axioms, but the multiaffine property is replaced by a much more complicated formula. Is there an analogue of the cancellation axiom that extends these blossoming theories in a canonical way? Can the analogue of the divided difference be incorporated into this extended blossoming theory, thus unifying the divided difference with the blossom in these more general settings?

Finally, although all our results here are derived only for functions of a single variable, there is a well known generalization of the blossom to polynomials in several variables [24]. There is also a notion of divided difference for functions of several variables [5]. Are these two theories compatible? Do they share a similar set of axioms and identities? Is there a natural generalization of the extended blossom to the multivariate setting, and if so does this extended blossom unify the theories of the multivariate blossom and the multivariate divided difference?

References

[1] Barry, P. J.: de Boor-Fix functionals and polar forms. Comput. Aided Geom. Des. *7*, 425–430 (1990).

[2] Barry, P. J.: de Boor-Fix functionals and algorithms for Tchebycheffian B-spline curves. Const. Approx. *12*, 385–408 (1996).

[3] Barry, P. J., Goldman, R. N.: Algorithms for progressive curves: Extending B-spline and blossoming techniques to the monomial, power and Newton dual bases. In: Knot insertion and deletion algorithms for B-spline curves and surfaces (Goldman, R., Lyche, T., eds.), pp. 11–63. Philadelphia: SIAM, 1993.

[4] de Boor, C.: A practical guide to splines. New York: Springer, 1978.

[5] de Boor, C.: A multivariate divided difference. Approx. Theory *8*, 1–10 (1995).

[6] de Boor, C., Fix, G.: Spline approximation by quasi-interpolants. J. Approx. Theory *8*, 19–45 (1973).

[7] de Casteljau, P.: Formes a Poles. Paris: Hermes, 1985.

[8] Dahmen, W., Micchelli, C. A., Seidel, H. P.: Blossoming begets B-splines built better by B-patches. Math. Comput. *59*, 97–115 (1992).

[9] Davis, P. J.: Interpolation and approximation. New York: Dover, 1975.

[10] Goldman, R. N.: Blossoming and knot insertion algorithms for B-spline curves. Comput. Aided Geom. Des. *7*, 69–81 (1990).

[11] Goldman, R. N.: The rational Bernstein bases and the multirational blossoms. Comput. Aided Geom. Des. *16*, 710–738 (1999a).

[12] Goldman, R. N.: Blossoming with cancellation. Comput. Aided Geom. Des. *16*, 671–689 (1999b).

[13] Goldman, R. N.: Rational B-splines and multirational blossoms (2000a) – in preparation.

[14] Goldman, R. N.: The multirational blossom: An axiomatic approach. (2000b) – in preparation.

[15] Goldman, R. N.: Axiomatic characterizations of divided difference. (2000c) – in preparation.

[16] Goldman, R. N., Barry, P. J.: Wonderful triangle. In: Mathematical methods in computer aided geometric design II (Lyche, T., Schumaker, L., eds.), pp. 297–320. San Diego: Academic Press, 1992.

[17] Lee E. T. Y.: A remark on divided difference. Am. Math. Monthly *96*, 618–622 (1989).

[18] Lyche, T., Schumaker, L., Stanley, S.: Quasi-interpolants based on trigonometric splines. J. Approx. Theory *95*, 280–309 (1998).

[19] Marsden, J. E.: Basic complex analysis. San Franscisco: W.H. Freeman, 1973.

[20] Marsden, M. J.: An identity for spline functions with applications to variation-diminishing spline approximation. J. Approx. Theory *3*, 7–49 (1970).

[21] Mazure, M.-L.: Blossoming of Chebyshev splines. In: Mathematical methods for curves and surfaces (Daehlen, M., Lyche, T., Schumaker, L., eds.), pp. 353–364. Nashville: Vanderbilt University Press, 1995.

[22] Pottmann, H.: The geometry of Tchebycheffian splines. Comput. Aided Geom. Des. *10*, 181–210 (1993).

[23] Ramshaw, L.: Blossoming: A Connect-the-Dots Approach to Splines. Digital Systems Research Center Technical Report *19*, Palo Alto (1987).

[24] Ramshaw, L.: Beziers and B-splines as multiaffine maps. In: Theoretical foundations of computer graphics and CAD (Earnshaw, R. A., ed.), pp. 757–776. NATO ASI Series F, Vol. 40, New York: Springer Verlag, 1988.

[25] Ramshaw, L.: Blossoms are polar forms. Comput. Aided Geom. Des. *6*, 323–358 (1989).

[26] Schumaker, L. L.: Spline functions: basic theory. New York: J. Wiley, 1981.

[27] Seidel, H. P.: A new multiaffine approach to B-splines. Comput. Aided Geom. Des. *6*, 23–32 (1989).
[28] Seidel, H. P.: Symmetric recursive algorithms for surfaces: B-patches and the de Boor algorithm for polynomials over triangles. Const. Approx. *7*, 257–279 (1991).
[29] Vegter, G.: The apolar bilinear form in geometric modeling. Math. Comput. *69*, 691–720 (1999).

R. Goldman
Department of Computer Science – MS-132
Rice University
6100 Main Street
Houston, TX 77005-1892
USA
e-mail: rng@cs.rice.edu

Computing [Suppl] 14, 185–198 (2001)

Computing

Localizing the 4-Split Method for G¹ Free-Form Surface Fitting

S. Hahmann, G.-P. Bonneau, and **R. Taleb,** Grenoble

Abstract

One common technique for modeling closed surfaces of arbitrary topological type is to define them by piecewise parametric triangular patches on an irregular mesh. This surface mesh serves as a control mesh which is either interpolated or approximated. A new method for smooth triangular mesh interpolation has been developed. It is based on a regular 4-split of the domain triangles in order to solve the vertex consistency problem. In this paper a generalization of the 4-split domain method is presented in that the method becomes completely local. It will further be shown how normal directions, i.e. tangent planes, can be prescribed at the patch vertices.

1. Introduction

Numerous areas of application such as geometric modeling, scientific visualization, medical imaging need to pass a surface through a set of data points. Not all of them need smooth surfaces. In geometric design, however, it is often desirable to produce visually smooth surfaces, i.e. surfaces with continuously defined tangent planes. Closed surfaces of arbitrary topological type can generally not be defined as a map of domain on $\mathbb{R}^2$ into $\mathbb{R}^3$ without introducing undesirable singularities. Defining a surface on a triangulated mesh where every patch is the image of one domain triangle allows for arbitrary topological types.

The problem of constructing a parametric triangular G^1 continuous surface interpolating an irregular mesh in space has been considered by many. All methods are local and can be classified depending on how they solve the vertex consistency problem, which occurs when joining with G^1 continuity an even number of C^2-patches around a vertex: there are Clough-Tocher domain splitting methods [2, 9, 15, 16], convex combination schemes [4–6, 13], boundary curve schemes [14, 10], algebraic methods [1], singular parameterizations [12], quasi G^1 interpolants [11].

Recently another type of triangular interpolation schemes has been developed [7] which can be called *triangular 4-split method.* A regular domain triangle 4-split leads to the construction of four quintic Bézier patches which form a macro-patch in one-to-one correspondence to a mesh face. They have one polynomial degree less than Loop's scheme [10] but one degree more than Piper's [15] or Shirman-Séquin's methods [16]. The triangle 4-split is a new approach in parametric triangular mesh interpolation and has several advantages, as explained in chapter 3.1.

It is also a local scheme in that changes of a vertex in the surface mesh only modify a small number of surface patches. But it is not *completely local.* Complete locality would mean that changes of a mesh vertex only affect the patches incident to this vertex. This is a very important property, because more the scheme is local, the more it is well adapted for use in an interactive design system. Real-time modifications of a complex object require minimum computation and display time.

The main aim of the present paper is to generalize the triangular 4- split method in order to make it completely local. A welcome side effect is that interpolation of tangential data will now be possible. In Section 2, the vertex consistency problem is described, and some notations are introduced. In Section 3, it is shown how the G^1 interpolation/approximation scheme introduced in [7] can be made completely local by using a virtual neighbourhood for each input vertex. Section 4 shows how the complete locality can be used to interpolate tangent planes, or to optimize the shape of the output G^1 surface. Eventually, Section 5 gives some examples.

2. The Vertex Consistency Problem

2.1. Notations

For the purpose of this paper, a *control mesh* $\mathcal{M}$ is a set of vertices, edges and triangular faces that describe an oriented 2-manifold in $\mathbb{R}^3$. The number of faces or edges incident to a vertex is referred to as the *order* of a vertex. The set of vertices sharing the edges incident to a vertex is called *vertex neighbourhood.*

In one-to-one correspondence to the mesh faces a collection of triangular patches joining each other with tangent plane continuity is constructed. They are called *macro patches* M^i and consist of four C^1 continuous triangular Bézier patches. The resulting piecewise polynomial surface $\mathcal{S}$ can either be an approximation to the control mesh or it can interpolate the vertices of $\mathcal{M}$. In both cases, it can optionally interpolate given normal directions at the patch vertices. The reader is supposed to be familiar with Bézier curves and triangular Bézier patches, otherwise details can be found in [3].

2.2. G^1-conditions

When constructing a network of polynomial patches with G^1 continuity special attention has to be paied to what happens at the patch vertices. For this reason, the parameterization of the macro-patches has been chosen as illustrated in Fig. 1. Each macro-patch M^i is the image of the unit triangle in $\mathbb{R}^2$.

The index $i = 1, \ldots, n$ is taken modulo n, where n is the order of the mesh vertex corresponding to M^i (0, 0).

Let $M^{i-1}(u_{i-1}, u_i)$ and $M^i(u_i, u_{i+1})$ be two adjacent patches that share a common boundary, i.e. $M^{i-1}(0, u_i) = M^i(u_i, 0)$ for $u_i \in [0, 1]$. M^{i-1} and M^i meet with G^1 continuity if there exists a scalar function Φ_i such that

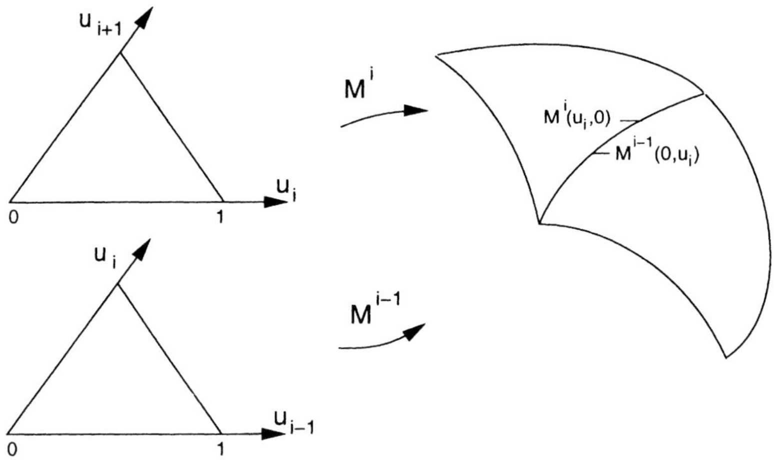

Figure 1. Parameterization

$$(C) \qquad \Phi_i(u_i)\frac{\partial M^i}{\partial u_i}(u_i,0) = \frac{1}{2}\frac{\partial M^i}{\partial u_{i+1}}(u_i,0) + \frac{1}{2}\frac{\partial M^{i-1}}{\partial u_{i-1}}(0,u_i), \quad i = 1,\ldots,n.$$

These simplified G^1-conditions are used in order to keep the scheme of as low degree as possible.

Difficulties can now arise when joining several polynomial patches together around a common vertex with G^1 continuity. This problem has been mentioned by several authors and can be called *vertex consistency problem* [14] or *twist compatibility problem* [17]. At a vertex, where n patches meet, G^1 continuity can generally not be achieved by simply solving the linear system of n equations (C). The G^1 continuity at such a vertex is directly related to the twists. For polynomial patches, which lie in the continuity class C^2, both twists are identical

$$\frac{\partial^2 M^i}{\partial u_i \partial u_{i+1}}(0,0) = \frac{\partial^2 M^i}{\partial u_{i+1}\partial u_i}(0,0), \quad i = 1,\ldots,n.$$

Therefore, additional conditions at the patch corner, which involve the twists, have to be satisfied for G^1 continuity of a network of patches:

$$\frac{1}{2}\frac{\partial^2 M^i}{\partial u_i \partial u_{i+1}}(0,0) + \frac{1}{2}\frac{\partial^2 M^{i-1}}{\partial u_{i-1}\partial u_i}(0,0) = \Phi^1\frac{\partial M^i}{\partial u_i}(0,0) + \Phi^0\frac{\partial^2 M^i}{\partial u_i \partial u_i}(0,0), \quad i = 1,\ldots,n \tag{1}$$

where $\Phi^0 := \Phi_i(0)$ and $\Phi^1 := \Phi_i'(0)$ are further simplifying assumptions to the G^1-conditions in the present paper. System (1) is obtained by differentiating (C) with respect to u_i taken at $u_i = 0$.

In matrix notation, the system (1) states as follows

$$(T) \qquad T\bar{\boldsymbol{t}} = \Phi^1 \bar{\boldsymbol{r}}^1 + \Phi^0 \bar{\boldsymbol{r}}^2,$$

where

$$T = \begin{bmatrix} \frac{1}{2} & 0 & \cdots & & \frac{1}{2} \\ \frac{1}{2} & \frac{1}{2} & & \cdots & 0 \\ & & \ddots & & \\ 0 & \cdots & \frac{1}{2} & \frac{1}{2} & 0 \\ 0 & \cdots & & \frac{1}{2} & \frac{1}{2} \end{bmatrix}, \quad \bar{\boldsymbol{r}}^1 = \begin{bmatrix} \frac{\partial M^1}{\partial u_1}(0,0) \\ \vdots \\ \frac{\partial M^n}{\partial u_n}(0,0) \end{bmatrix}, \quad \bar{\boldsymbol{r}}^2 = \begin{bmatrix} \frac{\partial^2 M^1}{\partial u_1 \partial u_1}(0,0) \\ \vdots \\ \frac{\partial^2 M^n}{\partial u_n \partial u_n}(0,0) \end{bmatrix},$$

and $\bar{\boldsymbol{t}}$ is the vector of the twists.

This system can have singularities when n is even, due to the circulant structure of the matrix T, when solving for the twists. In that case, special solutions of the first and second derivatives of the boundary curves at the vertex have to be found in order to get a solution of system (T). In the present paper a similar solution to [7, 10] is given. It consists of determining patch boundary curves, such that the first and second derivative vectors $\bar{\boldsymbol{r}}^1$, $\bar{\boldsymbol{r}}^2$ lie in the image space of the matrix T. More details in chapter 3.2.

3. Generalized Triangle 4-Split Interpolation Method

3.1. Domain 4-Split

Let us first introduce the basic idea of the present method as introduced in [7]. We aim to construct a piecewise polynomial surface interpolating or approximation the control mesh $\mathcal{M}$ by a *complete local method.* The vertex consistency problem will be solved by constructing first a network of boundary curves subject to (T). An obvious requirement on these curves is therefore, that their first and second derivatives at the vertices should be independent from each other. This implies that a boundary curve corresponding to an edge of $\mathcal{M}$ has to be of degree 5 at least. Another important requirement on the curve network is to keep the polynomial degree as low as possible.

To this intend a regular 4-split of the domain triangle is introduced as shown in Fig. 2. It enables us to take piecewise polynomial boundary curves of degree 3 instead of 5. The twist compatibility system (T) can now be solved for each vertex independently. This means, that for each vertex of $\mathcal{M}$ the cubic boundary curve pieces corresponding to the edges incident to that vertex can be constructed independently from the joining curves pieces.

In general, the 4-split allows to construct not only a low degree curve network but also low degree cross-boundary tangents, which finally leads to a piecewise

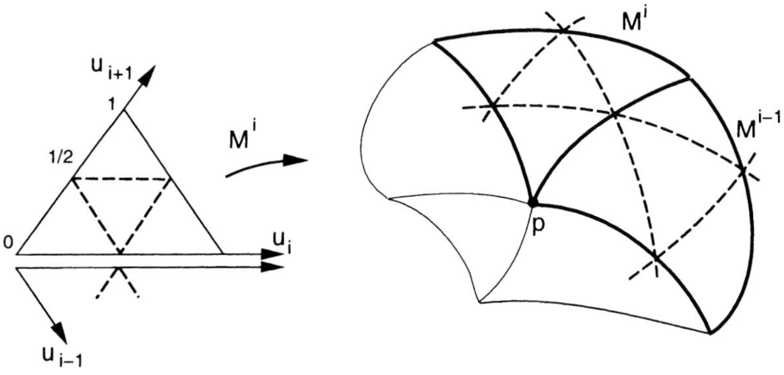

Figure 2. 4-split of the domain triangles

quintic surface spline. Each macro-patch M^i is composed of four quintic Bézier triangles.

3.2. Boundary Curve Network and Cross-Boundary Tangents

A network of curves being the boundary curves of the macro-patches is constructed in correspondence to the edges of $\mathcal{M}$. These curves are so-called *twist compatible* curves because they satisfy twist compatibility conditions (T).

Each boundary curve between two adjacent patches is a piecewise (2 pieces) cubic Bézier curve parameterized on $\{0, \frac{1}{2}, 1\}$. Around each vertex of $\mathcal{M}$ the control points $\boldsymbol{b}_0^i, \boldsymbol{b}_1^i, \boldsymbol{b}_2^i$, $i = 1, \ldots, n$, of all incident boundary curves are constructed independently from the joining curve piece of the opposite vertices. The "midpoints" $\boldsymbol{b}_3^i$ are then constructed in order to have C[1] boundary curves. See Fig. 3 for the notations.

At a vertex $\boldsymbol{v}$ the Φ_i-functions which are defined on the incident edges to $\boldsymbol{v}$, are first determined by calculating $\Phi_i(0)$ and $\Phi_i(1)$ from system (C) by solving it for $u_i = 0$

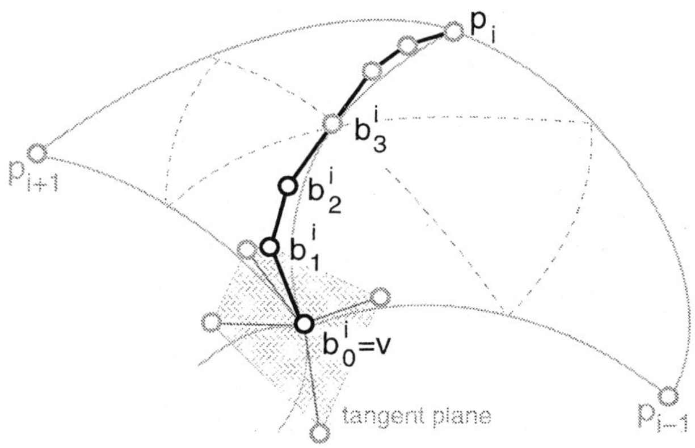

Figure 3. Control points of the boundary curves at vertex **v**

and $u_i = 1$ resp., which gives $\Phi^0 = \Phi_i(0) = \cos\left(\frac{2\pi}{n}\right)$ and $\Phi_i(1) = 1 - \cos\left(\frac{2\pi}{n_i}\right)$. The domain 4-split now enables to seperate vertex derivatives and to take the Φ_i-function piecewise linear.

$$\Phi_i(u_i) = \begin{cases} \cos\frac{2\pi}{n}(1 - 2u_i) + u_i & \text{for } u_i \in \left[0, \frac{1}{2}\right] \\ (1 - u_i) + \left(1 - \cos\frac{2\pi}{n_i}\right)(2u_i - 1) & \text{for } u_i \in \left[\frac{1}{2}, 1\right] \end{cases}$$

Let us now adopt a matrix notation for the boundary curve control points between $\boldsymbol{v}$ and $\boldsymbol{p}_i, i = 1, \ldots, n$:

$$\bar{\boldsymbol{b}}_0 := \begin{bmatrix} \boldsymbol{b}_0^1 \\ \vdots \\ \boldsymbol{b}_0^n \end{bmatrix}, \quad \bar{\boldsymbol{b}}_1 := \begin{bmatrix} \boldsymbol{b}_1^1 \\ \vdots \\ \boldsymbol{b}_1^n \end{bmatrix}, \quad \bar{\boldsymbol{b}}_2 := \begin{bmatrix} \boldsymbol{b}_2^1 \\ \vdots \\ \boldsymbol{b}_2^n \end{bmatrix}, \quad \bar{\boldsymbol{p}} := \begin{bmatrix} \boldsymbol{p}_1 \\ \vdots \\ \boldsymbol{p}_n \end{bmatrix}, \quad \bar{\boldsymbol{v}} := \begin{bmatrix} \boldsymbol{v} \\ \vdots \\ \boldsymbol{v} \end{bmatrix},$$

where $\bar{\boldsymbol{p}}$ is referred to as the *vertex neighborhood* of $\boldsymbol{v}$.

The following choice for the boundary curve Bézier points near the vertex $\boldsymbol{v}$ satisfies the G^1-conditions (C). Simultanously, these control points lie in the image space of matrix T, and therefore allow to solve the system (T) for the twists. See also [7, 10]:

$$\begin{aligned} \bar{\boldsymbol{b}}_0 &= \alpha\bar{\boldsymbol{v}} + B^0\bar{\boldsymbol{p}}, \\ \bar{\boldsymbol{b}}_1 &= \alpha\bar{\boldsymbol{v}} + B^1\bar{\boldsymbol{p}}, \\ \bar{\boldsymbol{b}}_2 &= \left[(\gamma_0 + \gamma_1)\alpha + \frac{\gamma_2}{3}\right]\bar{\boldsymbol{v}} + B^2\bar{\boldsymbol{p}}, \quad \gamma_0 + \gamma_1 + \gamma_2 = 1, \end{aligned} \tag{3}$$

where B^0, B^1, B^2 are $n \times n$ matrices defined by

$$\begin{aligned} B_{ij}^0 &= \frac{1-\alpha}{n} \\ B_{ij}^1 &= \frac{1 - \alpha + \beta\cos\left(\frac{2\pi(j-i)}{n}\right)}{n} \\ B_{ij}^2 &= \frac{(\gamma_0 + \gamma_1)(1-\alpha) + \gamma_1\beta\cos\left(\frac{2\pi(j-i)}{n}\right)}{n} + \gamma_2 \begin{cases} 1/6 & \text{if } j = i-1, i+1 \\ 1/3 & \text{if } j = i \\ 0 & \text{otherwise.} \end{cases} \end{aligned} \tag{4}$$

The free parameters $\alpha, \beta, \gamma_1, \gamma_2$ control the interpolation/approximation of mesh vertices, the first and second derivatives. In [8] it is shown how they can be set optimally. The control points of the joining curves pieces $\boldsymbol{b}_0^k$, $\boldsymbol{b}_1^k$, $\boldsymbol{b}_2^k$ and $\boldsymbol{b}_3^k = \boldsymbol{b}_3^i$ are found by applying the formulas (3) and (4) to the neighbouring mesh points $\boldsymbol{p}_i$ of $\boldsymbol{v}$. k is the index of $\boldsymbol{v}$ relative to the neighborhood of $\boldsymbol{p}_i$.

The cross-boundary tangents are subject to the G^1 conditions (C), the vertex consistency constrains (T) and the curve network and are set to be equal

$$\begin{aligned}\frac{\partial M^i}{\partial u_{i+1}}(u_i, 0) &= \Phi_i(u_i)\frac{\partial M^i}{\partial u_i}(u_i, 0) + \Psi_i(u_i)V_i(u_i),\\ \frac{\partial M^{i-1}}{\partial u_{i-1}}(0, u_i) &= \Phi_i(u_i)\frac{\partial M^i}{\partial u_i}(u_i, 0) - \Psi_i(u_i)V_i(u_i).\end{aligned} \tag{5}$$

The scalar function Ψ_i and the vector function V_i are built of minimal degree so as to interpolate the values of the cross-derivatives and the twists at the vertices v and p_i:

$$\begin{aligned}\Psi_i(u_i) &= \sin\frac{2\pi}{n}(1 - u_i) + \sin\frac{2\pi}{n_i}u_i & &(\textit{linear})\\ V_i(u_i) &= \sum_{k=1}^{n} v_i^k B_k^2(2u_i) \quad u_i \in \left[0, \frac{1}{2}\right], & &(\textit{piecewise quadratic})\end{aligned} \tag{6}$$

where

$$\bar{v}_0 = V^0\bar{p}, \quad \bar{v}_1 = V^1\bar{p}, \quad \bar{v}_2 = \frac{1}{2}\bar{v}_1 + \frac{1}{2}\bar{v}_1^k \, . \tag{7}$$

The $n \times n$ matrices V^0 and V^1 are given by

$$V_{ij}^0 = \frac{6\beta}{n}\sin\left(\frac{2\pi(j - i)}{n}\right), \quad i, j = 1, \ldots, n,$$

$$\begin{aligned}V_{ij}^1 &= \frac{1}{\psi_i^0}\left[(6\phi^1 - 48\phi^0 + 24\phi^0)\tan\left(\frac{\pi}{n}\right) - 6\psi_i^1\right]\frac{\beta}{n}\sin\left(\frac{2\pi(j - i)}{n}\right)\\ &+ \frac{4}{\psi_i^0}\gamma_2\phi^0\begin{cases}1 & \text{if } j = i + 1\\ -1 & \text{if } j = i - 1,\end{cases}\end{aligned} \tag{8}$$

where $\Phi^0 = \Phi_i(0)$, $\Phi^1 = \Phi_i'(0)$ and $\Psi_i^1 = \Psi_i'(0)$ are known from (2) and (6).

Although the boundary curves and the cross-boundary tangents are piecewise cubic, the macro-patches will be piecewise quintic. With quartic patches a vertex consistency problem could occur at the boundary mid-points which are supplementary vertices of order 6 (see domain triangle 4-split). This problem is automatically solved by the special choice of the cross-boundary tangents (7).

The explicit Bézier representation of the boundary curves is already known. In order to obtain quintic curves two degree elevations have to be performed of (3). Some further simple calculations combining (5)–(8) with (3) are necessary to get the first inner row of Bézier points of the macro-patches from the cross-boundary tangents. The formulas are explicitly given in [7].

3.3. Filling-in the Macro-Patches

Each macro-patch is composed of four quintic triangular Bézier patches. The boundary curves of a macro-patch are the twice degree elevated curves of Section 3.2. The cross-boundary tangents of Section 3.2 determine the first inner row of control points after one degree elevation. The remaining 15 inner control points, which are highlighted in Fig. 4, are used for joining the four inner patches with C^1 continuity. Six of them can be chosen arbitrarily.

3.4. Complete Locality

The present triangle 4-split method has several properties. It is an *affine invariant* scheme because only affine combinations of mesh vertices of $\mathcal{M}$ are used. An *explicit closed form* Bézier representation of the quintic patches is known. Several *shape parameters* and free control points are available for local shape modifications or shape optimizations. And the present method is *local* in that changes of a mesh vertex of $\mathcal{M}$ only affect a small number of patches.

Locality is a very important property for interpolation schemes. The advantages of the locality are obvious: the algorithms are generally numerically stable, no linear systems of equations have to be solved here. The algorithms are fast, because local modifications of the input data imply only local updates of the interpolating surface. Interactive real-time modeling of 3D objects with a large and complicate input mesh becomes possible.

It is easy to see, that the 4-split method is local: the algorithm works in two steps. First the boundary curves and cross-boundary tangents are constructed piecewise around each vertex and then joined together. For each vertex $\boldsymbol{v}$ the incoming curve pieces (Fig. 5) are calculated by using only the local neighbourhood points $\bar{\boldsymbol{p}} = [\bar{\boldsymbol{p}}_1, \ldots, \boldsymbol{p}_n]^T$ of $\boldsymbol{v}$, see equations (3) and (7) and the icosahedron example in Fig. 6(a). Once the boundary curves and cross-boundary tangents are fixed a second step of the algorithm consists of calculating the remaining inner control points for each macro-patches locally, i.e. independent from the neighbouring macro-patches.

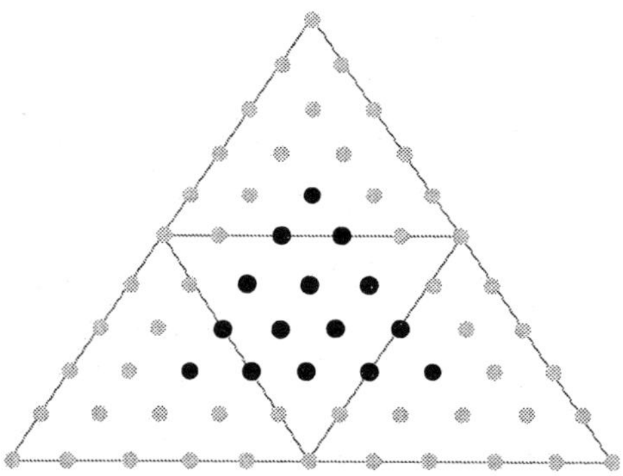

Figure 4. 15 free inner control points making macro-patches C^1

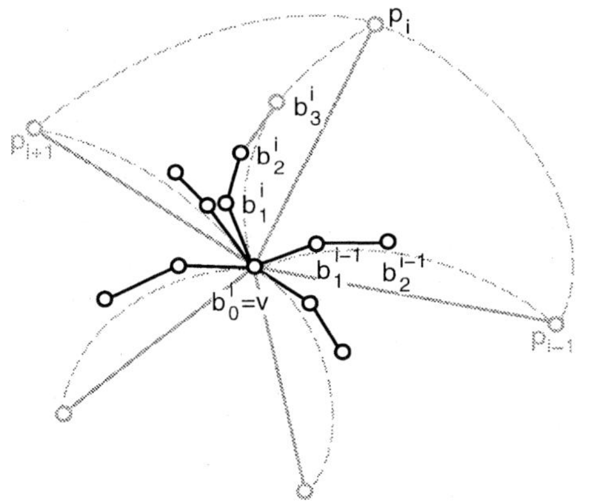

Figure 5. Boundary curves incident to $\boldsymbol{v}$. A first step of the algorithm consists of calculating these curve piece for each vertex, idem for the cross-boundary tangents, and of joining them together in the middle

Modification of a mesh vertex $\boldsymbol{v}$ has therefore influence on all the macro-patches having $\boldsymbol{v}$ in common, and on all the macro-patches having the neighbourhood points $\bar{\boldsymbol{p}} = [\boldsymbol{p}_1, \ldots, \boldsymbol{p}_n]^T$ of $\boldsymbol{v}$ in common. This is due to the boundary curve construction, as mentioned below. For the icosahedron example, Fig. 6b shows all boundary curves which are concerned, when vertex $\boldsymbol{v}$ is modified.

It turns out that this method is not as local as it is desirable and useful. It will now be shown, that it is possible to *generalize the 4-split method* in order to make it a complete local interpolation scheme. *Complete local* means here, that modifications of a mesh vertex $\boldsymbol{v}$ only influence the n (order of $\boldsymbol{v}$) macro-patches incident to $\boldsymbol{v}$, see fig. 6(c). To this end the control points of the boundary curves, $\bar{\boldsymbol{b}}_0$, $\bar{\boldsymbol{b}}_1$, $\bar{\boldsymbol{b}}_2$, and the control points of the cross-boundary tangents $\bar{\boldsymbol{v}}_0$, $\bar{\boldsymbol{v}}_1$ should be made independent of the vertex neighbourhood $\bar{\boldsymbol{p}}$ of $\boldsymbol{v}$. It can be observed that satisfying the G^1-condition for $\bar{\boldsymbol{b}}_0$, $\bar{\boldsymbol{b}}_1$, $\bar{\boldsymbol{b}}_2$ and $\bar{\boldsymbol{v}}_0$, $\bar{\boldsymbol{v}}_1$ doesn't depend on a particular choice of $\bar{\boldsymbol{p}}$. The curve control points $\boldsymbol{b}_1^i$ for example satisfy the G^1-conditions because they are the result of an first order Fourier approximation of n distinct points. Thus

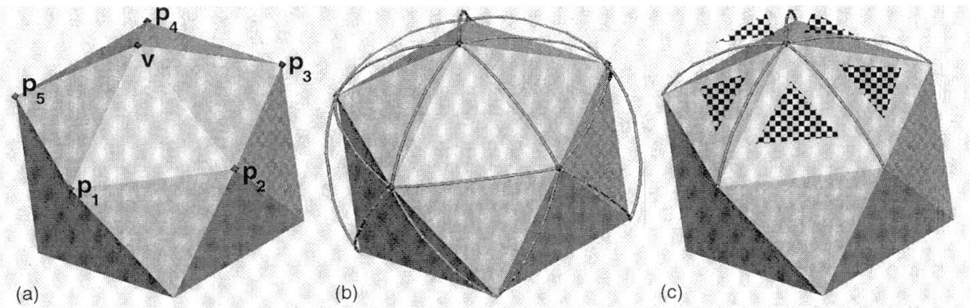

Figure 6. The input mesh is a regular polyhedra, an icosahedron. **a** Local neighbourhood points $\boldsymbol{p}_i$ of a mesh vertex $\boldsymbol{v}$. **b** Boundary curves which depend on vertex $\boldsymbol{v}$. The control polygons of the piecewise degree five curves are shown. **c** Macro-patches and boundary curves depending on vertex $\boldsymbol{v}$ when using the concept of virtual neighbourhoods in the algorithm

they make the first derivative of the boundary curves lying in the image space of (T). Similarly for the others. Furthermore, the construction of the boundary curve pieces (3) and the cross-boundary tangent pieces (7) is local around a mesh vertex $\boldsymbol{v}$. The vertex neighbourhood $\bar{\boldsymbol{p}}$ can therefore be replaced by another new "virtual" neighbourhood $\bar{\boldsymbol{p}}^* = [\boldsymbol{p}_1^*, \ldots, \boldsymbol{p}_n^*]^T$. The following equations replace (3) and (7) in the algorithm.

New boundary curve's Bézier points:

$$\begin{aligned} \bar{\boldsymbol{b}}_0 &= \alpha\bar{\boldsymbol{v}} + B^0\bar{\boldsymbol{p}}^*, \\ \bar{\boldsymbol{b}}_1 &= \alpha\bar{\boldsymbol{v}} + B^1\bar{\boldsymbol{p}}^*, \\ \bar{\boldsymbol{b}}_2 &= \left[(\gamma_0 + \gamma_1)\alpha + \frac{\gamma_2}{3}\right]\bar{\boldsymbol{v}} + B^2\bar{\boldsymbol{p}}^*, \quad \gamma_0 + \gamma_1 + \gamma_2 = 1, \end{aligned} \tag{9}$$

$$\bar{\boldsymbol{v}}_0 = V^0\bar{\boldsymbol{p}}^*, \quad \bar{\boldsymbol{v}}_1 = V^1\bar{\boldsymbol{p}}^*, \quad \bar{\boldsymbol{v}}_2 = \frac{1}{2}\bar{\boldsymbol{v}}_1 + \frac{1}{2}\bar{\boldsymbol{v}}_1^k, \tag{10}$$

where the matrices B^0, B^1, B^2, V^0, V^1 are given by (4) and (8). Doing this for all mesh vertices finally leads to a complete local mesh fitting scheme.

4. Choice of Virtual Vertex Neighbourhood

Up to now the 4-split method calculates first and second order derivative informations of the surface $\mathscr{S}$ at a vertex $\boldsymbol{v}$ by using the n neighbourhood points

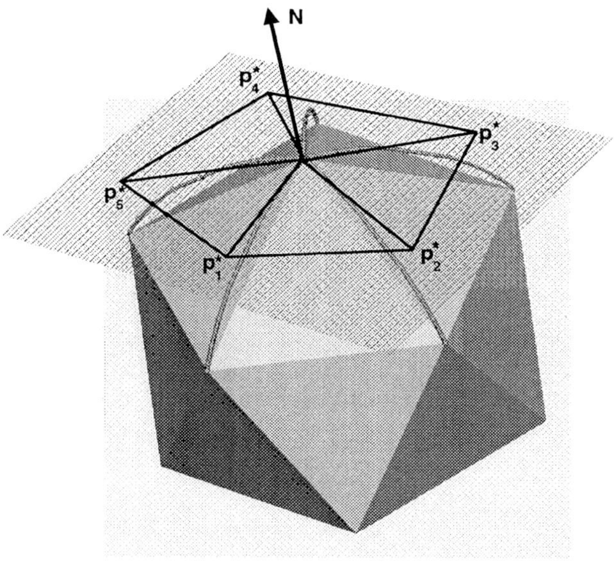

Figure 7. The virtual neighbourhood points $\boldsymbol{p}_i^*$ lie in a plane together with the vertex $\boldsymbol{v}$ orthogonal to **N** in order to make the surface interpolating the given normal vector **N**

$\boldsymbol{p}_1, \ldots, \boldsymbol{p}_n$ of $\boldsymbol{v}$. They are vertices of the input mesh and are therefore not free. In the generalized method, presented in the previous chapter, this set of n points can be chosen arbitrarily for each mesh vertex. How these novel degrees of freedom can be used in order to obtain pleasing shapes or in order to create shapes design handles is now shown in the following subsections.

4.1. How Many Degrees of Freedom?

By replacing the true neighbourhood points $\boldsymbol{p}_1, \ldots, \boldsymbol{p}_n$ by the virtual neighbourhood points $\boldsymbol{p}_1^*, \ldots, \boldsymbol{p}_n^*$ additional degrees of freedom are created at ewach vertex. They can be used either for normal vector interpolation at the mesh vertices or for surface fairing methods. The number of degrees of freedom depends on how the points $\boldsymbol{p}_i^*$ are combined in equation (9) and (10). The number of degrees of freedom for calculating the first derivative points in (9) for example is equal to the rank of the matrix B^1, which is equal 2. How these two vector valued degrees of freedom can be employed is subject of the next sections.

4.2. Interpolation of Normal Vector Input

The tangent plane of $\mathcal{S}$ at a vertex $\boldsymbol{v}$ is spanned by the points $\boldsymbol{b}_1^i, i = 1, \ldots, n$ and $\boldsymbol{v}$. These $n+1$ points are all lying in the same plane, since the boundary curve's first derivatives

$$\frac{\partial M^i}{\partial u_i}(0,0) = 6(\boldsymbol{b}_1^i - \boldsymbol{v})$$

satisfy the G¹ conditions at the vertex $\boldsymbol{v}$. The Bézier control points are obtained from a weighted averaging of the virtual neighbourhood points $\boldsymbol{p}_i^*$ given by:

$$\bar{\boldsymbol{b}}_1 = \begin{bmatrix} \boldsymbol{b}_1^1 \\ \vdots \\ \boldsymbol{b}_1^n \end{bmatrix} = \boldsymbol{v} + \frac{\beta}{n} \begin{bmatrix} \sum_{j=1}^n \cos\left(\frac{2\pi(j-1)}{n}\right) \boldsymbol{p}_j^* \\ \vdots \\ \sum_{j=1}^n \cos\left(\frac{2\pi(j-n)}{n}\right) \boldsymbol{p}_j^* \end{bmatrix}.$$

In other words, the normal vector of $\mathcal{S}$ at a vertex $\boldsymbol{v}$ is the weighted combination of the normal vectors of the n planes spanned by $\{\boldsymbol{v}, \boldsymbol{p}_i^*, \boldsymbol{p}_{i+1}^*\}$. The weights come from the G¹ conditions, but the points $\boldsymbol{p}_i^*$ are free. It is therefore possible to interpolate a given normal vector at the mesh vertices. To this end, the oriented points $\boldsymbol{p}_i^*$ have to lie in a plane together with $\boldsymbol{v}$, which is orthogonal to the given normal vector.

4.3. Shape Optimization

By replacing the true vertex neighbourhood of a mesh vertex $\boldsymbol{v}$ by the new points $\boldsymbol{p}_i^*$, $i = 1, \ldots, n$, which are free, extranous degrees of freedom have been created for the whole surface scheme. If normal direction interpolation is not desired, the points $\mathbf{p}_i^*$ can be determined by some optimization process on the curve network.

The shape of the resulting surface depends mainly on the shape of the boundary curves. A "well shaped" curve network should f.ex. avoid undulations. The free virtual neighbourhood $\bar{\boldsymbol{p}}^*$ and the free curve shape parameters $\beta, \gamma_1, \gamma_2$ are available for each mesh vertex. They can be determined by local or global optimization on the curve network by using some minimum norm criteria, like energy functionals. Based on this concept of virtual neighbourhood points, the paper [8] proposes and tests various appropriate criteria for shape optimizations.

5. Complete Local Fitting of Arbitrary Meshes

A first example, Fig. 8, simply illustrates the surface construction steps: input mesh, boundary curves, filling-in the macro-patches. The input mesh is a triangulated regular polyhedron with 12 vertices on the unit sphere, called icosahedron. The boundary curves are computed first, see Fig. 8a. The paramater α is set to one, the mesh vertices are therefore interpolated. The macro-patches are then filled-in with four quintic Bézier patches each. The resulting surface is shown in Fig. 8b. The shape parameters and free inner control points have been chosen in order to approximate the unit sphere. The L_∞-error between the surface and the

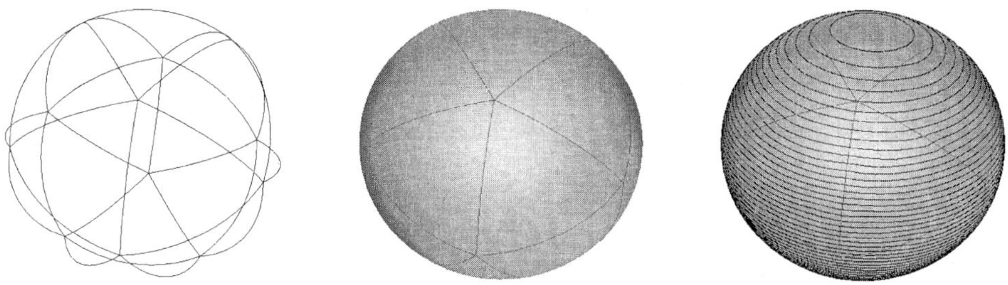

Figure 8. Interpolated icosahedron with isophote analysis

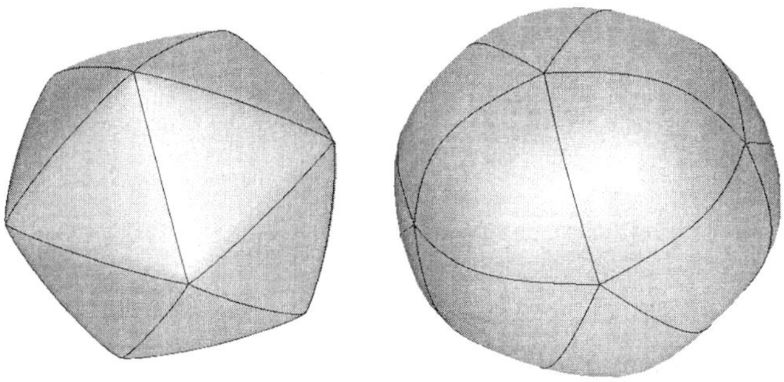

Figure 9. Interpolated icosahedron with different shape parameters

unit sphere is 0.0033. An isophote analysis in Fig. 8c shows the **global smoothness** of the spline surface.

It is then possible to choose other **shape parameters**, which are stretching the boundary curves and flattening the macro-patches see Fig. 9a or rounding out the curves and patches. see 9b.

The **complete locality** of the surface scheme is illustrated on the icosahedron example in Fig. 10. In both examples one mesh vertex has been modified and it can be observed that only the surface macro-patches incident to this vertex have been modified, see Fig. 10b,d. The left image of each example shows the control nets of the Bézier patches. The four patches of each macro patch are colored individually, see Fig. 10a,c.

The next example, Fig. 11, shows another surface with vertices of order 6 and 4. Additional to the input mesh normal directions are interpolated at the mesh

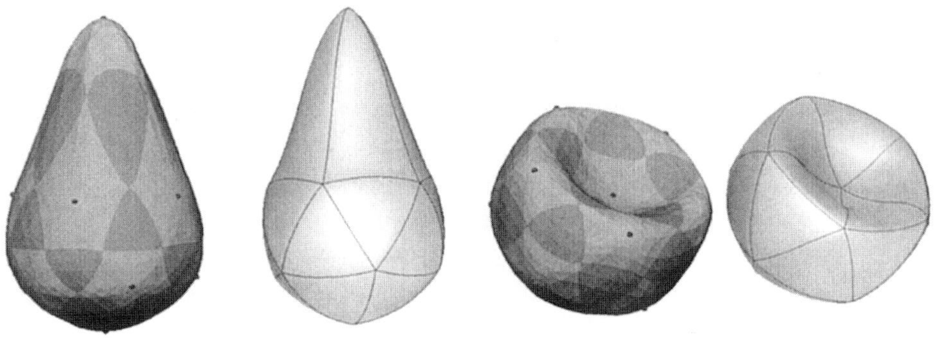

Figure 10. Locally modified icosahedron

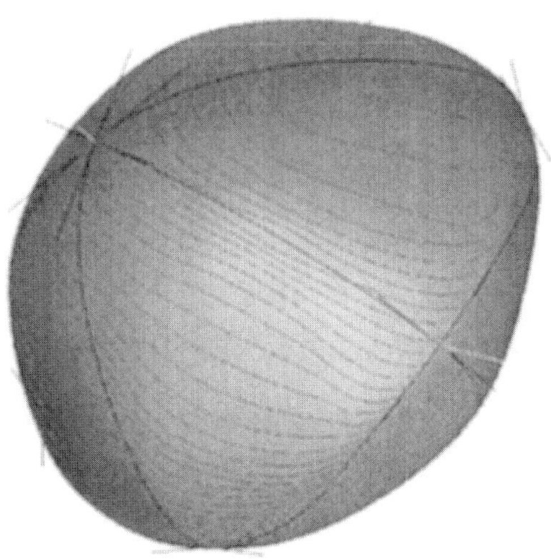

Figure 11. Normal interpolation, vertices of order 4 and 6

vertices. They are shown as gray lines in Fig. 11. The shape parameters are fixed automatically by a local form optimization method (Section 4.3).

References

[1] Bajaj, C.: Smoothing polyhedra using implicit algebraic splines. Comput. Graphics *26*, 79–88 (1992).
[2] Farin, G.: A construction for visual C^1 continuity of polynomial surface patches. Comput. Graphics Image Proc. *20*, 272–282 (1982).
[3] Farin, G.: Curves and surfaces for computer aided geometric design 4th ed. New York: Academic Press, 1997.
[4] Gregory, J. A.: N-sided surface patched. In: The mathematics of surfaces (Gregory, J. ed.), pp. 217–232. Oxford: Clarendon Press, 1986.
[5] Hagen, H.: Geometric surface patches without twist constraints. Comput. Aided Geom. Des. *3*, 179–184 (1986).
[6] Hagen, H., Pottmann, H.: Curvature continuous triangular interpolants. In: Mathematical methods in computer aided geometric design (Lyche, T., Schumaker, L. L. eds.), pp. 373–384. New York: Academic Press, 1989.
[7] Hahmann, S., Bonneau, G.-P.: Triangular G^1 interpolation by 4-splitting domain triangles. Comput. Aided Geom. Des. *17*, 731– 757 (2000).
[8] Hahmann, S., Bonneau, G.-P., Taleb, R.: Smooth irregular mesh interpolation. In: Curve and surface fitting: Saint-Malo 1999 (Cohen, A., Rabut, C., Schumaker, L. L. eds.), pp. 237–246. Nashville: Vanderbilt University Press, 2000.
[9] Jensen, T.: Assembling triangular and rectangular patches and multivariate splines. In: Geometric modeling: algorithms and new trends (Farin, G. ed.), pp. 203–220. Philadelphia: SIAM, 1987.
[10] Loop, C.: A G^1 triangular spline surface of arbitrary topological type. Comput. Aided Geom. Des. *11*, 303–330, (1994).
[11] Mann, S.: Surface approximation using geometric Hermite patches. PhD dissertation. University of Washington, 1992.
[12] Neamtu, M., Pluger, P.: Degenerate polynomial patches of degree 4 and 5 used for geometrically smooth interpolation in $\mathbb{R}^3$. Comput. Aided Geom. Des. *11*, 451–474 (1994).
[13] Nielson, G.: A transfinite, visually continuous, triangular interpolant. In: Geometric modeling: algorithms and new trends (Farin, G. ed.), pp. 235–246. Philadelphia: SIAM, 1987.
[14] Peters, J.: Smooth interpolation of a mesh of curves, Construct. Approx. *7*, 221–246 (1991).
[15] Piper, B. R.: Visually smooth interpolation with triangular Bézier patches. In: Geometric modeling: algorithms and new trends (Farin, G. ed.), pp. 221–233. Philadelphia: SIAM, 1987.
[16] Shirman, L. A., Séquin, C. H.: Local surface interpolation with Bézier patches. Comput. Aided Geom. Des. *4*, 279–295 (1987).
[17] Van Wijk, J. J.: Bicubic patches for approximating non-rectangular control meshes. Comput. Aided Geom. Des. *3*, 1–13 (1986).

S. Hahmann
G.-P. Bonneau
R. Taleb
Laboratoire LMC-CNRS
University of Grenoble
B.P. 53, F-38041 Grenoble cedex 9
France
e-mail: Stefanie.Hahmann@imag.fr

Computing [Suppl] 14, 199–218 (2001)

Computing

Surface Reconstruction Using Adaptive Clustering Methods

B. Heckel, Mountain View, CA, **A. E. Uva,** Bari, **B. Hamann,** and **K. I. Joy,** Davis, CA

Abstract

We present an automatic method for the generation of surface triangulations from sets of scattered points. Given a set of scattered points in three-dimensional space, without connectivity information, our method reconstructs a triangulated surface model in a two-step procedure. First, we apply an adaptive clustering technique to the given set of points, identifying point subsets in regions that are nearly planar. The output of this clustering step is a set of two-manifold "tiles" that locally approximate the underlying, unknown surface. Second, we construct a surface triangulation by triangulating the data within the individual tiles and the gaps between the tiles. This algorithm can generate multiresolution representations by applying the triangulation step to various resolution levels resulting from the hierarchical clustering step. We compute deviation measures for each cluster, and thus we can produce reconstructions with prescribed error bounds.

AMS Subject Classifications: 65D05, 65D07, 65D10, 65D17, 68U05, 68U07.

Key Words: Surface reconstruction, reverse engineering, clustering, multiresolution representation, triangulation, hierarchical reconstruction.

1. Introduction

Surface reconstruction is concerned with the generation of continuous models (triangulated or analytical) from scattered point sets. Often, these point sets are generated by scanning physical objects or by merging data from different sources. Consequently, they might be incomplete, contain noise or be redundant, which makes a general approach for reconstructing surfaces a challenging problem. In many instances, high complexity and varying level of detail characterize an underlying object. Multiple approximation levels are needed to allow rapid rendering of reconstructed surface approximations and interactive exploration. Surface reconstruction problems arise in a wide range of scientific and engineering applications, including reverse engineering, grid generation, and multiresolution rendering.

We introduce a surface reconstruction method that is based on *cluster analysis*. Our approach generates a surface reconstructed from arbitrary point sets, *i.e.*, scattered data without connectivity information. The reconstructed model is generated in two steps. First, we apply an adaptive clustering method to the point set, producing a set of almost flat shapes, so-called "tiles", that locally approximate the underlying surface. Each tile is associated with a cluster of points. Since each cluster is "nearly planar" we can assume that the data within a cluster can be represented as a height field with respect to the best-fit plane defined by the tile. We can either triangulate all data points in the tile to produce a high-resolution mesh locally representing the surface or we can choose to only triangulate the boundary points defining the polygon of the tile to create a low-resolution local surface approximation.

Second, we triangulate the gaps between the tiles by using a *constrained Delaunay triangulation*, producing a valid geometrical and topological model. We compute distance estimate for each cluster, which allows us to calculate an error measure for the resulting triangulated models. By considering a set of error tolerances, we can construct a hierarchy of reconstructions. Figure 1 illustrates the steps of the algorithm.

In Section 2, we review algorithms related to surface reconstruction that apply to our work. In Section 3, we discuss the mathematics of clustering based on *principal component analysis* (PCA) and the generation of tiles. In Section 4, we describe the triangulation procedure that uses tiles as input and produces a triangulation as output. This section discusses the triangulation of the tiles themselves as well as the method for triangulating the space between the tiles. Results of our algorithm are provided in Section 5. Conclusions and ideas for future work are provided in Section 6.

2. Related Work

Given a set of points $\{\mathbf{p}_i = (x_i, y_i, z_i)^T,\ i = 1, \ldots, n\}$ assumed to originate from a surface in three-dimensional space, the goal of surface reconstruction is to gen-

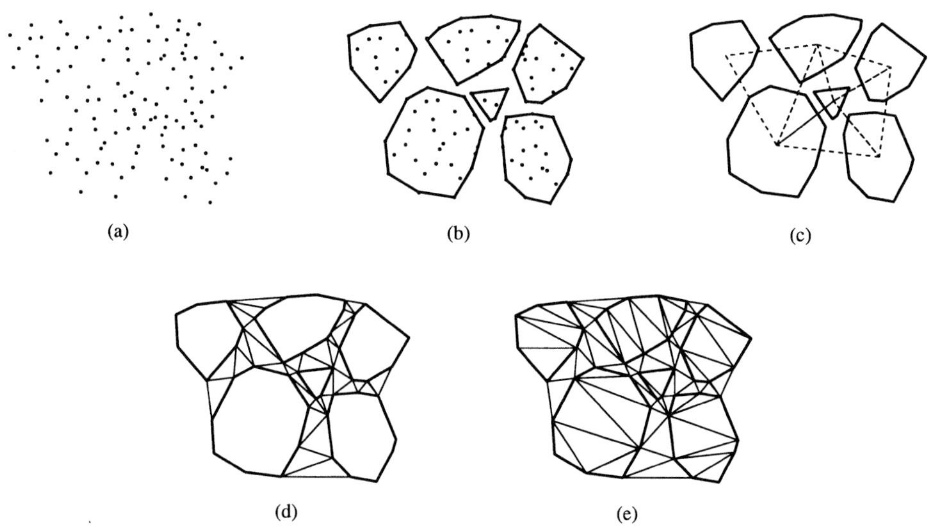

Figure 1. The major steps of the reconstruction algorithm. Given the scattered points in **a** we create the tiles shown in **b** using adaptive clustering. The connectivity graph of these tiles is superimposed in **c** and this graph is used to construct the triangulation of the area between the tiles, shown in **d**. By triangulating the tiles themselves we obtain the final triangulation, shown in **e**

erate a triangulated model approximating the unknown surface. The representation and reconstruction of three-dimensional shapes has been a significant problem in the computer graphics, computer vision, and mechanical engineering communities for several years. Most research has focused on providing a known data structure along with a set of heuristics that enable an approximating mesh to be constructed from the set of sample points.

Boissonnat [8] was one of the first to address the problem of surface reconstruction from a scattered point set. He uses a nearest neighbor criterion to produce an advancing front along the surface. From an initial point $\mathbf{p}_0$, an edge is generated between $\mathbf{p}_0$ and its nearest neighbor $\mathbf{p}_1$. An initial "contour" is generated by considering the two edges $\overline{\mathbf{p}_0\mathbf{p}_1}$ and $\overline{\mathbf{p}_1\mathbf{p}_0}$. This contour is then propagated by selecting a point $\mathbf{p}_2$ in the neighborhood of the edge (considering the k nearest neighbors of $\mathbf{p}_0$ and $\mathbf{p}_1$) such that the projection of $\mathbf{p}_2$ in the tangent plane T, generated by a least-squares method using the neighborhood about the edge, "sees" the projected edge under the largest angle. The point $\mathbf{p}_2$ is added to the contour, creating a triangle, and the algorithm continues with each edge of the contour. Under certain restrictive, non-folding conditions this algorithm is guaranteed to work.

Hoppe et al. [17] and Curless and Levoy [10] utilize a regular grid and produce a signed distance function on this grid. Hoppe et al.'s method [17] is based on a *zero-set* approach for reconstruction, using the given points to create a signed distance function d, and then triangulating the isosurface $d = 0$. They determine an approximate tangent plane at each point $\mathbf{p}$, using a least-squares approximation based on the k nearest neighbors of $\mathbf{p}$. Using adjacent points and tangent planes, they determine the normal to the tangent plane, which is then used to determine the signed distance function. The triangulation is then generated using the marching cubes algorithm of Lorensen et al. [23]. This algorithm produces an approximating triangulation. The approximation is treated as a global optimization problem with an energy function that directly measures deviation of the approximation from the original surface.

Curless and Levoy [10] present an approach to merge several range images by scan-converting each image to a weighted signed distance function in a regular three-dimensional grid. The zero-contour of this distance function, is then triangulated using a marching cubes algorithm [23]. This algorithm also produces an approximating mesh to the data points. The closeness of the approximation is determined by the size of the grid elements.

Boissonnat [8], Attali [3] and Amenta et al. [2] utilize the properties of the *Delaunay triangulation* [30] to assist in generating an interpolating mesh for a set of sample points. Boissonnat's second algorithm [8] first generates a *Delaunay tetrahedrization* $\mathscr{T}$ of the points as an intermediate structure. The boundary of this tetrahedral mesh defines the convex hull of the data points. The algorithm then progressively removes tetrahedra from $\mathscr{T}$, such that the boundary of the resulting set of tetrahedra remains a polyhedron. A drawback of this approach is that no change of the topology is allowed, and consequently, it is impossible to create a surface formed of several connected components and having holes.

Attali [3] utilizes a *normalized mesh*, a subset of the Delaunay triangulation, to approximate a surface represented by a set of scattered data points. When applied to "*r*-regular shapes" in two dimensions, this method is provably convergent. Unfortunately, in three dimensions, heuristics must be applied to complete a surface. The general idea is to construct the Delaunay mesh in two dimensions and remove those triangles that do not contribute to the normalized mesh. The boundary of the remaining triangles forms the boundary of the surface.

Amenta et al. [2] use a three-dimensional "Voronoi diagram" and an associated (dual) Delaunay triangulation to generate certain "crust triangles" on the surface that are used in the final triangulation. The output of their algorithm is guaranteed to be topologically correct and converges to the original surface as the sampling density increases.

The *alpha shapes* of Edelsbrunner et al. [12], which define a simplicial complex for an unorganized set of points, have been used by a number of researchers for surface reconstruction. Guo [14] describes a method for reconstructing an unknown surface of arbitrary topology, possibly with boundaries, from a set of scattered points. He uses three-dimensional alpha shapes to construct a simplified surface that captures the "topological structure" of a scattered data set and then computes a curvature-continuous surface based on this structure. Teichmann and Capps [33] also utilize alpha shapes to reconstruct a surface. They use a local density scaling of the alpha parameter, depending on the sampling density of the mesh. This algorithm requires the normal to the surface to be known at each point.

Bajaj et al. [4] use alpha shapes to compute a domain surface from which a signed distance function can be approximated. After decomposing a set of scattered points into tetrahedra, they fit algebraic surfaces to the scattered data. Bernardini and Bajaj [5] also utilize alpha shapes to construct the surface. This approach provides a formal characterization of the reconstruction problem and allows them to prove that the alpha shape is homeomorphic to the original object and that approximation within a specific error bound is possible. The method can produce artifacts and requires a local "sculpting step" to approximate sharp edges well.

The ball-pivoting algorithm of Bernardini et al. [6] utilizes a ball of a specified radius that pivots around an edge of a seed triangle. If it touches another point another triangle is formed, and the process continues. The algorithm continues until all reachable edges have been considered, and then it re-starts with another seed triangle. This algorithm is closely related to one using alpha shapes, but it computes a subset of the 2-faces of the alpha shape of the surface. This method has provable reconstruction guarantees under certain sampling assumptions, and it is a simply implemented.

Mencl [25] and Mencl and Müller [26] use a different approach. They use an algorithm that first generates a *Euclidean minimum spanning tree* for the point set. This spanning tree is a tree connecting all sample points with line segments so that the sum of the edge lengths is minimized. The authors extend and prune this tree depending on a set of heuristics that enable the algorithm to detect features, connected components and loops in the surface. The graph is then used as a guide to generate a set of triangles that approximate the surface.

The idea of generating clusters on surfaces is similar to the generation of "superfaces" as done by Kalvin and Taylor [21, 22]. This algorithm uses random seed faces, and develops "islands" on the surface that grow through an advancing front. Faces on the current superface boundary are merged into the evolving superface when they satisfy the required merging criteria. A superface stops growing when there are no more faces on the boundary that can be merged. These superfaces form islands that partition the surface and can be triangulated to form a low-resolution triangulation of the surface. Hinker and Hansen [16] have developed a similar algorithm.

The idea of stitching surfaces together has been used by Soucy and Laurendeau [32], who have designed a stitching algorithm to integrate a set of range views. They utilize "canonical subset triangulations" that are separated by a minimal parametric distance. They generate a parametric grid for the empty space between the non-redundant triangulations and utilize a constrained Delaunay triangulation, computed in parameter space, to fill empty spaces. Connecting these pieces allows them to get an integrated, connected surface model.

The algorithm we present is based on a new approach. We utilize an adaptive clustering method [24] to generate a set of "tiles" that represent the scattered data locally. The resulting disjoint tiles, together with the space between the tiles, can be triangulated. Several steps are necessary to implement this method. First, the tiles must be generated. We utilize principal component analysis (PCA) to determine clusters of points that are nearly coplanar. Each tile is generated from the boundary polygon of the convex hull of a cluster of points that have been projected into the best-fit plane. We use a hierarchical clustering scheme that splits clusters where their errors are too large. We determine a connectivity graph for the tiles by generating a Delaunay-like triangulation of the tile centers. Finally, we triangulate the tiles and the space between them by using a localized constrained Delaunay triangulation. By triangulating the original points within the tiles we can obtain a locally high-fidelity and high-resolution representation of the data. By triangulating only the boundary polygons of the tiles, we can also generate a low-fidelity and low-resolution representation of the data.

3. Hierarchical Clustering

Suppose we are given a set of distinct points

$$\mathcal{P} = \{\mathbf{p}_i = (x_i, y_i, z_i)^T, \quad i = 1, \ldots, n\},$$

where the points lie on or close to an unknown surface. We recursively partition this point set by separating it into subsets, or clusters, where each subset consists of nearly coplanar points. In this section, we describe a hierarchical clustering algorithm that utilizes PCA, see Hotelling [19], Jackson [20] or Manly [24], to establish best-fit planes for each cluster. These planes enable us to measure the distance between the original points in the clusters and the best-fit planes, and to establish the splitting conditions for the clusters.

3.1. Principal Component Analysis

Given a set of n points in three-dimensional space, the covariance matrix S of the point set is

$$S = \frac{1}{n-1}(D^T D),$$

where D is the matrix

$$D = \begin{pmatrix} x_1 - \bar{x} & y_1 - \bar{y} & z_1 - \bar{z} \\ \vdots & \vdots & \vdots \\ x_n - \bar{x} & y_n - \bar{y} & z_n - \bar{z} \end{pmatrix} \tag{1}$$

and

$$\bar{\mathbf{c}} = (\bar{x}, \bar{y}, \bar{z})^T = \left(\frac{1}{n}\sum_{i=1}^{n} x_i, \frac{1}{n}\sum_{i=1}^{n} y_i, \frac{1}{n}\sum_{i=1}^{n} z_i \right)^T \tag{2}$$

is the geometric mean of the n points.

The 3×3 matrix S can be factored as $U^T L U$, where L is diagonal and U is an orthonormal matrix. The diagonal elements of L are the eigenvalues $\lambda_{\max}, \lambda_{\mathrm{mid}}$, and $\lambda_{\min}$ of S (ordered by decreasing absolute values), and the columns of U are the corresponding normalized eigenvectors $\vec{\mathbf{e}}_{\max}$, $\vec{\mathbf{e}}_{\mathrm{mid}}$, and $\vec{\mathbf{e}}_{\min}$. These mutually perpendicular eigenvectors define the three axis directions of a local coordinate frame with center $\bar{\mathbf{c}}$.

We use the values of $\lambda_{\max}, \lambda_{\mathrm{mid}}$, and $\lambda_{\min}$ to determine the "degree of coplanarity" of a point set. Three cases are possible:

- Two eigenvalues – λ_{mid} and $\lambda_{\min}$ – are zero, and one eigenvalue – $\lambda_{\max}$ – has a finite, non-zero absolute value. This implies that the n points are collinear.
- One eigenvalue – $\lambda_{\min}$ – is zero, and the other two eigenvalues – $\lambda_{\max}$ and λ_{mid} – have finite, non-zero absolute values. This implies that the n points are coplanar.
- All three eigenvalues have finite, non-zero absolute values.

The eigenvector $\vec{\mathbf{e}}_{\max}$ defines the *orthogonal regression line*, which minimizes the sums of the squares of deviations perpendicular to the line itself. The eigenvectors $\vec{\mathbf{e}}_{\max}$ and $\vec{\mathbf{e}}_{\mathrm{mid}}$ describe the regression plane, which minimizes the sums of the squares of the deviations perpendicular to the plane. Figure 2 illustrates this local coordinate system.

We define the vectors

$$\vec{\mathbf{w}}_{\max} = \vec{\mathbf{e}}_{\max}/\sqrt{|\lambda_{\max}|},$$
$$\vec{\mathbf{w}}_{\text{mid}} = \vec{\mathbf{e}}_{\text{mid}}/\sqrt{|\lambda_{\text{mid}}|}, \quad \text{and}$$
$$\vec{\mathbf{w}}_{\min} = \vec{\mathbf{e}}_{\min}/\sqrt{|\lambda_{\min}|}$$

and let W be the matrix whose columns are $\vec{\mathbf{w}}_{\max}$, $\vec{\mathbf{w}}_{\text{mid}}$, and $\vec{\mathbf{w}}_{\min}$, respectively. The matrix W can be written as $W = UL^{-1/2}$, where

$$L^{-1/2} = \begin{pmatrix} \frac{1}{\sqrt{|\lambda_{\max}|}} & 0 & 0 \\ 0 & \frac{1}{\sqrt{|\lambda_{\text{mid}}|}} & 0 \\ 0 & 0 & \frac{1}{\sqrt{|\lambda_{\min}|}} \end{pmatrix}$$

and

$$WW^T = UL^{-1/2}L^{-1/2}U^T = UL^{-1}U^T = S^{-1}.$$

There is another way to look at this coordinate frame. Given a point $\mathbf{p} = (x, y, z)^T$, one can show that

$$\begin{aligned} \mathbf{p}^T S^{-1}\mathbf{p} &= \mathbf{p}^T WW^T\mathbf{p} \\ &= (W^T\mathbf{p})^T(W^T\mathbf{p}) \\ &= \mathbf{q}^T\mathbf{q}. \end{aligned}$$

The quadratic form $\mathbf{p}^T S^{-1}\mathbf{p}$ defines a norm in three-dimensional space. This affine-invariant norm, which we denote by $\|\cdot\|$, defines the square of the length of a vector $\mathbf{v} = (x, y, z)^T$ as

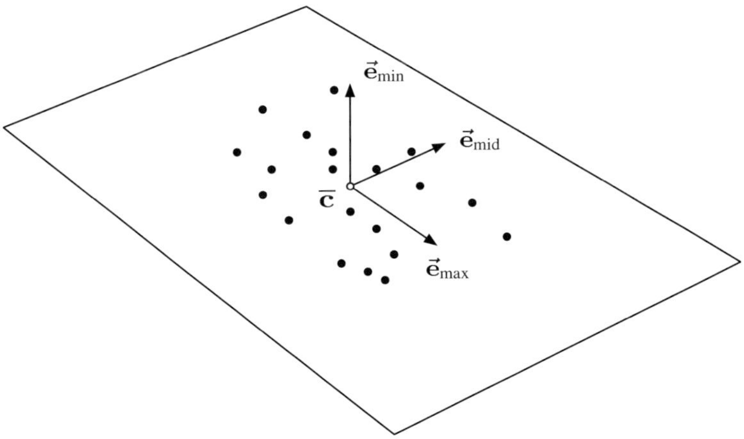

Figure 2. Principal component analysis (PCA) of a set of points in three-dimensional space. PCA yields three eigenvectors that form a local coordinate system with the geometric mean $\bar{\mathbf{c}}$ of the points as its local origin. The two eigenvectors $\vec{\mathbf{e}}_{\max}$ and $\vec{\mathbf{e}}_{\text{mid}}$, corresponding to the two largest eigenvalues, define a plane that represents the best-fit plane for the points. The eigenvector $\vec{\mathbf{e}}_{\min}$ represents the direction in which we measure the error

$$\|\mathbf{v}\|^2 = \mathbf{v}^T S^{-1} \mathbf{v}, \tag{3}$$

see [28, 29]. The "unit sphere" in this norm is the ellipsoid defined by the set of points $\mathbf{p}$ satisfying the quadratic equation $\mathbf{p}^T S^{-1} \mathbf{p} = 1$. This ellipsoid has its major axis in the direction of $\vec{\mathbf{e}}_{\max}$. The length of the major axis is $\sqrt{|\lambda_{\max}|}$. The other two axes of this ellipsoid are in the directions of $\vec{\mathbf{e}}_{\text{mid}}$ and $\vec{\mathbf{e}}_{\min}$, respectively, with corresponding lengths $\sqrt{|\lambda_{\text{mid}}|}$ and $\sqrt{|\lambda_{\min}|}$. We utilize this ellipsoid in the clustering step.

We consider a point set as "nearly coplanar" when $\sqrt{|\lambda_{\min}|}$ is small compared to $\sqrt{|\lambda_{\text{mid}}|}$ and $\sqrt{|\lambda_{\max}|}$. If our planarity condition is not satisfied, we recursively subdivide the point set and continue this subdivision process until all point subsets meet the required planarity condition. We define the *error* of a cluster as $\sqrt{|\lambda_{\min}|}$, which is the maximum distance from the least-squares plane.[1]

The PCA calculation is linear in the number of points in the point set. The essential cost of the operation is the calculation of the covariance matrix. The calculation of the eigenvalues and eigenvectors is a fixed-cost operation, as it is performed for a 3×3 matrix.

3.2. Splitting Clusters

We use PCA to construct a set of clusters for a given point set $\mathscr{P}$. In general, the eigenvalues implied by the original point set $\mathscr{P}$ are non-zero and finite, unless the given points are collinear or coplanar. The eigenvalue $\lambda_{\min}$ measures, in some sense, the deviation of the point set from the plane that passes through $\bar{\mathbf{c}}$ and is spanned by the two eigenvectors $\vec{\mathbf{e}}_{\max}$ and $\vec{\mathbf{e}}_{\text{mid}}$.

If the error of a cluster $\mathscr{C}$ is greater than a certain threshold, we split the cluster into two subsets along the plane passing through $\bar{\mathbf{c}}$ and containing the two vectors $\vec{\mathbf{e}}_{\text{mid}}$ and $\vec{\mathbf{e}}_{\min}$. This bisecting plane splits the data set into two subsets. The general idea is to perform the splitting of point subsets recursively until the maximum of all clusters errors has a value less than a prescribed threshold, *i.e.*, a planarity condition holds for all the clusters generated. For any given error tolerance, the splitting of subsets always terminates when each cluster consists of less than four points.

This method can fail to orient clusters correctly if the density of the surface samples is not sufficient. For example, in areas where two components of a surface are separated by a small distance, the algorithm may produce one cluster consisting of points from both components, see Figure 3. This fact causes the algorithm to produce an incorrect triangulation. However, if the sample density is high in these areas, the splitting algorithm will eventually define correctly oriented clusters.

[1] Potential outliers in a data set are removed in the scanning process. If outliers exist in the data, an "average" error of $\sqrt{\frac{|\lambda_{\min}|}{n}}$, where n is the number of points in the cluster, produces better results.

This method is also useful when the density of sample points is highly varying. In these regions, the algorithm correctly builds large clusters with low error. The triangulation step can thus create a triangulation correctly in areas that have few or no samples, see [32].

3.3. Reclassification of Points during Splitting

Generating clusters based only on splitting planes can generate irregular clusters of points, where many points can be separated by long distances. Since a bisecting plane may not be the ideal place to best separate the cluster, the algorithm may produce irregular triangulations. To remedy this we utilize a reclassification step to adjust clusters locally.

Initially, we place all points in one cluster. During each iteration of the cluster splitting algorithm, the cluster with the highest internal error is split. After splitting this cluster, a local *reclassification step* is used to improve the "quality" of the clusters. This reclassification step is illustrated for a planar curve reconstruction in Fig. 4.

Suppose that cluster $\mathscr{C}$ is to be split. To split $\mathscr{C}$ into two subsets $\mathscr{C}_1$ and $\mathscr{C}_2$, we define the two points $\mathbf{p}_1 = \bar{\mathbf{c}} - \vec{\mathbf{v}}_{\max}$ and $\mathbf{p}_2 = \bar{\mathbf{c}} + \vec{\mathbf{v}}_{\max}$, where $\vec{\mathbf{v}}_{\max} = \sqrt{|\lambda_{\max}|}\vec{\mathbf{e}}_{\max}$. These points are on the orthogonal regression line and the ellipsoid $\mathbf{p}^T S^{-1}\mathbf{p} = 1$ associated with $\mathscr{C}$.

Let $\mathscr{C}_3, \mathscr{C}_4, \ldots, \mathscr{C}_k$ be the "neighboring clusters" of $\mathscr{C}$, and let $\bar{\mathbf{c}}_3, \bar{\mathbf{c}}_4, \ldots, \bar{\mathbf{c}}_k$ be their respective cluster centers. Using the points $\bar{\mathbf{c}}_1 = \mathbf{p}_1, \bar{\mathbf{c}}_2 = \mathbf{p}_2, \bar{\mathbf{c}}_3, \ldots,$ and $\bar{\mathbf{c}}_k$, we determine k new clusters $\mathscr{C}'_1, \mathscr{C}'_2, \ldots,$ and $\mathscr{C}'_k$, where a point $\mathbf{p}$ is an element of a cluster $\mathscr{C}'_i$ if the distance between $\mathbf{p}$ and $\bar{\mathbf{c}}_i$ is the minimum of all distances $\|\mathbf{p} - \bar{\mathbf{c}}_j\|, j = 1, \ldots, k$. The new clusters obtained after this step replace the original cluster $\mathscr{C}$ and the clusters in the neighborhood clusters of $\mathscr{C}$.

The neighbor clusters of a cluster $\mathscr{C}$ are defined by a cluster connectivity graph. Section 3.5 details the construction of this graph. This graph is also used to determine the triangulation of the area between the clusters, as described in Section 4.

(a) (b)

Figure 3. Principal component analysis requires a sufficient sampling density when two components of a surface are separated by a relatively small distance. In **a** the number of samples in the indicated region is not sufficient for the cluster generation algorithm to generate two separate clusters on the different components. In **b** the sampling density is sufficient for the splitting algorithm to orient two clusters correctly

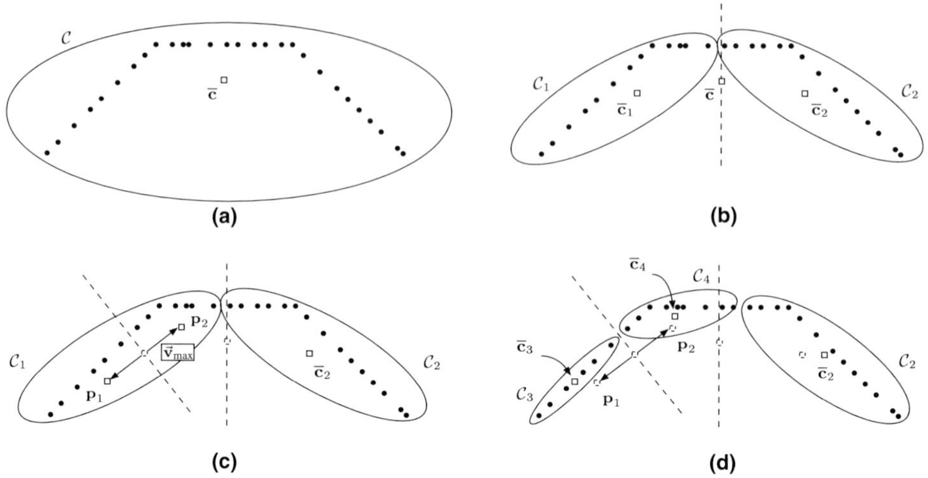

Figure 4. Planar example of reclassification. Given the set of points shown in **a** forming a single cluster $\mathscr{C}$, the algorithm splits this cluster, forming the clusters $\mathscr{C}_1$ and $\mathscr{C}_2$ shown in **b**. To split cluster $\mathscr{C}_1$ with center $\bar{\mathbf{c}}_1$, two new points, $\mathbf{p}_1 = \bar{\mathbf{c}}_1 - \vec{\mathbf{v}}_{\max}$ and $\mathbf{p}_2 = \bar{\mathbf{c}}_1 + \vec{\mathbf{v}}_{\max}$ are defined, as shown in **c**. All points are then reclassified considering $\mathbf{p}_1, \mathbf{p}_2$ and $\bar{\mathbf{c}}_2$, producing the new clusters $\mathscr{C}_2$, $\mathscr{C}_3$ and $\mathscr{C}_4$, shown in **d**. This process may be repeated with the new clusters, defining $\bar{\mathbf{c}}_2, \bar{\mathbf{c}}_3$, and $\bar{\mathbf{c}}_4$ as the geometric means of the respective clusters, forming yet another set of clusters that better approximates the data

The reclassification step is potentially the most time-consuming step per iteration, since its time complexity depends on the number of clusters in the local neighborhood. The average number of neighbors in the *cluster connectivity graph* can be assumed to be a constant, which means that the complexity of the reclassification is linear in the number of points contained in the neighboring clusters. We limit this reclassification step to the clusters in the neighborhood to keep it a local process. The time needed for the reclassification step decreases as the cluster sizes shrink.

3.4. Tile Generation

The set of clusters partitions the original data set. The resulting clusters all satisfy a coplanarity condition. For each cluster $\mathscr{C}$, the cluster center $\bar{\mathbf{c}}$ and the two eigenvectors $\vec{\mathbf{e}}_{\max}$ and $\vec{\mathbf{e}}_{\text{mid}}$ define a plane P that minimizes the sum of the squares of the plane-to-point distances for the associated points. We project all points associated with cluster $\mathscr{C}$ into the plane P and compute the convex hull of the projected points in P, see Fig. 5. We determine the boundary polygon H of this convex hull and generate the boundary polygon T of the cluster by "lifting" the points defining H back to their original positions in three-dimensional space. We call T the "tile" associated with cluster $\mathscr{C}$, and H the "planar tile" associated with $\mathscr{C}$. The principal orientation of T is implied by the cluster's associated eigenvector $\vec{\mathbf{e}}_{\min}$. Figure 6 illustrates the tile generation process for a model used in Eck et al. [11].

3.5. The Connectivity Graph

To accurately calculate the neighbors of a cluster, we require that a connectivity graph for the clusters be maintained. This graph can be generated from the tiles, as they form a Voronoi-like partition of the cluster set.

The set of tiles implies an approximation of the underlying surface. We generate the connectivity graph by generating a Delaunay graph of the cluster centers along the surface implied by the tiles, see Mount [27]. To simplify the task we use the planar tiles to approximate geodesic distances on the surface, as shown in Fig. 7.

This graph is generated by a second step of the algorithm. If a Delaunay graph cannot be generated in a certain area, we continue to split clusters in this area until the graph can be completed. In areas where two surface components are separated by a small distance, the Delaunay graph cannot be generated.

The graph can also be used to generate surface boundaries. An edge of the graph can be mapped to three line segments, one which represents the distance between the clusters, see Figure 7. If this distance is greater than a given threshold, the edge can be eliminated from the graph. We can detect these "boundary clusters" in the

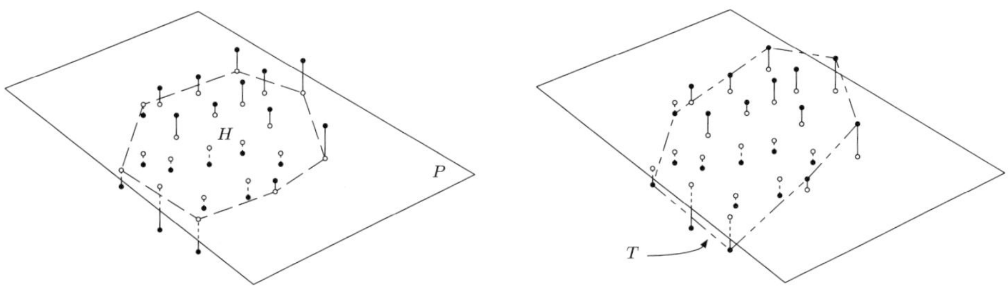

Figure 5. Given a cluster of points, the points are projected onto the regression plane P. The boundary polygon of the convex hull H of the projected points is generated. "Lifting" the points defining the convex-hull boundary polygon back to their original position in three-dimensional space defines the non-planar tile boundary polygon T

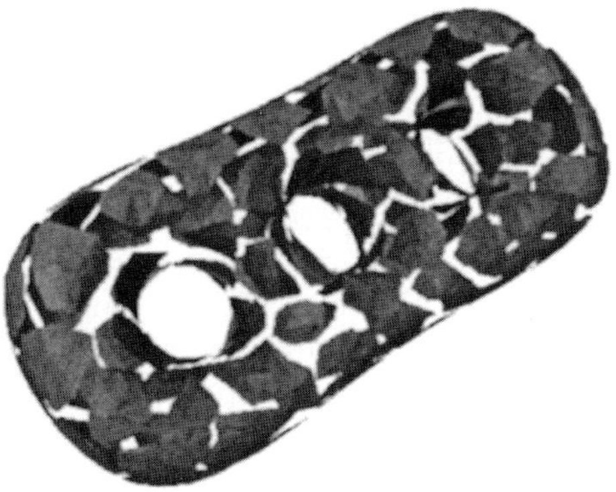

Figure 6. Tiles generated for the "three-holes" data set. The initial data set consists of 4000 points. The initial tiling of the data set consists of 120 tiles

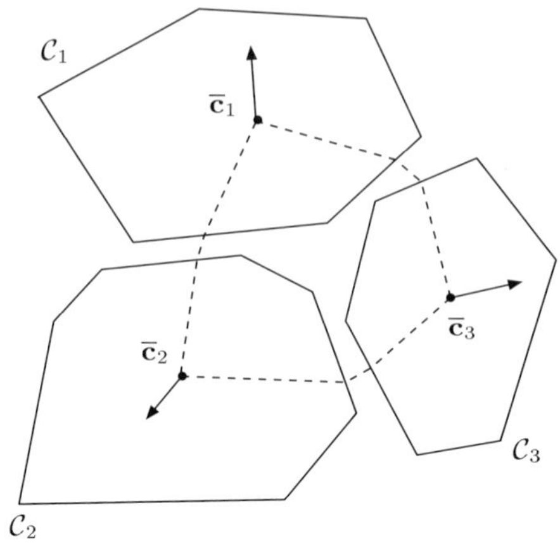

Figure 7. Distance measured on the tiles approximates the geodesic distances on the underlying unknown surface. These distances are used to generate the Delaunay-like triangulation of the cluster centers

triangulation step and modify the triangulation between the clusters to create surface boundaries.

4. Generating the Triangular Mesh

Since each cluster is "nearly planar" we can assume that the data within the cluster can be represented as a height field with respect to the best-fit plane. Thus, we can project the data onto the best-fit plane and triangulate the data using a two-dimensional Delaunay triangulation. The result triangulates the area within the convex hull of the projected points. This triangulation can be "lifted" to a triangulation of the tile associated with the cluster by using the points' original locations in three-dimensional space.

A high-resolution triangulation of the points in a cluster is obtained by considering all points of the cluster. To obtain a lower-resolution triangulation, we consider only the points of the boundary polygon of the convex hull of the projected points. A Delaunay triangulation of these points can also be lifted to form a triangulation of the tile. Since we know the maximal deviation of the points of the cluster from the best-fit plane, we can measure the deviation of the lower-resolution triangulation from the high-resolution one.

To generate a triangulation of the space between the tiles, we utilize the connectivity graph generated in the clustering step. Here, we consider a "triangle" T in the Delaunay-like graph and the three clusters $\mathscr{C}_1, \mathscr{C}_2$, and $\mathscr{C}_3$ whose centers define the vertices of this triangle, as shown in Fig. 8. We determine a plane $\mathscr{P}$ on which the three clusters can be bijectively mapped.[2] The normal of this plane can

[2]There are cases where a bijective map cannot be constructed. In these cases, we split clusters recursively until the construction of such a map is possible for all clusters. Even if this strategy fails, and this has never been the case with our models, the triangulation cannot be generated automatically in this area.

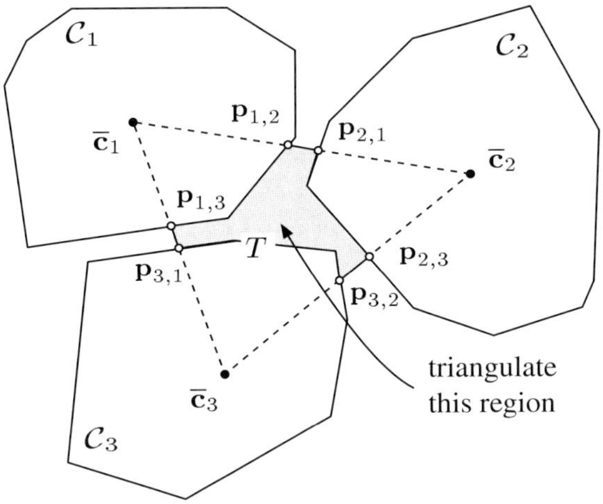

Figure 8. Three tiles projected onto a plane. The intersection points $\mathbf{p}_{i,j}$ between the edges of the tiles C_i and the edges of the triangle T are added to the set of tile boundary vertices. This enables us to triangulate the area of the triangle using a constrained two-dimensional Delaunay triangulation that preserves the boundaries of the tiles

be obtained by selecting one of the normals of the best-fit planes of one of the three clusters or by averaging the normals of the best-fit planes of the three clusters connected by the triangle T.

Considering Fig. 8, we operate on the area bounded by the triangle and the data set containing the vertices $\bar{\mathbf{c}}_1, \bar{\mathbf{c}}_2$, and $\bar{\mathbf{c}}_3$ of the triangle T, the points of the tiles contained in T, and the six additional points $\mathbf{p}_{1,2}, \mathbf{p}_{2,1}, \mathbf{p}_{1,3}, \mathbf{p}_{3,1}, \mathbf{p}_{2,3}$, and $\mathbf{p}_{3,2}$, *i.e.*, the points where the edges of the triangle intersect the tile boundary polygons. We apply a constrained Delaunay triangulation step, see Okabe et al. [30], to this point set, which preserves the edges of the tile boundary polygons.

Figure 9 illustrates this process. The region to be triangulated (shaded) is bounded by three convex curves (segments of the tile boundaries) and three line segments. A Delaunay triangulation does not provide a triangulation such that the segments of the tile boundary polygons are preserved in the triangulation. By identifying the missing edges we can perform edge-flipping to obtain the required constrained Delaunay triangulation. The final triangulation in the area of the triangle T is generated by "lifting" all vertices back to their original positions.

This triangulation procedure adds additional points to the tile boundary polygons. These points can be eliminated by identifying the triangles that share these points. A constrained Delaunay triangulation applied to such areas generates triangles that fill the same area, but do not contain the additional points $\mathbf{p}_{i,j}$. Figure 10 illustrates this process, and Fig. 11 shows the three-holes data set using a low-resolution representation of the tiles, together with a triangulation of the space between the tiles.

This algorithm can also be adapted for situations where tiles lie on the boundary of a surface. Given two planar tiles $\mathscr{C}_1$ and $\mathscr{C}_2$ that have been projected onto a

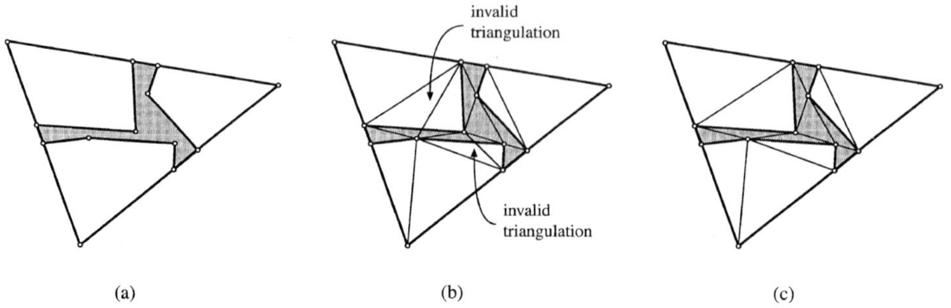

Figure 9. Triangulating the region inside a triangle T. The points to be triangulated are shown as circles in **a**; in **b** a Delaunay triangulation has been generated; and in **c** edge-flipping operations have been used to construct a correct triangulation. By removing the triangles that lie within the tiles, we obtain a triangulation of the shaded area

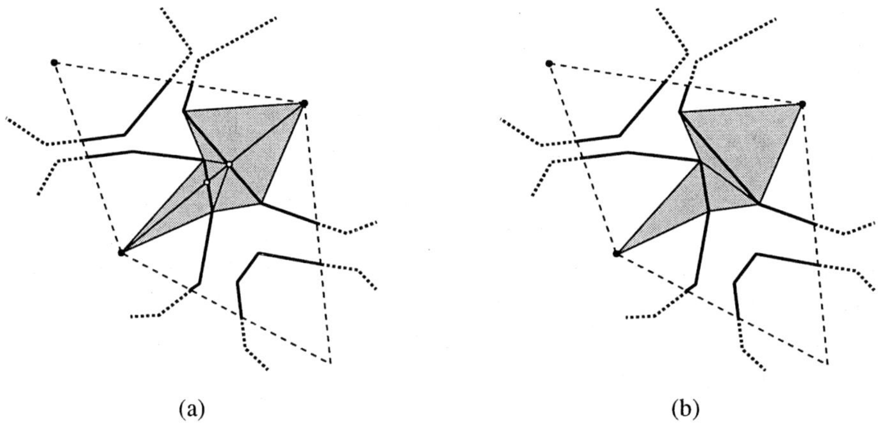

Figure 10. Eliminating unnecessary intersections points on tile boundaries. By considering those triangles that have additional points (shown as circles) among their vertices, shown in **a** we can ignore those points and locally apply a constrained Delaunay triangulation to this area, creating the desired triangulation in **b**

plane, the area to be triangulated lies outside the two tiles and inside the area defined by the line joining the centers of the triangles and a line segment on the boundary of the convex hull of the planar tiles. Generating a constrained Delaunay triangulation of this area produces the required triangulation, see Fig. 12.

5. Results

We have used this algorithm to produce reconstructions for a variety of data sets. The input to the algorithm is based either upon the desired error tolerance associated with the clusters, or the total number of clusters generated by the adaptive splitting algorithm.

Figures 13 and 14 show a reconstruction of a data set representing a car body. The original data set contains 20,621 points, and it is represented by 400 tiles. Figure

Figure 11. Reconstruction of the three-holes data set. The triangulation is formed by generating triangles from edges of the tile boundary polygons and the tile centers. The triangulation between the tiles is shown

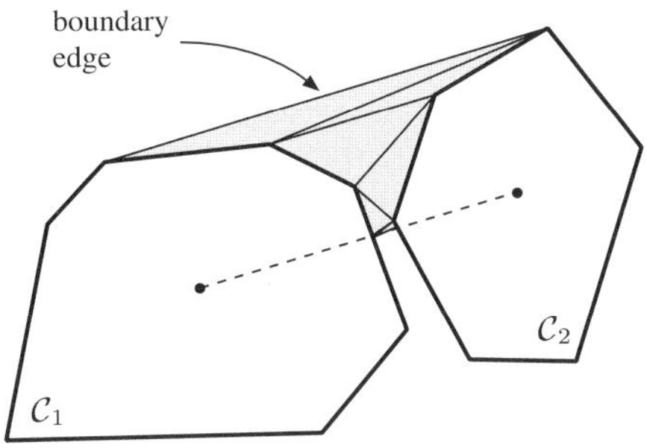

Figure 12. Triangulating the region between a boundary edge and the line joining the centers of two boundary tiles. The boundary edge is part of the convex hull of the two tiles

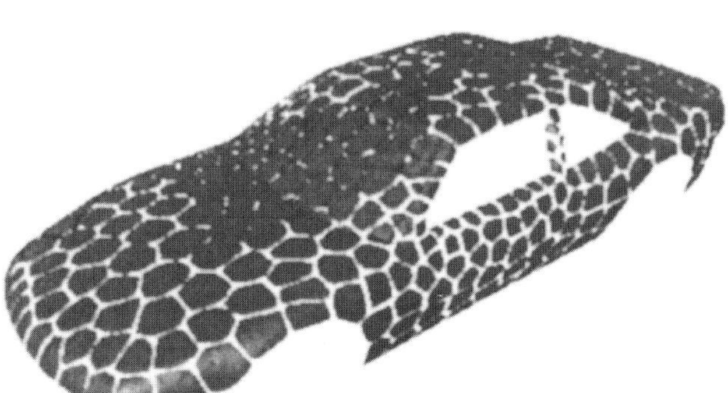

Figure 13. Tiles generated for a car body data set. The original data set contains 20,621 data points. This reconstruction contains 400 tiles

Figure 14. Complete reconstruction of the car body

Figure 15. Reconstruction of the hypersheet data set. The original data set contains 6,752 points, and 200 clusters were generated

13 shows the triangulation of the tiles generated from the first step of the algorithm. Figure 14 shows the complete triangulation of the data set. For this data set, we have identified the boundaries by modifying the connectivity graph. Edges of the final connectivity graph were deleted whenever the length between the clusters exceeded a certain threshold. Thus, the windows and the bottom of the car are not triangulated in this example.

Figure 16. Dragon data set (tiles only). The original data set contains 100,250 points, and 5,000 tiles were generated

Figure 17. Low-resolution approximating mesh of the dragon data set

Figure 15 shows a reconstruction of the "hypersheet" data set used in Hoppe et al. [18]. This data set contains 6,752 points, and the reconstruction is based on 200 clusters.

Figure 16–18 show reconstructions of the "Stanford dragon" data set. The original data set contains 100,250 points and is represented here with 5,000 tiles. Figure 16 shows the tiles generated from the first step of the algorithm, and Fig. 17 shows a low-resolution triangulation. Here, we have triangulated the vertices of the tile boundary polygons and have added the triangles in the space between the tiles. Figure 18 shows a complete triangulation of the data set.

Figure 18. High-resolution reconstruction of the dragon data set

Table 1. Statistics for the models. The triangulation time depends primarily on the number of tiles

Data set	Number of points	Number of tiles	Cluster generation time in seconds	Triangulation time in seconds
Three-holes	4,000	120	6	119
Hypersheet	6,752	200	12	240
Automobile body	20,621	400	17	182
Dragon	100,250	5,000	375	1,860

All models were generated using PCA to analyze the clusters. The reconstructions are therefore affine-invariant. Table 1 provides timing statistics for the reconstructions of the models shown in Fig. 11 and Figs. 13–18. These models were generated on an SGI Onyx2 using a single 195MHz R10000 processor.

6. Conclusions

We have presented a new algorithm that allows the generation of triangulated surface models from discrete point sets without connectivity information. This algorithm uses an adaptive clustering approach to generate a set of two-manifold tiles that locally approximate the under-lying unknown surface. We construct a triangulation of the surface by triangulating the data within the individual tiles and triangulating the gaps between the tiles. Approximating meshes can be generated by directly triangulating the boundary polygons of the tiles. Since the deviation from the point set is known for each cluster, we can produce approximate reconstructions with prescribed error bounds.

If a given data set has connectivity information, then our algorithm can be viewed as a generalization of the vertex-removal algorithm of Schroeder et al. [31]. Instead of removing a vertex and re-triangulating the resulting hole, we remove clusters of nearly coplanar points and re-triangulate the hole generated by removing the cluster. This is an immediate extension of our approach. We also plan to extend our algorithm to reconstruct surfaces with sharp edges and vertices.

We plan to extend our approach to the clustering of more general scattered data sets representing scalar and vector fields, defined over two-dimensional and three-dimensional domains. These are challenging problems as faster algorithms for the generation of data hierarchies for scientific visualization are becoming increasingly important due to our ability to generate ever larger data sets.

Acknowledgements

This work was supported by the National Science Foundation under contracts ACI 9624034 (CAREER Award), through the Large Scientific and Software Data Set Visualization (LSSDSV) program under contract ACI 9982251, and through the National Partnership for Advanced Computational Infrastructure (NPACI); the Office of Naval Research under contract N00014-97-1-0222; the Army Research Office under contract ARO 36598-MA-RIP; the NASA Ames Research Center through an NRA award under contract NAG2-1216; the Lawrence Livermore National Laboratory under ASCI ASAP Level-2 Memorandum Agreement B347878 and under Memorandum Agreement B503159; and the North Atlantic Treaty Organization (NATO) under contract CRG 971628 awarded to the University of California, Davis. We also acknowledge the support of ALSTOM Schilling Robotics and SGI. We thank the members of the Visualization Group at the Center for Image Processing and Integrated Computing (CIPIC) at the University of California, Davis.

We would like to thank the reviewers of this paper. Their comments have improved the paper greatly.

References

[1] Algorri, M.-E., Schmitt, F.: Surface reconstruction from unstructured 3D data. Comput. Graphics Forum *15*, 47–60 (1996).

[2] Amenta, N., Bern, M., Kamvysselis, M.: A new Voronoi-based surface reconstruction algorithm. In: SIGGRAPH 98 Conference Proceedings (Cohen, M., ed.), pp. 415–422. Annual Conference Series, ACM SIGGRAPH. New York: ACM Press, 1998.

[3] Attali, D.: r-regular shape reconstruction from unorganized points. Computational Geometry Theory and Applications *10*, 239–247 (1998).

[4] Bajaj, C. L., Bernardini, F., Xu, G.: Automatic reconstruction of surfaces and scalar fields from 3D scans. Comput. Graphics *29*, Annual Conference Series 109–118 (1995).

[5] Bernardini, F., Bajaj, C. L.: Sampling and reconstructing manifolds using alpha-shapes. In: Proc. 9th Canadian Conf. Computational Geometry, pp. 193–198 (1997).

[6] Bernardini, F., Mittleman, J., Rushmeier, H., Silva, C., Taubin, G.: The ball-pivoting algorithm for surface reconstruction. IEEE Trans. Visual. Comput. Graphics *5*, 145–161 (1999).

[7] Bittar, E., Tsingos, N., Gascuel, M.-P.: Automatic reconstruction of unstructured 3D data: Combining medial axis and implicit surfaces. Comput. Graphics Forum *14*, C/457–C/468 (1995).

[8] Boissonnat, J.-D.: Geometric structures for three-dimensional shape representation. ACM Trans. Graphics *3*, 266–286 (1984).

[9] Bolle, R. M., Vemuri, B. C.: On three-dimensional surface reconstruction methods. IEEE Trans. Pattern Anal. Mach. Intell. PAMI-13, *1*, 1–13 (1991).

[10] Curless, B., Levoy, M.: A volumetric method for building complex models from range images. Comput Graphics *30*, Annual Conference Series 303–312 (1996).

[11] Eck, M., DeRose, T., Duchamp, T., Hoppe, H., Lounsbery, M., Stuetzle, W.: Multiresolution analysis of arbitrary meshes. In: SIGGRAPH 95 Conference Proceedings (Cook, R., ed.), pp. 173–182. Annual Conference Series, ACM SIGGRAPH. New York: ACM Press, 1995.

[12] Edelsbrunner, H., Mücke, E. P.: Three-dimensional alpha shapes. ACM Trans. Graphics *13*, 43–72 (1994).

[13] Gordon, A. D.: Hierarchical classification. In: Clustering and classification (Arabie, R., Hubert, L., DeSoete, G., eds.), pp. 65–105. Singapore: World Scientific, 1996.

[14] Guo, B.: Surface reconstruction: from points to splines. Comput. Aided Des. *29*, 269–277 (1997).

[15] Heckel, B., Uva, A., Hamann, B.: Clustering-based generation of hierarchical surface models. In: Proceedings of Visualization 1998 (Late Breaking Hot Topics) (Wittenbrink, C., Varshney, A., eds.), pp. 50–55. Los Alamitos: IEEE Computer Society Press, 1998.
[16] Hinker, P., Hansen, C.: Geometric optimization. In: Proceedings of the Visualization '93 Conference (San Jose, CA, Oct. 1993) (Nielson, G. M., Bergeron, D., eds.), pp. 189–195. Los Alamitos: IEEE Computer Society Press, 1993.
[17] Hoppe, H., DeRose, T., Duchamp, T., McDonald, J., Stuetzle, W.: Surface reconstruction from unorganized points. Comput. Graphics *26*, 71–78 (1992).
[18] Hoppe, H., DeRose, T., Duchamp, T., McDonald, J., Stuetzle, W.: Mesh optimization. Comput. Graphics *27*, 19–26 (1993).
[19] Hotelling, H.: Analysis of a complex of statistical variables into principal components. J. Educat. Psychol. *24*, 417–441, 498–520 (1993).
[20] Jackson, J. E.: A user's guide to principal components. New York: Wiley, 1991.
[21] Kalvin, A. D., Taylor, R. H.: Superfaces: Polyhedral approximation with bounded error. In: Medical Imaging: Image Capture Formatt. Display, *2164*, 2–13 (1994).
[22] Kalvin, A. D., Taylor, R. H.: Superfaces: Polygonal mesh simplification with bounded error. IEEE Comput. Graphics Appl. *16*, 64–77 (1996).
[23] Lorensen, W. E., Cline, H. E.: Marching cubes: a high resolution 3D surface construction algorithm. Comput. Graphics *21*, 163–170 (1987).
[24] Manly, B.: Multivariate statistical methods, A primer. New York: Chapman & Hall, 1994.
[25] Mencl, R.: A graph-based approach to surface reconstruction. Comput. Graphics Forum *14*, C/445–C/456 (1995).
[26] Mencl, R., Müller, H.: Graph-based surface reconstruction using structures in scattered point sets. In: Proceedings of the Conference on Computer Graphics International 1998 (CGI-98) (Los Alamitos, California, June 22–26 1998) (Wolter, F.-E., Patrikalakis, N. M., eds.), pp. 298–311. Los Alamitos: IEEE Computer Society Press, 1998.
[27] Mount, D. M.: Voronoi diagrams on the surface of a polyhedron. Technical Report CAR-TR-121, CS-TR-1496, Department of Computer Science, University of Maryland, College Park, MD, May 1985.
[28] Nielson, G. M.: Coordinate-free scattered data interpolation. In: Topics in multivariate approximation (Schumaker, L., Chui, C., Utreras, F., eds.), pp. 175–184. New York: Academic Press, 1987.
[29] Nielson, G. M., Foley, T.: A survey of applications of an affine invariant norm. In: Mathematical methods in computer aided geometric design (Lyche, T., Schumaker, L., eds.), pp. 445–467. San Diego: Academic Press, 1989.
[30] Okabe, A., Boots, B., Sugihara, K.: Spatial tesselations – concepts and applications of Voronoi diagrams. Chichester: Wiley, 1992.
[31] Schroeder, W. J., Zarge, J. A., Lorensen, W. E.: Decimation of triangle meshes. Comput. Graphics *26*, 65–70 (1992).
[32] Soucy, M., Laurendeau, D.: A general surface approach to the integration of a set of range views. IEEE Trans. Pattern Anal. Mach. Intell. *17*, 344–358 (1995).
[33] Teichmann, M., Capps, M.: Surface reconstruction with anisotropic density-scaled alpha shapes. In: Proceedings of Visualization 98 (Oct. 1998), (Ebert, D., Hagen, H., Rushmeier, H., eds.), pp. 67–72. Los Alamitos: IEEE Computer Society Press, 1998.

B. Heckel
PurpleYogi.com, Inc.
201 Ravendale
Mountain View, CA 94043
USA
e-mail: heckel@PurpleYogi.com

A. E. Uva
Dipartimento di Progettazione
e Produzione Industriale
Politecnico di Bari
Viale Japigia 182
70126 Bari
Italy
e-mail: uva@dppi.poliba.it

B. Hamann
K. I. Joy
Center for Image Processing and Integrated Computing (CIPIC)
Department of Computer Science
University of California
Davis, CA 95616-8562 USA
e-mails: hamann@cs.ucdavis.edu, joy@cs.ucdavis.edu

Computing [Suppl] 14, 219–232 (2001)

Computing

An Algorithm to Triangulate Surfaces in 3D Using Unorganised Point Clouds

G. Kós, Budapest

Abstract

Reconstructing surfaces from a set of unorganised sample points in the 3D space is a very important problem in reverse engineering. Most algorithms first build a triangular mesh to obtain an approximate surface representation. In this paper we describe an algorithm which works by creating and merging local triangular complexes to obtain an unambiguous 2D-manifold triangulation. We use all the given sample points as vertices, which is a natural requirement. Our method is able to handle open boundaries and holes, different geni (for example tori) and unoriented surfaces in a computationally efficient way.

1. Introduction

Given a set of unorganized points, which lie approximately on the boundary surface of a three-dimensional object, for which there is no a priori information about the topology of the given points. Our goal is to reconstruct the topology of the surface by building a triangular mesh using the given points. This problem is well-known in computer vision and computer graphics, and also a key issue in reverse engineering of shapes (see [10]), where complete and accurate CAD models need to be built based on measured data.

There are several special considerations concerning the measured data sets. Physical measurements always superimpose some noise on the ideal data points; the point density is often very uneven due to curvature variations, and undesirable, outlying elements may also occur. Typically the point set is formed by merging multiple measurements, which creates very inhomogeneous distributions for the united point clouds. The point cloud may contain holes due to occlusion, i.e. there may be surface portions which cannot be measured from any of the viewing directions. It is also typical that the point set represents not a complete volumetric object, but only certain surface portions of the boundary, and only these parts need to be reconstructed.

The goal of most approaches is to build a 3D triangular mesh based on the data points. In some cases only the given sample points are used as vertices, but in other cases artificial vertices are used. The approaches differ also in the assumptions concerning the surface topology.

Many algorithms are based on the Delaunay tessellation of the sample point set, or an α-shape of the points. The concept of α- shapes is also strongly related to discovering the topology of a given point cloud [6]. The α-shape is a subset of the 3-2-1-0 dimensional simplices – i.e. tetrahedra, triangles, edges and vertices respectively – of the Delaunay tessellation. Only those elements are kept which lie on a sphere of radius less than α which has no points in its interior. If the sample points are uniformly distributed and the curvature of the surface is lower than the sampling density, then α can be chosen in such a way that those triangles will be kept which contribute to the external surface of the object. A generalization of α-shapes – using local weights based on the local point densities – was also suggested in [7].

In the early work of Boissonnat [4], two different approaches were presented. The first one builds a triangulation in an incremental manner by always adding a "close" point to the current structure. The second one removes tetrahedra from the Delaunay triangulation of the convex hull of the points and thus performs sculpting step by step until the final volume of the polyhedron is obtained. These methods are somewhat limited for disconnected surface portions and objects with holes.

Choi [5] also suggested an incremental technique for triangulation; however, the points were assumed to be projectable to a given plane.

Veltkamp [11] suggested a generalization of Boissonnat's second algorithm. He creates the so-called γ-neighbourhood graph, which is a superset of the set of triangles in the Delaunay triangulation. He then selects a subset of it to obtain a closed, genus-0 triangle mesh. The selection method starts with the convex hull again.

The problem of building the actual boundary surface from the α-shape is also difficult from an algorithmic point of view, because often non-connected, non-manifold sets of elements need to be processed. Related problems and solutions were reported by Guo et al. [8].

The concept of the weighted α-shapes was pursued in the works of Bajaj and Bernardini et al. [2, 3], where Boissonnat's sculpting technique is applied on a so-called α-solid. This method is efficient in reconstructing sharp features.

Amenta, Bern and Kamvysselis in [1] recently published another approach based on Voronoi diagrams and Delaunay triangulation. They add some artificial vertices to the original points, then compute the Delaunay triangulation of the set, and lastly by removing every object which has at least one artificial vertex. This algorithm may not work very well if noise is present.

One of the most important works is due to Hoppe et al. (see [9]). In this paper a piecewise linear function $f : \Re^3 \to \Re$ is created to estimate the signed distance from the boundary of the object. Then the zero-set of this function is extracted by a special *marching cubes* algorithm. This algorithm can be used for surfaces with arbitrary topology, and it can detect open boundaries. A disadvantage of Hoppe's method is that the marching cubes algorithm requires a huge amount of memory, and is time consuming. Another minor problem is that the implicit function is not continuous everywhere, and special care is required in the marching cubes algorithm to preserve consistency.

Our algorithm in this paper attempts to get rid of the limitations of the above approaches. The basic principle is to merge locally defined triangulations which leads to a consistent global triangular mesh at the end.

The following basic requirements need to be satisfied:

- We would like to handle arbitrary, unorganized point clouds with uneven distributions. The mesh should connect all the data points (if possible).
- The surface boundary is allowed to be open or the union of several components. The algorithm must properly handle holes, recognize open boundaries and reconstruct disjoint components.
- Our only assumption on the surface topology is that it is a 2D-manifold. It may contain arbitrary number of holes and handles.
- Our method should be able to reconstruct un-oriented meshes (for example, the Moebius strip) as well.
- The method should be computationally efficient and robust.

In the following sections the basic steps of our triangulation algorithm will be discussed, followed by a few examples and suggestions for further improvements.

2. Description of the Algorithm

The triangulation algorithm can be divided into three phases:

1. *Pre-processing*. Here multiple points are removed and some attributes of the vertices, for example, the surface normals are computed.
2. *Triangulation*. This is the main step, when a consistent triangulation is built.
3. *Post-processing*. This includes optimising the surface such as smoothing the triangulation.

2.1. Pre-Processing Steps

2.1.1. Clustering Points

First we put the sample points into an octree-like structure of clusters to speed up the search for points within an arbitrary ball.

The points are put in *boxes*, and the boxes are linked to each other to form a tree. Each box is a bounding box and there is a root box, which contains the whole dataset. If a box contains more than a given number of points, it is divided to two smaller boxes. (In our experiments the maximum number of points was set to 20.) The direction of the division is across the longest edge of the bounding box.

2.1.2. Removing Coincident Points

Some data sets may contain multiple points, which may cause problems in the course of creating triangles, or – if the multiplicity of a point is greater than the limit for the boxes – clustering fails. Thus, multiple points must be deleted, i.e. after clustering, coincident ones are detected and removed.

2.1.3. Building a Neighbourhood Graph

For efficiency reasons, we introduce a special structure to find close points. We build a graph whose vertices are the sample points.

We start with a graph with an empty edge set. For each point P, we take the n closest points, $Q_1, \cdots Q_n$ and give a subset of the edges $PQ_1 < \cdots < PQ_n$ to the graph. The elements of the subset are chosen one by one in this increasing order with the following constraint: if the angle $\angle Q_iPQ_j$ is less than a given threshold φ, the edges PQ_i and PQ_j are not chosen at the same time. (See Fig. 1).

Doing this for each vertex, we obtain the neighbourhood graph (see Fig. 2).

In our tests we used $n = 25$ and $\varphi = 30°$. If the local density of the sample points are approximately the same in each direction, n can be chosen smaller, for ex-

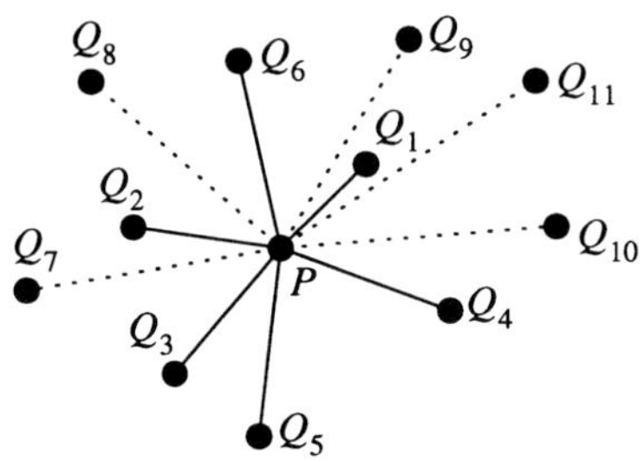

Figure 1. Inserting edges $PQ_1, PQ_2, PQ_3, PQ_4, PQ_5$ and PQ_6 into the graph

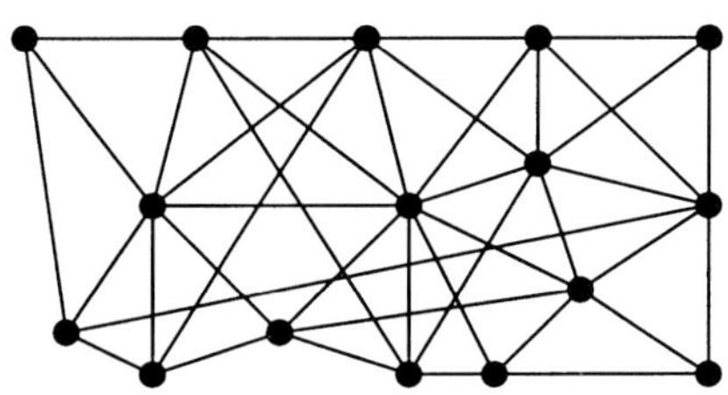

Figure 2. Neighbourhood graph for 16 points

ample to 10. If the pointset is uneven, – for example, it containts very long scanlines – n must be greater.

If $\varphi < 60°$, the following statement can be proven easily. For any $1 \leq m \leq n$ there exists a set of indices $1 \leq i_1 \leq \cdots \leq i_k = m$ such that $PQ_{i_1}, Q_{i_1}Q_{i_2}, \ldots, Q_{i_{k-1}}Q_{i_k}$ are all edges of the final graph.

2.1.4. Computing Maximum Triangle Sizes

We do not want to triangulate large holes or the outer boundary of the surface. For this reason, we only process triangles that are not larger than a given size.

The *size* of a triangle is the diameter of the smallest circle which contains the triangle. If all the angles of the triangle are less than 90 degrees, it is the diameter of the circumcircle. Otherwise, it is the length of the longest side.

In our current implementation, at each vertex P, a local upper limit – the so-called *local maximum triangle size* – is computed at each vertex. Triangle sizes are constrained not to exceed all the corresponding maximum triangle sizes at the three vertices.

If the closest n points to P are again $Q_1, \ldots, Q_n$, the maximum triangle size at P is chosen as $3 \cdot PQ_n$. (In our experiments $n = 50$.)

2.1.5. Normal Estimation

To estimate the surface normal at some point P, we take the closest n neighbouring points (chosen as $n = 25, 50$ and 100 in our experiments, denoted by $Q_1, \ldots, Q_n$). Then we build a local co-ordinate system with the origin at P, and find the implicit quadric of the form

$$F(\mathbf{x}) = \mathbf{x}^t A\mathbf{x} + \mathbf{n} \cdot \mathbf{x}$$

satisfying the condition $|\mathbf{n}| = 1$, for which

$$\sum_i F^2(\overrightarrow{PQ_i})$$

is minimal. Then the normal of the surface $F(\mathbf{x} - P) = 0$ at P is simply $\mathbf{n}$. After some elimination, this leads to a very simple eigenvalue-problem.

The more usual method used in the literature is to fit a least squares plane (equivalent to setting $A = 0$) and optimising only $\mathbf{n}$. If the curvature is high, the error in this method can be very large at points near the surface boundary.

Even for oriented surfaces, it is very difficult to guarantee the consistency of the normals over the whole surface. Our algorithm does not use the orientation of the estimated normals.

2.2. Triangulation

The motivation behind our algorithm is the generalised α-shape of the sample points *on the surface*. First we define this structure. For a surface $\mathcal{S}$ let the *distance* between points A and B be the minimum length of the arcs on the surface which connect A and B. Then for some points $P_1, \ldots, P_n$ on the surface we define the *Voronoi cells* $\mathcal{C}_1, \ldots, \mathcal{C}_n \subset S$. For a given $\mathcal{C}_k, 1 \leq k \leq n; \mathcal{C}_k$ contains those points Q of S for which point P_k is the closest to Q amongst $\{P_1, \ldots, P_n\}$.

If the cells $\mathcal{C}_i$ and $\mathcal{C}_j$ are adjacent (i.e. they have a common boundary arc), we connect points P_i and P_j with the shortest arc in $\mathcal{S}$. If the points $P_1, \ldots P_n$ are dense enough in $\mathcal{S}$, these arcs will divide $\mathcal{S}$ into triangles. In singular cases – when there are at least four points on the same circle – more sided polygons may also occur, which can be further divided into triangles. It is natural to call the triangles obtained the *generalised, curved Delaunay triangulation* (see Fig. 3).

After removing the triangles that have greater size than the maximum, we may call the remaining set of triangles the generalised α-shape of the points $P_1, \ldots, P_n$.

This generalisation keeps many properties of the Delaunay triangulation. For example, the interiors of the circumcircles of the triangles contain none of the points $P_1, \ldots, P_n$.

2.2.1. The Angle Criterion

Assume that we have a quadrilateral $ABCD$ on $\mathcal{S}$ and we have to decide whether to choose the triangles ABC and CDA or the triangles BCD and DAB.

On surfaces with constant curvature, the decision is easily made by comparing the sum of the opposite angle pairs of the quadrilateral. If $\angle ABC + \angle CDA < \angle BCD + \angle DAB$, the circumcircles of ABC and CDA contain the points D and B respectively, therefore we use the triangles ABC and CDA; otherwise we take the triangles BCD and DAB (see Fig. 4).

In our method we ignore that we are dealing with curved surfaces, and always use the criterion above.

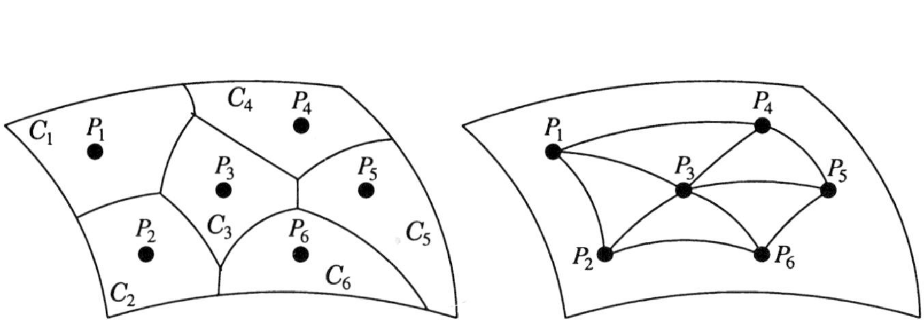

Figure 3. Generalised Voronoi cells and Delaunay triangulation

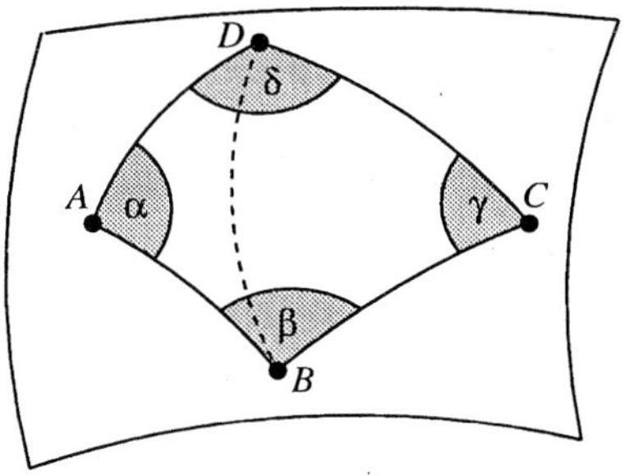

Figure 4. If $\alpha + \gamma < \beta + \delta$, connect B and D rather than A and C

To compute an angle $\angle$BAC, we project points B and C to points B' and C' in the tangent plane at point A, then take the angle $\angle B'AC'$ (see Fig. 5). The goal of this step is to eliminate the effect of the change in normal direction.

Generally, for arbitrary four points A, B, C, D we say that A and C are connectable if $\angle ABC + \angle CDA < \angle BCD + \angle DAB$; conversely, B and D are connectable if $\angle ABC + \angle CDA > \angle BCD + \angle DAB$. (These angles are projected angles).

2.2.2. Creating Candidate Triangles

Now, after demonstrating the heuristic background, we describe the detailed algorithm to create the triangulation. In the first step a loop of triangles is generated around each sample point. These triangles are candidates to become elements of the final triangulation.

At an arbitrary point A, the algorithm locally creates Delaunay triangulation of the surface and chooses the triangles which have A as a vertex. This means that we want to build a list $B_1, \ldots B_k$ of points with the following properties:

- From the normal direction at A, the points $B_1, \ldots, B_k$ are indexed in counter-clockwise direction.
- For any point C in the angle sector B_iAB_{i+1}, in the quadrilateral AB_iCB_{i+1} the points A and C should not be connectable.

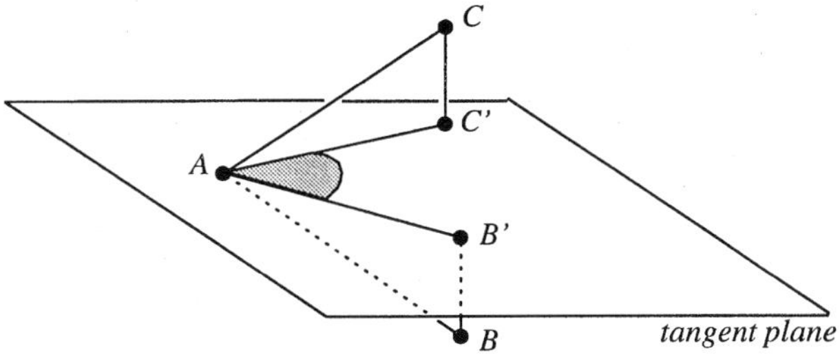

Figure 5. Projected angle BAC

After building this point list, we take all the triangles AB_iB_{i+1}, which contain A and say that the point A has a *vote* for these triangles (see the explanation later).

To generate points B_i, the algorithm works by inserting and deleting points dynamically. We use the following structures: the list of the inserted points, and a queue to store the candidate points. Initially the point list is empty, and the queue contains the neighbours of A *obtained from the neighbourhood graph.*

In each step we take the current point C from the queue which is the closest to A. If C lies in the (projected) angle sector B_iAB_{i+1}, we test whether A and C are connectable in the quadrilateral AB_iCB_{i+1}. If the criterion fails, we discard the point C.

If A and C are connectable, we insert C into the list between B_i and B_{i+1}. After inserting C, some of the points $B_1, \ldots, B_k$ may need to be deleted. The point B_j must be deleted if the points A and B_j are not connectable in the quadrilateral $AB_{j-1}B_jC$ or ACB_jB_{j+1} (see Fig. 6).

If we insert C in the list, we put its neighbours, (in the neighbourhood graph) in the queue. To avoid multiple storing, we mark the stored points, and only insert unmarked ones.

This iteration is repeated until the queue becomes empty.

2.2.3. Create a Consistent Triangulation

In most cases the set of all candidate triangles does not form a manifold structure. For this reason, after creating the candidate triangles, a consistent triangle set needs to be selected.

We register a triangle (i.e. put it into the final triangulation) if

- the triangle mesh remains manifold;
- the triangle mesh remains oriented (if it is constrained to be oriented);
- the triangle does not overlap any registered triangle.

If any of these conditions fails, the triangle is deleted.

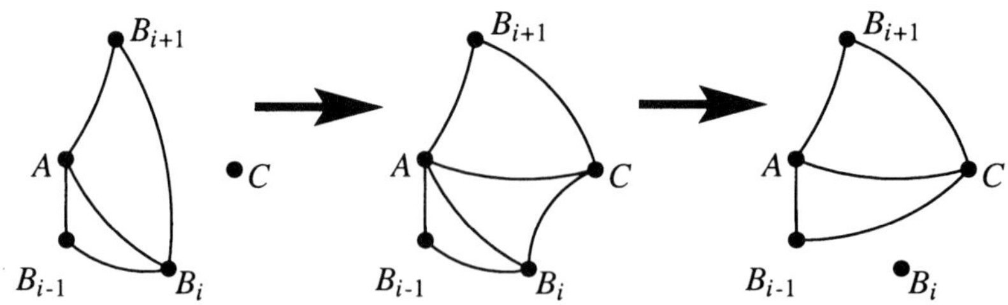

Figure 6. Inserting point C and deleting B_i

Two triangles overlap, if they have a common vertex, and their orthogonal projections to the tangent plane at that vertex have a common interior point. (see Fig. 7).

To register triangles, we sort them. We call some triangles *better* than others and try to register these before the others.

Each triangle has two properties. The most important property is the number of votes of its vertices (see Section 2.2.2). The *best* triangles have three votes; these were chosen as candidate triangles three times. The good triangles have two votes; they were chosen twice, but for the third vertices different candidate triangles were created. The remaining ones have only a single vote.

For each triangle we compute the three angles between the normal vector of the triangle and the estimated normals at the vertices. The maximum error is called the smoothness error of the triangle.

We say that an arbitrary triangle is better than another one, if it has more votes, or has the same number of votes, but a smaller smoothness error.

2.2.4. Filling Holes

After registering topologically consistent triangles, some holes may remain. These holes must be filled.

In the current implementation we use a simple heuristic method to fill holes. The origin of the method is a basic trick to find the 2D Delaunay triangulation of convex polygons.

Suppose that we have an arbitrary convex polygon $P_1P_2 \cdots P_n$ (in the plane) and want to find an index i for which the triangle $P_1P_2P_i$ is an element of the Delaunay triangulation of the polygon. By convexity, the points $P_3, \ldots, P_{i-1}, P_{i+1}, \ldots, P_n$ lie on the same side of the line P_1P_2, and – by the empty circumcircle property of the Delaunay triangulation – they are outside the circumcircle of triangle $P_1P_2P_i$. This implies that angles $\angle P_1P_3P_2, \ldots, \angle P_1P_{i-1}P_2, \angle P_1P_{i+1}P_2, \ldots, \angle P_1P_nP_2$ are all smaller than $\angle P_1P_iP_2$.

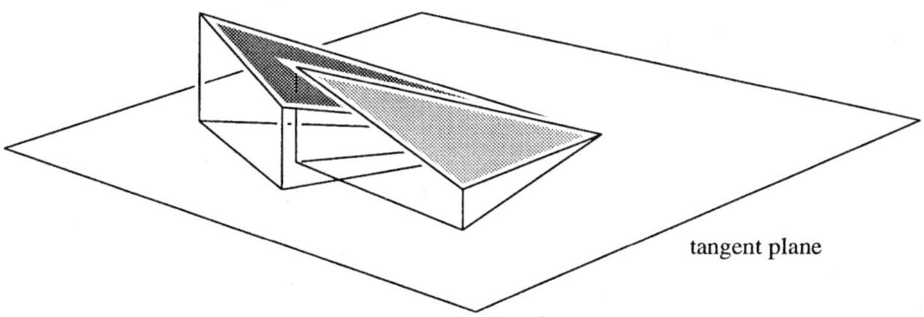

Figure 7. Overlapping triangles

Thus the answer is very simple: choose i such that the angle $P_1P_iP_2$ is the largest (see Fig. 8).

Suppose that we have a hole $P_1P_2\cdots P_n$; it is bounded by the triangles $P_1P_2Q_1$, $P_2P_3Q_2, \dots, P_nP_1Q_n$, and P_1P_2 is the shortest edge of the polygon $P_1P_2 \cdot P_n$. We set $2 < k \leq n$ in the following way:

- the angle between the triangles $P_1P_2Q_1$ and $P_1P_2P_k$ should be greater than 90 degrees, if it is possible;
- the angle $P_1P_kP_2$ should be maximal.

Then we try to register the triangle $P_1P_2P_k$, and fill holes $P_2P_3\cdots P_k$ and $P_kP_{k+1}\cdots P_nP_1$ (see Fig. 9).

Of course, the triangles used for hole filling must satisfy the maximum size criterion. Thus small holes are filled in this step, but large holes remain open.

2.2.5. Insert Non-Processed points

Some points may occur which are not vertices of any registered triangle. These points are simply inserted into the closest triangle, dividing it into three parts.

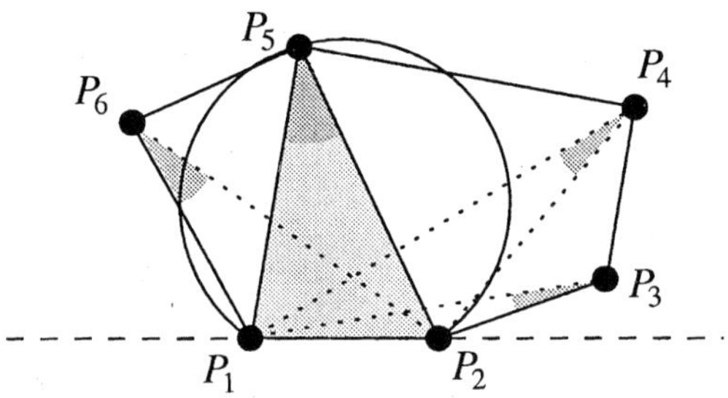

Figure 8. Finding the Delanuay triangulation of a convex polygon by choosing the largest angle

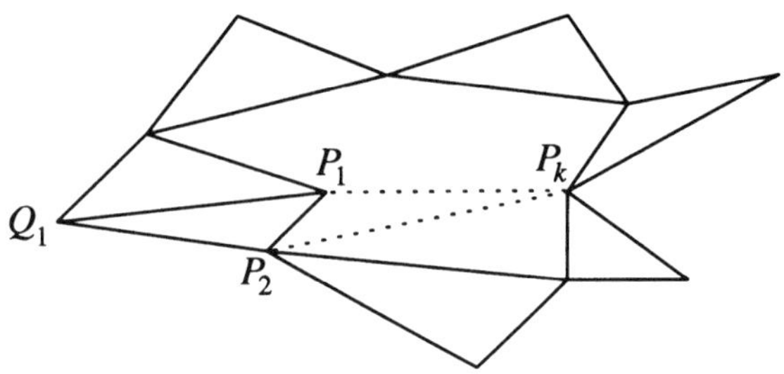

Figure 9. Filling a hole

2.3. Post-Processing

After creating a consistent triangulation an optimising step is performed, using simple edge swapping (see Fig. 10), keeping the original vertices.

2.3.1. Smoothing

There are many smoothing algorithms published in the literature, based on various optimizing principles, for example, minimising curvature integrals.

We prefer a different method. For any three points P_1, P_2 and P_3 of the point set we define the *error* of the triangle $P_1P_2P_3$, and minimize the sum of these errors. The definition is based on the difference between the estimated normals at the vertices and the normal of the triangle.

Denote the estimated normal at vertex P_i by $N_i (i = 1, 2, 3)$ and the normal of the triangle by N_t. We compute the angles between N_t and N_i. The error of the triangle $P_1P_2P_3$ is defined as the minimum of these angles.

The smoothing process is a loop which runs until there is no possible edge flipping which decreases the sum of triangle errors. In any state there is a set of candidate edges which have to be checked. Though the set of candidates may grow – each flipping makes the four neighbouring edges candidate – the algorithm cannot go to an infinite loop, because the sum of triangle errors strictly decreases.

In the beginning of the process all edges are candidates. Then the edges are checked one by one until the set of candidates becomes empty. In the current implementation there is no definite sorting in the set of candidate edges.

2.4. Examples

We have implemented the algorithm described above in C++. In this section we show some examples and results (see Fig. 11–13). We ran these examples on a 400 MHz Pentium II PC with 128MB RAM, under Linux.

To visualise the data sets the Visualisation Toolkit (VTK) was used.

For Klein's bottle some points were discarded to obtain a hole and avoid self-intersection. This hole was large enough that the algorithm did not fill it. For this test, generating an unoriented mesh was allowed. In Fig. 13, the picture on the left

Figure 10. Simple edge swapping

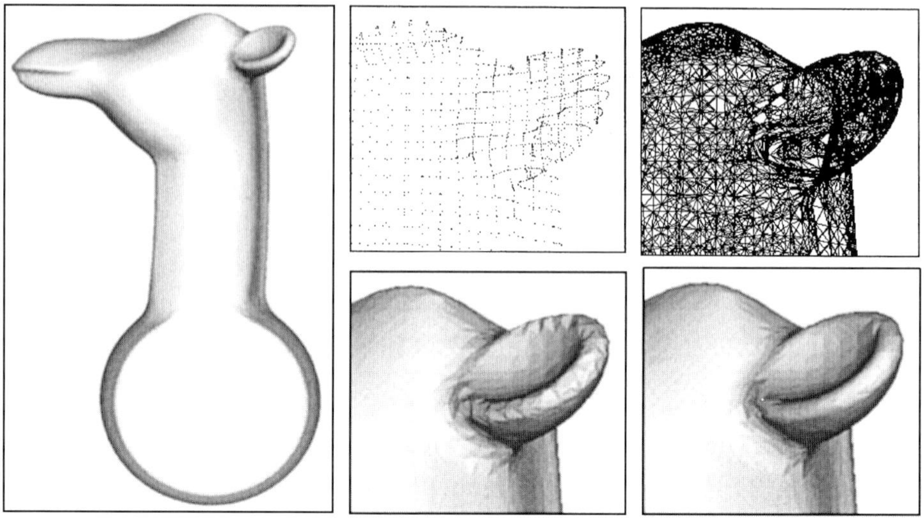

Figure 11. Giraffe (measured data from METROCAD GmbH, Saarbruecken). **a** A mesh with 6611 points and 13048 triangles; **b** points around the ear; **c** neighbourhood graph; **d** triangles without smoothing; **e** smoothed mesh. Elapsed time: 8.5 seconds

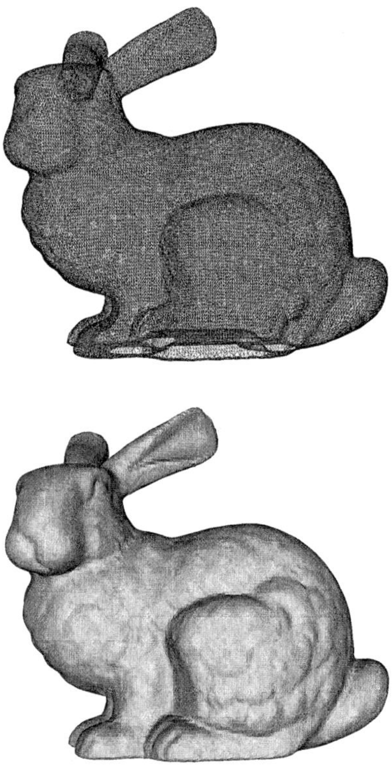

Figure 12. The Stanford bunny (measured data). 35947 points and 69451 triangles. Elapsed time: 69.9 seconds

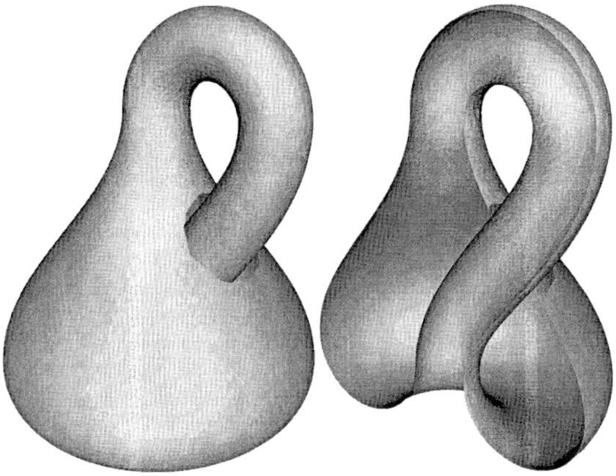

Figure 13. Klein's bottle (synthetic data). 8853 points and 17695 triangles. Elapsed time: 9.1 second

side shows the whole triangle mesh. The set on the right side is the same, but it was cut into half.

3. Conclusion and Future Work

A general 3D triangulation algorithm capable of processing unevenly distributed dense point clouds has been presented. The algorithm is based on reconstructing the relative Delaunay triangulation of the sample points on the surface.

The algorithm has been tested on several dozens of real and synthetic data sets. It works reliably if the distribution of the sample points is dense enough and not extremely uneven.

There are many aspects where the algorithm can be improved. Some of them are collected in the following list.

- *Building global orientation before creating triangles.* The sign of the normals should be set before creating any triangle, using a minimum weighted tree. This can help in many cases when there are two surface segments close together, but with opposite orientations.
- *Re-computing normals before smoothing.* The quality of smoothing depends on the accuracy of normal estimation. After triangulation, the normals should be re-computed.
- *Improving the hole filling step.* The current hole filling method is based on simple heuristics. For complicated holes it may fail.

Acknowledgements

This project started within the framework of an EU supported COPERNICUS project (RECCAD no. 1068) in 1997 and has also been supported by the National Science Foundation of the Hungarian

Academy of Sciences (OTKA no. 26203). Special thanks are due to Dr. Tamás Várady for directing my attention to this research area and for useful suggestions concerning this manuscript.

References

[1] Amenta, N., Bern, M., Kamvysselis, M.: A New Voronoi-based surface reconstruction algorithm. Comput. Graphics 415–421 (1998).
[2] Bajaj C., Bernardini, F., Chen, J., Schikore, D.: Automatic reconstruction of 3D CAD models. Proc. of the Int. J. Conf. on Theory and Practice of Geometric Modeling, Blaubeuren, Germany, October 1996.
[3] Bernardini, F.: Automatic reconstruction of CAD Models and properties from digital scans. Ph.D. Thesis, Purdue University, 1997.
[4] Boissonnat, J.-D.: Geometric structures for three-dimensional shape representation. ACM Trans. Graphics *3*, 266–286 (1984).
[5] Choi, B. K., Shin, H. Y., Yoon, Y. I., Lee, J. W.: Triangulation of scattered data in 3D space. Comput. Aided Des. *20*, 239–248 (1988).
[6] Edelsbrunner, H., Mücke, E. P.: Three-dimensional alpha shapes. ACM Trans. Graphics *13*, 43–72 (1994).
[7] Edelsbrunner, H.: Weighted alpha shapes. Technical Report UIUCDCD-R- 92-1760. Comp. Sci. Dept., Univ. Illinois, Urbana, Ill, 1992.
[8] Guo, B., Mennon, J., Willette, B.: Surface reconstruction using alpha shapes. Comput. Graphics *16*, 177–190 (1997).
[9] Hoppe, H., et al.: Surface reconstruction from unorganised points. Comput. Graphics, 71–76 (1992).
[10] Várady, T., Martin, R. R., Cox, J.: Reverse engineering of geometric models – an introduction. Comput. Aided Des. *29*, 255–268 (1997).
[11] Veltkamp, R. C.: Boundaries through scattered points of unknown density. Graph. Models Image Proc. *57*, 441–452 (1995).

G. Kós
Computer and Automation Research Institute
Budapest
Kende u. 13-17
H-1111 Budapest
Hungary
e-mail: kosgeza@sztaki.hu

Computing [Suppl] 14, 233–248 (2001)

Computing

Cylindrical Surface Pasting

S. Mann and **T. Yeung,** Waterloo, Ontario

Abstract

In the paper, we present cylindrical surface pasting, an extension of standard surface pasting that uses the surface pasting technique to blend two surfaces. The major issues discussed here are the domain mappings and the mapping of the feature control points. There are two types of domain mappings, depending on whether we paste as cylinder on a NUBS sheet or another NUBS cylinder. The mapping of the feature control points has to address both continuity and shape issues.

AMS Subject Classification: 68U07.

Key Words: Hierarchial Modelling, tensor product, β-splines, blending.

1. Introduction

Hierarchical modeling is an important research topic. Many surfaces have varying levels of detail, and modeling techniques that explicitly represent these levels of detail are useful in terms of reduced storage and in interactive modeling paradigms where users want to interact with their models at different levels of detail.

There are several methods for hierarchical modeling, including Hierarchical B-splines [6], various wavelet techniques, and LeSS [7]. Surface pasting is another hierarchical modeling method that has a couple of advantages over most other techniques. In particular, with surface pasting, the user can create a library of features, allowing for reuse of features. Further, unlike many techniques, the features, once pasted, can be reoriented in any direction on the base surface, and do not have to align with parametric directions.

Current surface pasting methods allow the user to paste one surface atop another. However, they do not allow for a single feature to connect two surfaces. Blending or filleting operations need to be employed to connect surfaces together. While there are many filleting methods, with the inspiration of standard surface pasting, we propose new a blending method in this paper, cylindrical pasting, that elaborates the domain mapping and displacement schemes of surface pasting, and applies it to place cylinders on NUBS base surfaces.

Our goal in cylindrical surface pasting is to extend the standard surface pasting method to a wider variety of modeling situations. Thus, while our method can be thought of as a blending method, we will treat it instead as a modeling technique, and in this paper we will focus on the mathematical details behind these operations rather than the user interface for modeling with these blends.

In the next section, we will state the relationship of cylindrical surface pasting to blending techniques. Then in Section 3, we will briefly review the standard surface pasting process. Section 4 is the heart of our paper, where we describe in detail the cylindrical surface pasting process. We conclude with some sample pasted surfaces and directions for future work.

2. Blending

Blending is an operation of creating smooth transitions between a pair of adjacent surfaces. Accordingly, the transition surface is simply called a *blend* or a *blending surface*. Blending methods that use parametric surfaces are the most popular techniques. Martin, Vida, and Varady have published a survey of different blending methods using parametric surfaces that clarifies the nature of blending and the relationships between various parametric blending methods [10].

Using the Martin-Vida-Varady terminology, the cylindrical surface pasting method described in this paper can be though of as a *local parametric-blending* method. In particular, we use a *trimline-based blend* as the basic idea for *Cylindrical Pasting*. In the following, a brief summary of the most important ideas in parametric blending is given. Figure 1 can be used as a guide to the different terms used in blending literature.

The surfaces to be joined smoothly (the surfaces being blended) are called *base surfaces*. The curve that forms the common boundary of a base surface and the blend surface is called a *trimline*. The base surfaces are trimmed at these curves. In

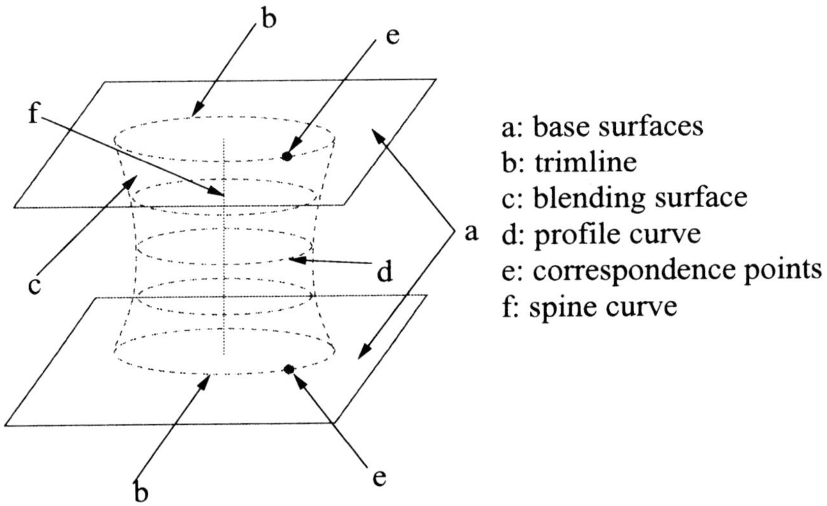

Figure 1. Terminology

general, the blending surface is created as a surface or volume swept along a given longitudinal trajectory, which is called the *spine curve*. At each point of the spine, a cross-sectional *profile curve* is associated with it that locally defines the shape of the blend. A profile can be constant or varying along the spine, and can be symmetric or asymmetric, and can be defined as a circular or free-form arc. Having two trimlines, a corresponding point pair (one point from each trimline) can be joined by a profile curve. Correspondences between these pairs of points need to be established by the *assignment* process.

Cylindrical Pasting is similar to trimline-based methods, which are a class of techniques where an auxiliary spine is generated from the two trimlines, mainly for the purposes of assignment and the creation of profile curves. Since we know that blending replaces parts of the base surfaces with blending surfaces, one obvious way of specifying such an operation is to decide explicitly which parts are to be substituted by choosing where the trimlines should lie on the base surfaces. Once a pair of trimlines has been chosen, a spine curve is used to choose corresponding points on the trimlines to be assigned together. The final important phase of trimline-based methods is a method of generating profile information that makes it possible to define the profile curves that connect assigned pairs of trimline points and contribute to the blending surface.

3. Surface Pasting Process

Tensor product B-spline surfaces play an important role in current surface design [5], especially in surface pasting. Since standard surface pasting is the starting point for the technique described in this paper, we give a quick review of how it works. For details, see any of the earlier works on the subject [1–3].

The pasting process is illustrated in Fig. 2. In surface pasting, we have both a base surface and a feature surface, each of which is in tensor-product B-spline form.

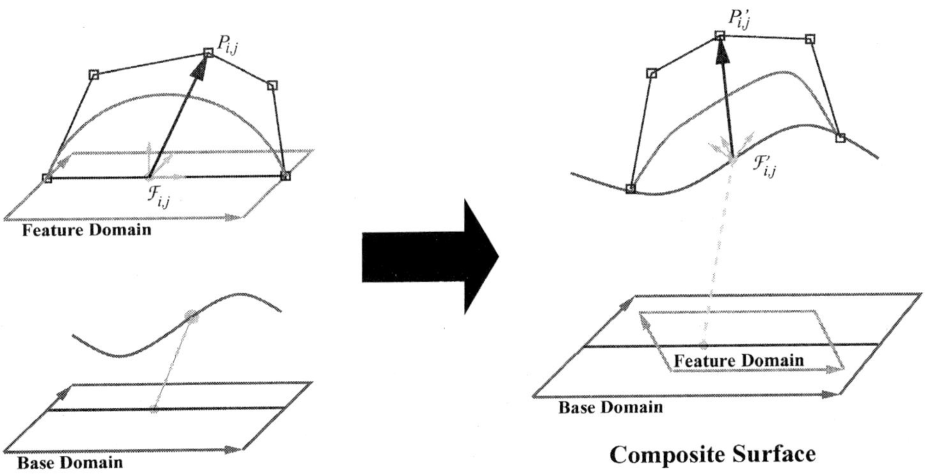

Figure 2. Pasting process

The basic idea is to adjust the feature's control points in such a way that the boundary of the pasted feature lies on or near the base surface, and the shape of the pasted feature reflects the original shape of the feature as well as the shape of the base surface on to which it is pasted.

To map the feature's control points, we first embed the feature's domain in the features's range (upper left of Fig. 2); i.e., we make the feature's domain be a subspace of the feature's range. Typically, we construct the feature surface to allow for an embedding of the domain that places the boundary control points of the feature at the Greville points of the embedded domain. Next, we construct a local coordinate frame $\mathscr{F}_{i,j} = \{u_{i,j}, v_{i,j}, w_{i,j}, \mathcal{O}\}$ for each feature control point $P_{i,j}$ with the origin $\mathcal{O}_{i,j}$ of each frame being the Greville point corresponding to the control point, with two of the frames's basis vectors being the parametric domain directions and the third basis vector being the direction perpendicular to the domain. Each control point $P_{i,j}$ is then expressed relative to its local coordinate frame $\mathscr{F}_{i,j}$ as $P_{i,j} = \alpha_{i,j}u_{i,j} + \beta_{i,j}v_{i,j} + \gamma_{i,j}w_{i,j} + \mathcal{O}_{i,j}$.

Next, we associate the feature's domain in the base's domain (right half of Fig. 2). This gives us the location on the base surface on to which we want to locate the feature. We now map each coordinate frame $\mathscr{F}_{i,j}$ on to the base surface, giving a new coordinate frame $\mathscr{F}'_{i,j} = \{u'_{i,j}, v'_{i,j}, w'_{i,j}, \mathcal{O}'_{i,j}\}$, whose origin $\mathcal{O}'_{i,j}$ is the evaluation of the base surface at $\mathcal{O}_{i,j}$, and two of its basis vectors lie in the tangent plane of the base surface at that point, the third being perpendicular to the tangent plane. We then use the coordinates of each feature control point $P_{i,j}$ relative to $\mathscr{F}_{i,j}$ to weight the elements of the frame $\mathscr{F}'_{i,j}$. This gives us the location of the pasted feature control point, $P'_{i,j} = \alpha_{i,j}u'_{i,j} + \beta_{i,j}v'_{i,j} + \gamma_{i,j}w'_{i,j} + \mathcal{O}'_{i,j}$.

4. Cylindrical Surface Pasting

Cylindrical Pasting is a surface modeling tool that integrates the techniques of parametric trimline-based blending into surface pasting, and creates a smooth transition cylinder between two base surfaces. We made modifications to the major pasting techniques (domain mapping and control points displacement) to adapt standard surface pasting to the cylindrical pasting environment.

In this paper, we give details of the final technique that we used for cylindrical surface pasting. A more complete description and a description of other ideas we tried that did not work can be found in a technical report [9]. However, we will describe one approach that we discarded and replaced with a better method, since this improved technique works both in standard pasting and cylindrical pasting.

The main idea in surface pasting is the mapping of the feature control points to get the feature surface to lie in the appropriate place relative to the base surface. There are three types of control points to map, each of which requires a different mapping technique. The first control points are those along the boundary of the feature. These should be mapped to achieve approximate C^0 continuity. The second layer of control points are mapped to achieve approximate C^1 continuity. And the remaining interior control points are mapped to achieve the desired feature shape. In this paper, we focus on the mapping of the first two layers, as their mapping is the pasting process; for completeness, we also give the mapping of the remaining control points, although they could be mapped using any standard extrusion method.

We will begin by stating the representation of the cylindrical feature used in our system. Next, we describe the first step to mapping the feature boundary control points, which is to associate the feature domain with the base domain. We then give the mapping of the first and second layers of control points. We then discuss our mapping of the remaining interior points. In Section 5, we give a brief overview of our user interface, and show some results of the cylindrical pasting process.

4.1. Representing a Cylinder Using NUBS

Our work deals with non-uniform B-splines (NUBS). To represent a cylindrical shape using a NUBS surface, we use the standard trick of identifying one of the edges of the domain rectangle with the opposite edge of the domain rectangle. The cross section of a cylinder is a circular curve. Although circles cannot be represented exactly using a NUBS curve, a NUBS curve can represent a closed curve that is a good approximation of a circle [4], although for our implementation we constructed the circular approximation by hand. To represent a closed curve with a cubic NUBS, we set the last three control points to be the first three control points, with an appropriate setting of the knot vector.

Mathematically, if we have a cubic B-spline with a knot vector $\{v_0, \ldots, v_N\}$ and control points $P_0, \ldots, P_{N-3}$, the following conditions must hold to get a closed curve:

$$P_0 = P_{N-5}, \quad P_1 = P_{N-4}, \quad P_2 = P_{N-3}, \tag{1}$$

and

$$\begin{aligned} v_1 - v_0 &= v_{N-3} - v_{N-4}, & v_2 - v_1 &= v_{N-2} - v_{N-3}, \\ v_3 - v_2 &= v_{N-1} - v_{N-2}, & v_4 - v_3 &= v_N - v_{N-1}. \end{aligned} \tag{2}$$

(Note: we are using the knot vector typically used with the blossoming variant of the B-spline; other forms of B-splines will typically put an additional knot at each end of the knot vector.)

A tensor B-spline surface has a two-dimensional domain defined in two parametric directions, U and V. We represent our cylinders by a rectangular domain where the V direction joins itself as in Eqs. 1 and 2, and U aligns with the axis of the cylinder. We will use a knot vector with full end knot multiplicity in the U direction.

4.2. Mapping of Domain for Cylinders

The first step in performing standard surface pasting is to map the domain of the feature surface to the domain of the base surface. The same procedure has to be done for Cylindrical Pasting. The rectangular domain of the feature cylinder has to be first mapped inside the domain of the base surface.

A cylinder can be pasted on two types of NUBS base surfaces: a normal NUBS surface, or a cylindrical NUBS surface. Depending on the type of the base surface, the rectangular domain of the feature cylinder will be transformed to the base domain in two different ways.

In the first case, the base surface is a normal NUBS surface with a rectangular domain. Only one of the two edges of the feature cylinder will be pasted on the base, as shown in Fig. 3a. We locate the position of the edge of the feature on the base surface through a domain association. The edge of the feature domain corresponding to the edge of the feature surface that is to lie on the base surface is mapped to circle in the base domain as shown in Fig. 3b. By default, we initially locate the domain for the feature cylinder at the center of the base domain with a predefined radius; the user may scale and translate this circle within the base domain. The second circle (dotted) in this figure is used for mapping the derivatives, as discussed in the next section.

In the second case, both the base and features surfaces are NUBS cylinders. Again, only one of the feature cylinder's edges is pasted on the base as illustrated Fig. 4a, with the top cylinder as the base. To locate the edge of the feature surface on the base surface, we again map an edge of the feature's domain into the base domain. As shown in Fig. 4b, the mapping of this edge is different. Since the base is a cylinder, we map the edge of the domain to a line that spans the base domain. Since the two sides of the base domain represent the seam of cylinder, we have mapped the closed curve of the edge of the feature surface to a closed curve on the base surface. The arrow in this figure is used to map the derivatives, as discussed in the next section.

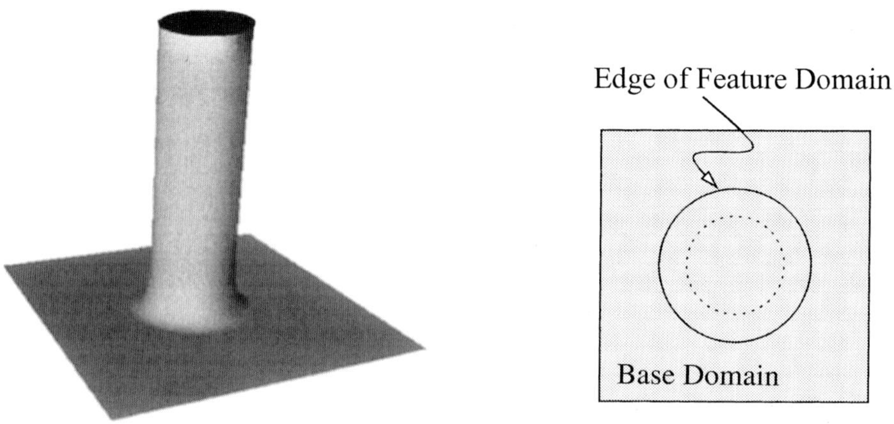

(a) World space (b) Domain space

Figure 3. A blending cylinder on a normal NUBS surface

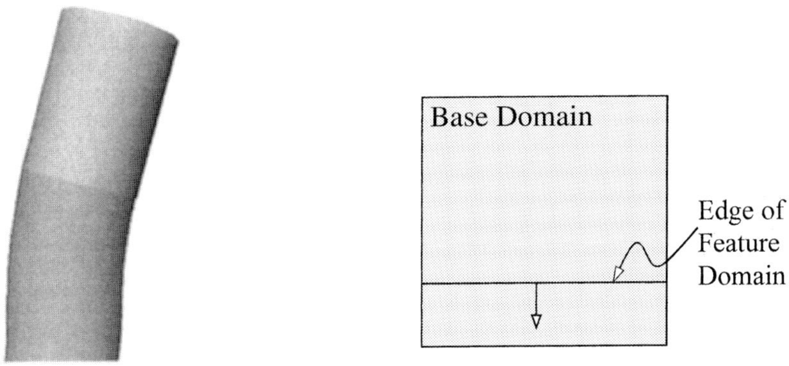

(a) World space (b) Domain space

Figure 4. A blending cylinder on a cylindrical NUBS surface

4.3. Control Points Displacement Scheme

As the mechanism of Cylindrical Pasting is based on the idea of trimline-based blending, the major issue is to determine the trimlines on the base surfaces. The body of the blend is constructed from the spine curve to be defined from the trimlines, and the profile curves along the spine curve.

In standard pasting, each control point of the feature surface is represented relative to a local coordinate frame. These frames are then mapped on to the base surface, and each feature control point is placed relative to the image of its local frame. For cylindrical pasting, we only have an identification of the first two rows of the feature control points with the base surface. These two layers can be mapped using the standard pasting method, although we will present a better mapping of the second layer of control points. The remaining feature control points must be mapped in a different manner. In this section, we discuss the mapping of the first two layers of control points, and in the next section we discuss the mapping of the remaining layers of control points.

The first two layers of control points at either end of the cylinder determine the position and first derivatives along the boundary. We call these two rows of control points the 0th and 1st layers in the world space, denoted as L_0 and L_1, respectively. L_0 is analogous to the trimline in the trimline-based blending method.

We mapped the L_0 layer of the feature cylinder in the same manner as standard pasting. Each L_0 point is located at the Greville point in the embedded feature domain, giving a zero displacement vector relative to its local frame, and the standard pasting procedure is used to paste these points. In other words, the boundary control points of the feature cylinder map to points on the base surface. This procedure is done for both edges of the feature cylinder. The result is that the boundary of the pasted cylinder will lie close, but not directly on, the base surface. If the C^0 discontinuity is too high, it can be reduced by performing knot insertion on the feature cylinder, as is done for standard pasting.

Next, we map the L_1 layer to approximate C^1 continuity, which in standard pasting required the displacement vectors for the L_1 layer control points to be set to zero. With cylindrical pasting, we only have the boundaries of the cylinder's domain associated with the base domain. So in our initial attempt, for pasting a cylinder on a NUBS surface we placed a second circle in the base domain, with the same center but smaller radius than the first circle (the dotted circle of Fig. 3b). And to paste a cylinder on another cylinder, we associated the L_1 layer with the base domain by using a vector perpendicular to the location of the feature domain edge within the base domain (the arrow in Fig. 4b). In both cases, once the identification of the L_1 layer of the feature with the base domain was made, we mapped the L_1 layer of the feature in the same manner as standard pasting.

To get a better feel for the C^0 and C^1 discontinuities, we looked at both the trimmed and untrimmed surfaces, and discovered two problems. The first problem is that while the boundary of the feature lies near the base, it is still unacceptably far from it as seen in Fig. 5b. This problem can be handled by inserting more knots in the V parametric direction. However, using the standard pasting method for the second layer of control points was problematic: the C^1 discontinuity was still unacceptably high Fig. 5a (in this figure, the feature cylinder is rendered as partially transparent). Adding knots in the V parametric direction has no effect on this discontinuity, and adding knots in the U parametric direction yields only small improvements.

To reduce the C^1 discontinuity, we changed the method for pasting the L_1 layer of control points. For cylindrical pasting, we still paste the L_0 layer of control points by using the method of standard pasting using zero length displacement vectors. Then, at each feature control point $C_{0,j}$ of L_0, we construct a coordinate frame $\mathscr{F}_{0,j}$, where $C_{0,j}$ is the origin, the unit derivative vector in the V direction is one coordinate vector, the derivative vector in the U direction is the second coordinate vector, with their cross-product forming the third derivative vector (Fig. 6a).

Next, we map the frames for the L_0 layer to the embedded domain of the base surface (Fig. 6b). Here, the two frame vectors that are tangent to the unpasted feature surface are mapped to the domain plane so that the basis vector tangent to

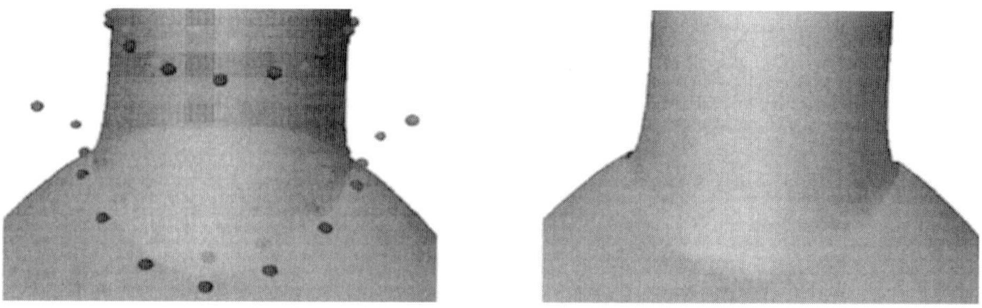

(a) Untrimmed base view (b) Trimmed base view

Figure 5. Pasting with C^1 continuity

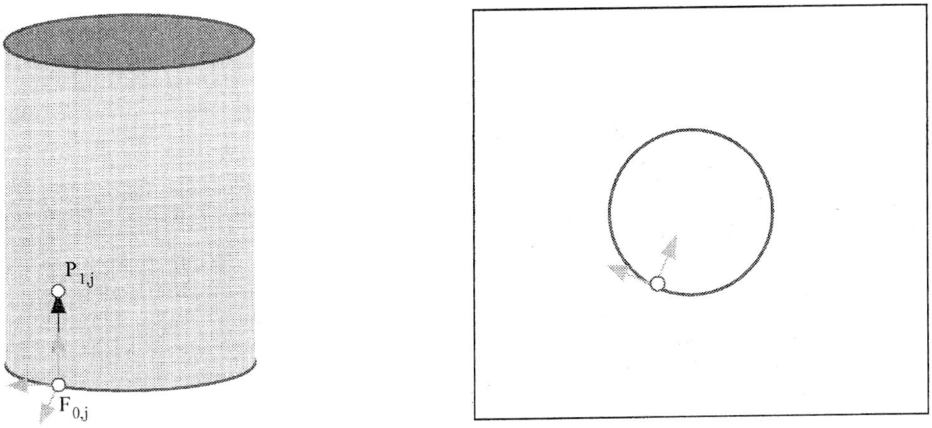

(a) Feature cylinder (b) Base domain

Figure 6. Cylindrical displacement mapping

the circle along the edge of the cylinder maps to be tangent to the circle in the base domain, and other tangent vector that lies in the tangent plane of the cylinder maps to be perpendicular to the circle, pointing inside the circle (the third basis vector is mapped parallel to the z-axis). We then map the frames on to the base surface and construct the $\mathscr{F}'_{0,j}$ frames. Each L_1 layer control point $P_{1,j}$ is then expressed as a displacement relative to frame $\mathscr{F}_{0,j}$, and as with standard pasting these values are used to weight the elements of $\mathscr{F}'_{0,j}$ to get the location of $P'_{1,j}$.

The net effect of the new method is to map differences of control points on the L_1 and L_0 layers (e.g., $P_{1,j} - P_{0,j}$) to cross-boundary derivatives of the base surface. With the new scheme both the C_0 and C_1 discontinuities are decreased as we insert knots in the V parametric direction of the feature.

This new method of mapping the L_1 layer has a lower C^1 discontinuity than the original method for mapping this layer as can be seen in the other images in this paper. Although devised for cylindrical pasting, this method for mapping the second layer of control points could easily be incorporated into standard pasting, and should give a reduction in C^1 discontinuity with no increase in computational cost.

4.4. The Remaining Control Points

Once we have mapped the L_0 and L_1 layers of both ends of the feature on to two bases surfaces, we need to map the remaining layers of the feature's control points. This is a standard extrusion problem; since our focus was on the pasting process (i.e., the mapping of the boundary control points), we implemented a simple technique using a spine curve, reparameterized to a near arc length parameterization, and then mapped the L_1 layers along this curve.

We considered using cubic Hermite splines to connect the L_i layers as Kim and Elber did [8]. Had we done this, then the mapping of the L_0 and L_1 layers described in the previous section would complete the mapping of our cylinder, and our method would essentially be identical to that of Kim and Elber. However, we intend to use our method for both blending and for longer connecting pieces, and we found that using only four layers of control points gave poor results for longer connecting pieces.

If we have more than four layers of control points, after mapping both pairs of L_0 and L_1 layers, we need to map the remaining interior control points of the feature cylinder. Initially we tried some simple linear interpolation techniques of the L_1 layers to locate the remaining interior control points. However, we found that these techniques gave us sharp creases and/or skews in our connecting cylinder, as illustrated in Fig. 7.

Instead, we decided to use a spine curve to specify the approximate path of the feature cylinder, and construct the remaining interior feature cylinder control points by mapping the L_1 layers to lie roughly perpendicular to this spine curve. The rest of this section gives the details of the construction of this spine curve and the mapping of the L_1 layers.

To get a well-shaped blending cylinder, we constructed the interior control points around a spine curve. This spine curve plays the role of the skeleton for the cylinder. It is a simple cubic Bézier curve defined by four control points: C_0, C_1, C_2 and C_3. Each of the two end points, C_0 and C_3, is the average point of the corresponding L_1 layers of control points. We then construct vectors $\vec{n}_0$ and $\vec{n}_3$ at C_0 and C_3 by summing the crossproducts with the surrounding points in the layer:

$$n_0 = \sum_j (C_{1,j} - C_0) \times (C_{1,j+1} - C_0)$$

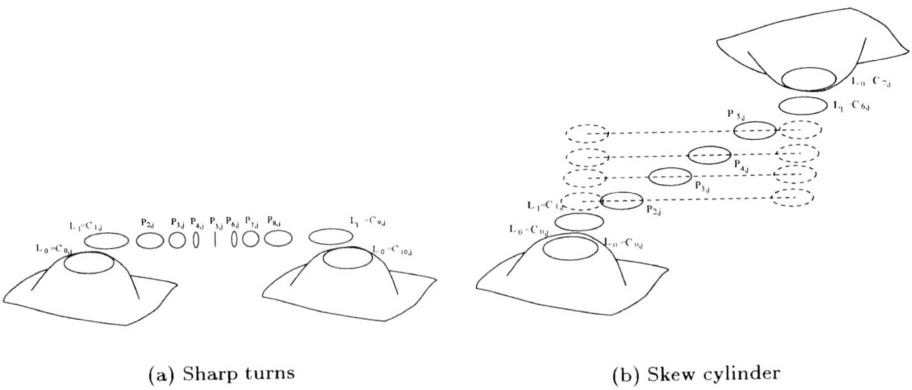

(a) Sharp turns (b) Skew cylinder

Figure 7. Linear interpolation leads to poor blends

$$n_3 = \sum_j (C_{n-1,j} - C_3) \times (C_{n-1,j+1} - C_0)$$

where $j + 1$ is taken modulo $m + 1$, with $n + 1$ and $m + 1$ being the number of control points in the two parametric directions. The orientations of the $\vec{n}_i$ are set to point away from the corresponding base surfaces.

Once we have $C_0, C_3, \vec{n}_0$ and $\vec{n}_3$, we locate the other two control points, C_1 and C_2 by placing C_1 on the line given by C_0 and $\vec{n}_0$, and C_2 on the line given by C_3 and $\vec{n}_3$. The C_i are now the control points of a simple cubic Bézier curve as shown in Fig. 8. The distance along the $\vec{n}_i$ that we place C_1 and C_2 is a curvature parameter that determines the bending of the cylindrical body, and is made available as a shape parameter for the user.

To define the cylindrical body, each remaining layer of control points P is built as a ring of points along the spine. We will build each of these curves as a linear interpolation of the mapped images of the two L_1 layers at both ends. These mapped images are denoted as L'_1 in Fig. 8.

The effect we want is for the left L_1 layer to sweep along the curve, gradually transforming into the right L_1. If we have $n + 4$ layers of control points in our blending cylinder, then four layers are L_i layers, and we need n profile curves. To construct the L'_1 and P layers, we first build a local coordinate frame at C_0 and one at C_3, and represent each of the control points in the L_1 layers relative to the corresponding frame. We then map these coordinate frames along the curve so that they are centered at a point $O(t)$ on the spine obtained by evaluating $O(t) = \sum C_i B_i^3(t)$ at some set of t values; the mapping of the basis vectors of the frames is described below.

While the values $t = \frac{i}{n+1}$, for $i = 1 \ldots n$, might seem like appropriate choices for sampling O, this Bézier curve is not arc- length parameterized. Thus, with uniform samplings of t, we get non-uniform samples on O, resulting in a blend surface with twists. To address that problem, we made an approximate arc-length parameterization of O by sampling O uniformly, computing the distance between these

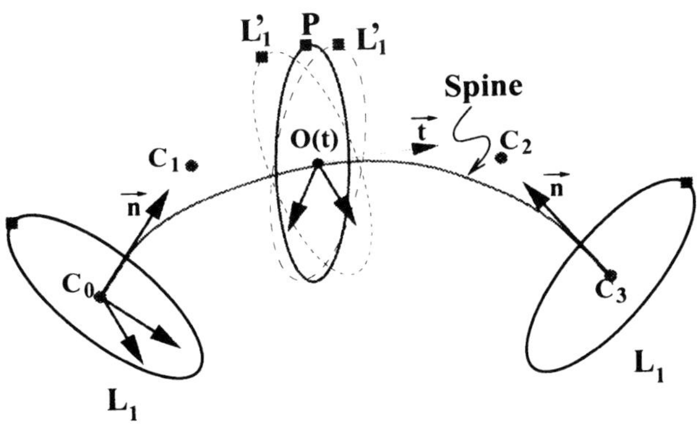

Figure 8. Method to determine an inbetween layer of control points

sample points, and using these distances to reparameterize the curve. The result is a close-to-arc-length parameterization, and rings of control points that are uniformly spaced over O.

Once we have the sample points on O, we need to map the L_1 layers to these sample points. We initially considered the idea of rotating L_1 along the spine curve with progressive degrees to get mapped images L_1' has been considered. Unfortunately, it is unclear how to find the appropriate degree variations for how much each L_1 should rotate to give the final profile to best represent the geometry of its base. Instead, we used a geometric transformation of $\vec{n}$, mapping $\vec{n}$ to the vector $\vec{t}$ tangent to the spine curve at $O(t)$. This gives the direction for locating the mapped coordinate frame derived from C_0, hence, the mapped control points can be used to locate L_1'. Applying the same process to L_1 at C_3, two mapped curves L_1' are obtained at $O(t)$. To obtain the final profile curve P that reflects the transition between the base surfaces, we applied linear interpolation (in the layer number) on the generated L_1's.

4.5. Correspondence Assignment Process

The cylindrical pasting process constructs the first two layers of control points on either end of the pasted cylinder by associating boundary control points with two edges of the cylinder domain, and then mapping these two edges into the base domain. The interior control points are found by transforming the second layer at either end of the cylinder along a curve, and then blending corresponding points, as described earlier in this chapter. However, one question that we have left unanswered is how to set up a correspondence between the two layers of blended control points.

The correspondence process is a non-trivial one. If we make a poor match between the two layers, then we introduce a twist in our blending surface. In our prototype for cylindrical pasting, we used the following process (which we note is inadequate in general):

1. Using the Cartesian coordinate system for the range space and the normal to the plane approximating the each second layer of pasted control points, find the coordinate direction most perpendicular to each normal.
2. Select the control point within the layer whose coordinate relative to the selected axis is a maximum.
3. Repeat the selection process with the second layer at the other end of the pasted cylinder.
4. Associate the two selected control points, and associate the remaining control points within the layers starting from the selected control points.

As stated, the last step still leaves unspecified the direction around the ring of control points. We chose this direction arbitrarily, and allow the user to change the direction for either ring of control points.

Note that this method works reasonably well if the two layers have relative locations similar to that on the right of Fig. 9, but is a poor choice if the layers have relative locations similar to that on the left of Fig. 9. Using a better method for solving the correspondence problem has been left for future work.

For more details on our method, see the technical report [9]; see the Vida-Martin-Varady survey [10] for references to other extrusion methods.

5. Results

We tested our cylindrical pasting method by blending two surfaces. Two examples were shown earlier (Figs. 3 and 4). A third example is shown in Fig. 10. In this figure, the bottom surface is a plane, while the top surface is a curved surface. The plane provides a useful test case since the pasting method for the boundary control points will result in the boundary of the feature meeting the plane with C^1

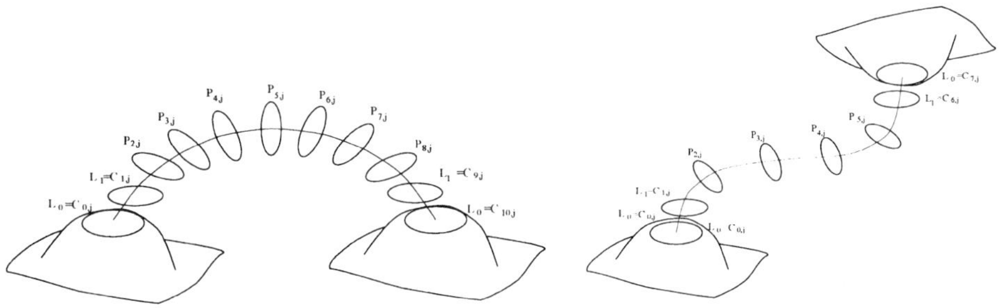

(a) Type 1 (b) Type 2

Figure 9. Good blend

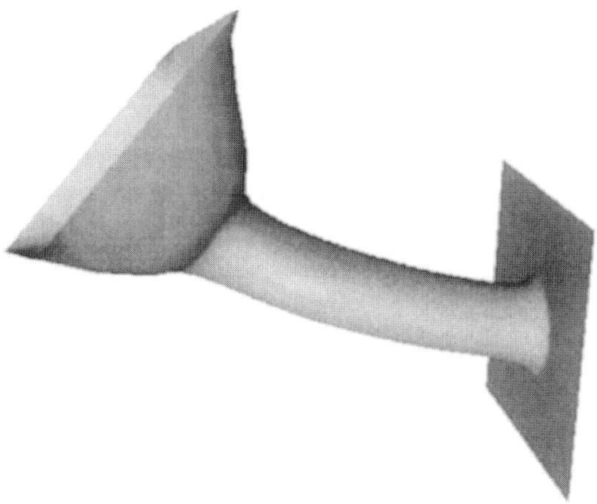

Figure 10. Cylindrical pasting example

continuity. Note, however, that once we trim the base, we will not have a C^0 join, since the feature boundary is not the trim curve. In any case, in this image we see that cylindrical pasting has the desired effect.

Our system was designed to test the mathematical ideas, and was implemented with a simple user interface. The following is an overview of the system. The user selects two surfaces to be joined with the mouse. The system places the boundary of the feature's domain in each of the base domains. The L_0 and L_1 layers of the feature's control points are mapped on to the base, an initial spine curve is created, and the remaining layers are set using the method described in Section 4.4.

The user can adjust both domains using sliders to adjust the radius of the circles, and can drag the circles/lines representing the feature domains in a pop-up domain window. The user also can adjust the spine curve using sliders to adjust the curvature of the spine curve. In the current system, the user has to visually inspect the joins of the features to the base, and tell the system to insert knots in the feature if the discontinuities are too high. After any adjustment, the system recomputes the blending feature.

Using our surface editor, we were able to drag the ends of the cylinder along the base surfaces and adjust other parameters at interactive speeds. The C^0 discontinuity was only visible when a small number of control points were used in the V parametric direction. This gap would disappear after performing knot insertion in the V direction, although some pixel drop-out was still visible due to the mismatch in tessellations between the base and feature surfaces.

The C^1 discontinuity was not visible, although if the cross-boundary tangents are too short then the sharp curvature at the join was visible.

6. Conclusions and Future Work

The goal of this work was to extend the surface pasting method to allow us to use the feature surfaces to connect two base surfaces. The results in this paper show how to modify the pasting process to achieve this goal. We can now use the surface pasting method to connect two surfaces, similar to other blending schemes. The advantages of cylindrical surface pasting are low degree connecting pieces that can be modified at interactive speeds.

Further, we found an improvement to the standard pasting method for mapping the cross-boundary control points. Our new method of mapping these control points results in a lower C^1 discontinuity, and knot insertion in the V direction reduces both the C^0 and C^1 discontinuities (for standard surface pasting, knot insertion in the V direction only reduces the C^0 discontinuity).

At this point, several major issues still need to be explored:

1. Domain curves. We used a NUBS approximation to a circle to map the edge of the feature domain into the base domain. However, we have no guarantee that a

circle in the base domain will map to a nice curve on the base surface. An ideal user interface would either allow the user to specify a curve on the base surface, or the system would find automatically a good first guess of a curve on the base surface, and map this curve backwards into the base domain.

2. Hierarchical Modeling. The goal of surface pasting is to provide a hierarchical modelling method that allows reuse of feature surfaces. The current version of cylindrical surface pasting is non-hierarchical. While some aspects of extending cylindrical surface pasting to be a hierarchial method are straightforward, other aspects will be more difficult. In particular, if you paste both ends of cylinder on to the same surface, then the resulting surface will be of higher genus than the original base surface[1]. Such topological issues will complicate the hierarhcal cylindrical surface pasting technique.

Recently, Gozonlez-Ochoa and Peters [7] have developed on offset method similar to surface pasting. Their method works on top of a winged-edge data structure, and readily solves these topology issues. Hierarchical modeling with cylindrical surface pasting will probably need to take a similar approach.

3. Fine tuning. The current system was a proof-of-concept implementation. The user interface is low level, with the user directly adjusting various parameters through sliders. Further, parts of what is now directly controlled by the user could be automated, such as automatically inserting knots to reduce the C^1 and C^0 discontinuities to the user specified tolerance.

Finally, our construction of the interior control points was an ad hoc construction intended to test the feasibility of cylindrical surface pasting. Instead, the shape of the cylindrical blend could be automatically set to achieve various goals (closest fit to a cylinder, minimize maximal curvature, etc.).

Acknowledgements

Many thanks to Richard Bartels and Kirk Haller, whose discussions of many of these issues proved invaluable. This work was supported by NSERC.

References

[1] Barghiel, C., Bartels, R., Forsey, D.: Pasting spline surfaces. In: Mathematical methods for curves and surfaces (Schumaker, L., Daehlen, M., Lyche, T., eds.), pp. 31–40. Vanderbilt: Vanderbilt University Press, 1995.

[2] Bartels, R., Forsey, D.: Spline overlay surfaces. Technical Report CS-92-08, University of Waterloo, Waterloo, Ontario, Canada N2L 3G1, 1991.

[3] Chan, L.: World space user interface for surface pasting. Master's thesis, University of Waterloo, Waterloo, Ontario, Canada N2L 3G1, 1996. Available as Computer Science Department Technical Report CS-96-32, ftp://cs-archive.uwaterloo.ca/cs-archive/CS-96-32/.

[4] Dokken, T., Dæhlen, M., Lyche, T., Mørken, K.: Good approximation of circles by curvature continuous bezier curves. Comput. Aided Geom. Des. *7*, 33–41 (1990).

[5] Farin, G.: Curves and surfaces for computer aided geometric design, 3rd ed. New York: Academic Press, 1994.

[6] Forsey, D., Bartels, R.: Hierarchical B-spline refinement. Comput. Graphics *22*, 205–212 (1988).

[1]Thanks to Jorg Peters for pointing this out.

[7] Gonzalez-Ochoa, C., Peters, J.: Localized-hierarchy surface splines (LeSS). In: ACM Symposium on Interactive 3D Graphics, 1999. Available as http://www.cise.ufl.edu/~jorg/jmisc/3dInteractive.ps.gz.
[8] Kim, K., Elber, G.: New approaches to freeform surface fillets. J. Visualization Comput. Anim. *8*, 69–80 (1997).
[9] Mann, S., Yeung, T.: Cylindrical surface pasting. Technical Report CS-99-13, University of Waterloo, Waterloo, Ontario, CANADA N2L 3G1, 1999. ftp://cs-archive.uwaterloo.ca/cs-archive/CS-99-13/.
[10] Vida, J., Martin, R. R., Varady, T.: A survey of blending methods that use parametric surfaces. Comput. Aided Des. *26*, 341–365 (1994).

S. Mann
T. Yeung
Computer Science Department
University of Waterloo
Waterloo, Ontario, N2L 3G1 CANADA
e-mail: smann@cgl.uwaterloo.ca

Computing [Suppl] 14, 249–265 (2001)

Computing

A Constraint-Based Method for Sculpting Free-Form Surfaces

P. Michalik and **B. Bruderlin**, Ilmenau

Abstract

We discuss the problem of creating editable features for free-form surfaces. The manipulation tool is a user-defined curve on the surface. The surface automatically follows changes of the curve keeping a predefined set of constraints satisfied, specifically the incidence and tangency along one or several surface-curves. We review and update our approach presented earlier [18] and show how the curve-surface composition can be expressed as a linear transformation. In this context, we also describe the so-called "aliasing" problem caused by an incompatibility of a general curve on a surface with the rectangular mesh of degrees of freedom of a tensor product surface. The proposed solution is a local reparametrization in accordance with the feature.

1. Introduction

Relational geometry is a very powerful paradigm which allows designers to create geometric models without exact a-priori knowledge of all coordinates. A designer sketches a basic form of a model and adds features, as needed later. In general, some kind of relations (constraints) between new and existing features may be defined, and will be maintained during the design process.

There are several approaches known for solving the constraints among "simple" geometric elements such as points, lines in 2D or planes in 3D, see e.g. [7]. However, the difficulties in these systems increase substantially when polynomial curves and surfaces are involved.

Assume a 3D point is constrained to be incident on a surface. If the position of the point is changed, the surface has to "follow" due to the defined incidence constraint. In the case of a plane or cylinder, the choices are usually obvious; we expect that the plane is rotated and/or translated into a new position, such that the incidence constraint is satisfied. A cylinder has an additional degree of freedom (the radius may change).

In the case of B-Spline surfaces the result of such an operation is not that obvious. We could use only the degrees of freedom associated with the affine transformation; this would reduce the changes of the surface to rotation, translation and scaling. Although this is useful, in some cases it might be too restrictive. If no constraints are defined, a piecewise bi-polynomial B-Spline has as many independent degrees of freedom as the number of control point coordinates. Defining a point-surface incidence constraint may only "consume" some of them, while the others are not influenced. For instance, if the constrained point is changed, only the dependent control points react locally. The surface exhibits a bumb around the position of the point. In principle, the same applies to restricting a curve to be incident on a surface, but the identification of influenced control points is substantially more complicated. We need to deform a surface according to a given point or curve, such that the associated constraints are satisfied, which may mean that the new surface exhibits a local change "along" the given curve.

1.1. Related Work

Free-from surface sculpting or deformation techniques originate from a need for more sophisticated editing of free-form surfaces. In the context of this article, two fundamental approaches are of importance: the free-form deformations technique (FFD) [22] and the variational methods, examined for example by Gossard, Celniker and Welch in [3] or [24].

Our previous work [18] is closely related to the variational methods and is briefly reviewed in the next section. The initial problem of that article is stated as follows: The user marks points or curves on the surface which will be edited to meet certain design criteria. The points or curves may be modified in 3D, while the prescribed constraints, particularly the incidence relation, are maintained. Thus, the design parameters of the curves define parameters of the model. The usage of other kinds of constraints, such as prescribed continuity along a curve or angles between surfaces meeting at a curve, are also possible.

The Extended Free-Form Deformation (EFFD) [4] or axial-FFD [17] pursues a similar goal. All FFD methods utilize the following principle: An existing free-form model A is embedded in an auxillary free-form primitive B. A functional dependency between degrees of freedom of A and B is found, such that changes of B are carried over to A: $A = f(B)$.

While the traditional FFD method embeds a free-form surface in a free-form volume, the axial-FFD technique [17] realizes a functional dependency between a free-form surface and a 3D curve – an "axis". The DOFs (control points) of the surface are attached to control points of the curve.

Once the auxialiary free-form primitive B has been found (not a trivial task at all), the problem of all FFD methods is to find a "good" embedding of A in B. Particularly, the EFFD technique requires solving non-linear equation systems. The axial FFD has an additional problem: the embedding of the surface in the "axis" is not unique, and some intuitive heuristics must be chosen. The other problem is the simultaneous satisfaction of additional constraints. A space of all deformations not violating the fixed constraints must be found, which can become difficult.

Finally, the method of "wires" developed by Singh and Fiume [23] should be mentioned. The wires are curves, which serve as the editing tool for surface sculpting. Although conceptually similar to axial-FFD, it utilizes an intuitive heuristics for embedding the DOFs of the model in the wires. The axial FFD and the wires-method do not guarantee incidence of the edited curve on the surface. In both methods, the surface only mimics the changes of the edited curve.

In the following, our previous approach [18] is reviewed and improved. Some changes have been made, which increase the efficiency and numerical stability of the method. We describe the "aliasing" problem which occurs in some cases. An alternative solution is proposed, utilizing the extended curve network inerpolation technique [11], solving the aliasing problem without the necessity of global constraints on the smoothness of the surface.

2. Revising the Previous Approach

Using the variational method, the incidence of a curve on a surface can be formulated in a very convenient way – as a linear system of equations, which allows combination of arbitary constraints of this kind, see e.g. [10] and the discussion below.

In [18] we formulated a continuous function approximation problem: Assuming the pre-image $u(t), v(t)$ of the curve $C(t)$ in the domain of the surface $S(u, v)$ is known (and constant), an error functional minimizing the distance square between the curve and a surface is:

$$\int_a^b \frac{1}{2} [S(u(t), v(t)) - C(t)]^2 dt \longrightarrow \min$$

The main contribution of [18] is an efficient algorithm based on functional composition using blossoms ([6], [21]) for setting up the Gaussian equations and efficient data structures for accessing the parameters (coefficients). We formulate an algebraic relation between the parameters (control points of the curve) $\mathbf{y}$ and the DOFs (control points of the surface) $\mathbf{x}$, which allows obtaining the surface directly, once the values of $\mathbf{y}$ are known. This is expressed as a linear system of equations $\mathbf{A} \cdot \mathbf{x} = \mathbf{Q} \cdot \mathbf{y}$ which is usually underdetermined. Solving the above equations for $\mathbf{x}$ yields:

$$\mathbf{A}' \cdot \mathbf{x} = \mathbf{Q}' \cdot \mathbf{y} \Longrightarrow \mathbf{x} = \mathbf{Q}' \cdot \mathbf{y} - \mathbf{N} \cdot \mathbf{p} \tag{1}$$

where the primes denote matrices resulting from Gaussian elimination, $\mathbf{N}$ denotes the null space of $\mathbf{A}$ and $\mathbf{p}$ denotes a vector of free variables. Eq. (1) defines a hyperplane in a space of the unknowns – the resulting DOFs of the surface. In general, the matrix $\mathbf{A}$ will be singular, and the surface retains some unconstrained degrees of freedom, which can be set (at least theoretically) to arbitrary values. In practice, the editing process is an ordered sequence of steps, with an intermediate surface in each: $S_1 \rightarrow S_2 \rightarrow \cdots \rightarrow S_n$, hence the values of the independent DOFs in the current step maybe assigned the values from the previous one.

The symbolic computations of the integrals and the Gaussian elimination have been shown to be the weak parts of the previous approach. Even after using all the

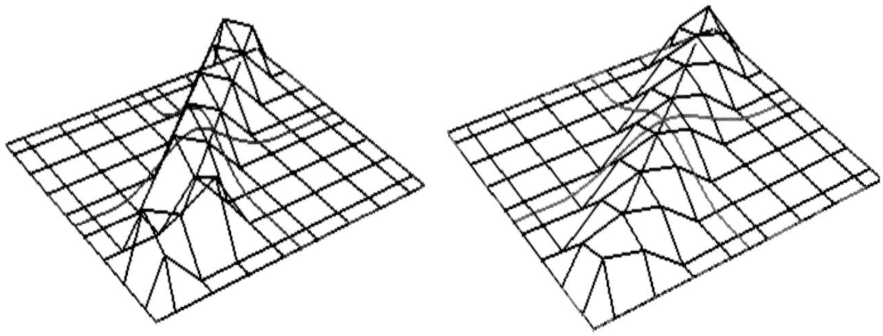

Figure 1. Control mesh of a 11×11, degree 2×2 surface deformed by a diagonal line segment; left: using Gaussian elimination; right: using SVD

speed up methods described in [18], the efficiency was not yet optimal, and the results were prone to numerical instability.

2.1. The SVD

Even with full pivoting (as originally proposed in [3]), the Gaussian elimination appeared to be insufficient for solving the highly singular sets of equations. The alternative proposed here is the use of SVD[1] algorithm using Givens-rotations in the intermediate elimination steps, as implemented in LAPACK [16].

However, in contrast to SVD, Gaussian elimination directly benefits from the knowledge of which of the variables $\mathbf{x}$ are independent (Eq. 1). This, of course, depends on the order of the eliminations steps. In general, the elimination is controlled by some kind of balancing criterion; usually the greatest element in each elimination step has priority. This can lead to catastrophic results, as can be seen in Fig. 1, on the left. Although the problem is symmetric, the result is very arbitrary, and depends on the order of the elimination steps. The pivoting is only an algebraic criterion. When applied without special knowledge of the problem structure, it leads to isolation of variables which solve the algebraic problem exactly, but are not satisfactory with regard to the geometric result.

2.1.1. Computing the SVD

The SVD algorithm performs a decomposition of the matrix $\mathbf{A}$ of size $m \times n$ into three orthogonal matrices: $\mathbf{A} = \mathbf{U} \cdot \Sigma \cdot \mathbf{V}^T$. The solution of the linear system $\mathbf{A} \cdot \mathbf{x} = \mathbf{y}$ can be obtained by solving the transformed system

$$\Sigma \cdot \xi = \beta \tag{2}$$

[1] Singular Value Decomposition

for transformed variables $\xi = \mathbf{V}^T \cdot \mathbf{x}$, $\beta = \mathbf{U}^T \cdot \mathbf{y}$ and resubstituting $\mathbf{x} = \mathbf{V} \cdot \xi$. More details and the algebraic background of SVD are given for instance in [15], [16].

Whenever singularity in the system matrix is expected, one needs to set a threshold value, up to which the singular values as delivered by the SVD are set to zero. Then the transformed system (Eq. 2) decomposes into r equations:

$$\sigma_i \cdot \xi_i = \beta_i, \quad (i = 0, \ldots, r) \tag{3}$$

and $m - r$ conditions:

$$0 = \beta_i, \quad (i = r + 1, \ldots, m) \tag{4}$$

Now the generalized solution of the original system $\mathbf{x} = \mathbf{x}^N + \hat{\mathbf{x}}$ can be obtained, with $\mathbf{x}^N$ being the smallest Euclidian normal solution, and $\hat{\mathbf{x}}$ the translation factor from the null space of $\mathbf{A}$:

$$\begin{aligned} \mathbf{x}^N &= \mathbf{V} \cdot \xi^N, \quad \xi_j^N = \beta_j / \sigma_j, \quad (j = 0, \ldots, r) \\ \hat{\mathbf{x}} &= \mathbf{V} \cdot \hat{\xi}, \quad (\hat{\xi}_j \text{ arbitrary}) \end{aligned}$$

The standard usage of SVD ignores the $\hat{\mathbf{x}}$ values (which are set to zero). For surface editing, we obviously do not want the surface to collapse into a small strip somewhere around the curve. Therefore, we set the values $\hat{\xi} = \mathbf{V}^T \cdot \mathbf{x}_p$ instead. This utilizes the solution $\mathbf{x}_p$ from the previous editing step and results in smooth changes of dependent DOFs (Fig. 1, right).

2.1.2. Additional SVD Benefits

A very useful side effect of the SVD is the explict knowledge of solvability conditions of the system (Eq. 4). If there are more equations than the determined unknowns (the number of control points of the curve might be greater than the number of determined DOFs), the solvability conditions define an orthogonal basis for a vector space $\mathcal{Y}$, such that whenever $\mathbf{y} \in \mathcal{Y}$, an exact solution exists. In practice, this means that the user makes a request on the curve ($\mathbf{y}'$). If $\mathbf{y}' \notin \mathcal{Y}$, the system automatically projects $\mathbf{y}'$ to $\mathcal{Y}$, such that $\|\mathbf{y}' - \mathbf{y}\| \rightarrow \min$.

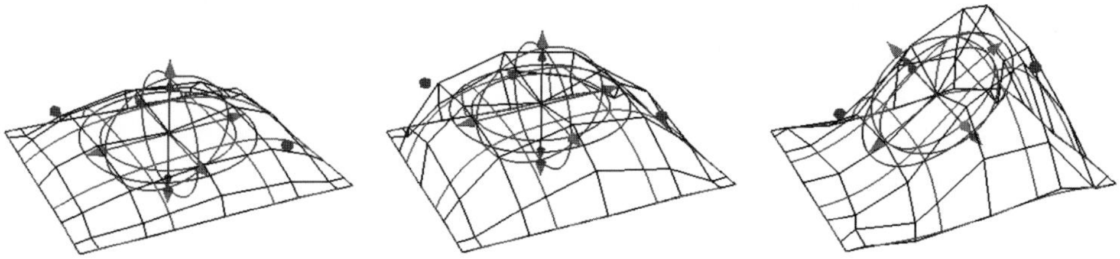

Figure 2. A circle-shaped curve on a 9×9 bi-cubic surface (only the control mesh is shown). The relation between the curve and the surface is computed by the direct method

Another consideration is the presence of simultaneous constraints. If they are all expressed by one equation system and the SVD is computed, the existence of a common solution can be determined immediately. A better way would be to sequentially solve for each constraint (or groups of constraints) and using the results of one step as a fixed set of conditions for the next step. However, this is still to be verified.

2.2. The Composition Matrix

Given the $u(t), v(t)$ representation of a curve in the domain of a parametric surface $S(u, v)$, the 3D curve incident on the surface is $C(t) = S(u(t), v(t))$, which can be expanded into:

$$\sum_p y_p B_p^\pi = \sum_i \sum_j x_{ij} \underbrace{B_i(u(t))B_j(v(t))}_{\Psi_{ij}(t)}$$

As long as the terms Ψ_{ij} remain constant, it is possible to express the resulting curve as a linear transformation $\mathbf{y} = \mathbf{C} \cdot \mathbf{x}$, with $\mathbf{y}$ being the control points of the resulting curve and $\mathbf{x}$ the DOFs of the surface. Indeed, the terms Ψ_{ij} only depend on $u(t), v(t)$ and the basis functions of the surface, not on x_{ij}. The B_p^π are known, since they also depend solely on $u(t), v(t)$ and the basis of the surface. These terms can be collected in a matrix. Once the composition algorithm is coded, the most efficient way is to collect the appropriate terms during the evaluation. Which terms should be compared and collected can be derived from the blossom-based composition algorithm (see [6]). The algorithm for computing the products of B-Spline basis functions is described in [19] or [9] and our '99 paper [18].

Thus, the control points of the curve $S(u(t), v(t))$ can be computed by applying the linear transformation expressed by the *composition matrix* to a vector of control points of the surface $\mathbf{x}$:

$$\begin{bmatrix} y_0 \\ y_1 \\ \vdots \\ y_p \end{bmatrix} = \begin{bmatrix} c_{0,0} & c_{0,1} & \cdots & c_{0,m\cdot n} \\ \vdots & \vdots & \ddots & \vdots \\ c_{p,0} & c_{p,1} & \cdots & c_{p,m\cdot n} \end{bmatrix} \begin{bmatrix} x_{k=f(0,0)} \\ x_{10} \\ \vdots \\ x_{mn} \end{bmatrix} \tag{5}$$

We choose the "rows-first" ordering of the indices of the tensor product: $k = i + m \cdot j$. It turns out that the columns of the matrix $\mathbf{C}$ created this way are exactly the coefficients of the free-form representation of $B_i(u(t))B_j(v(t)) = \Sigma_p c_{p,i+m\cdot j} B_p^\pi$. The ordering depends only on column-first or row-first treatment of the tensor product.

Finally, we compute the inverse transformation to $\mathbf{y} = \mathbf{C} \cdot \mathbf{x}$. Applying the concept of pseudo inverse (see [15]) ($\mathbf{C}$ is generally not square and contains singularities), this yields $\mathbf{x} = \mathbf{C}^+ \cdot \mathbf{y}$.

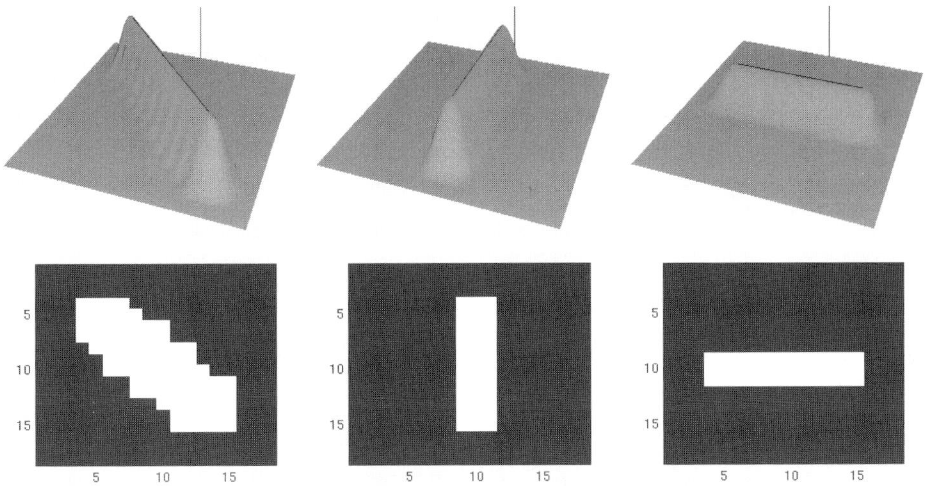

Figure 3. An attempt to constrain a diagonal, vertical and horizontal line on a 22×22 bi-quadratic B-Spline surface. The lower part of the figure shows the distribution of influenced control points in the domain of the surface. No aliasing can be observed on the middle and right surfaces

Compared to [18], this "direct" method dramatically reduces the complexity of setting up the equations. The complexity is only as high as the complexity of the composition algorithm (for complete analysis, see [6], [18]). The direct method bypasses the computation of the integrals of the variational equations. The blossom-based composition algorithm is known for its numerical stability. The elements of the matrix **C** are results of convex combinations of values which arise as convex combinations of the input values themselves. This guarantees an overall high precision of the matrix. In [10], a simpler version of this method was used to solve a similar problem for Bézier surfaces. The $u(t), v(t)$ domain curves were always assumed to be linear, and setting up the matrix **C** was hard-coded for this case.[2] In [18], we have shown how the blossom-based composition algorithm can be extended to B-Splines, without conversion to a Bézier basis. We introduced the "Multi-index tree" structure which allows quick access to precomputed data, particularly the B-Spline basis functions products and generalized basis function coefficients. We also analyzed the complexity of the composition algorithm modified for our purpose. The data structures and basic algorithms are all reused in the new approach described in Section 4. For B-Spline representation and visualization we used the IRIT system [8].

3. The "Aliasing" Effect

An effect, not that obvious at first sight, occurs when a surface is deformed along curves that cross the knot lines in arbitrary ways. The surface exhibits "bumps" at the border of the deformed region, as demonstrated in Fig. 3. The figure shows the result of constraining the incidence of a diagonal, vertical and horizontal line

[2] This work did not use the blossom-based composition algorithm.

on a 22×22 bi-quadratic B-Spline surface. The distribution of the dependent control points in the domain of the surface is shown in the bottom part of the figure. The example on the left is very bumpy, whereas in the middle and right examples, no bumps can be observed. A comparison with the staircase effect when drawing a line on a screen by assigning color values to a discrete grid of pixels comes to mind immediately. What is the reason for this "aliasing" and is there a way to remedy it?

The first question can be easily answered: the DOFs of a tensor product surface are aligned on a rectangular grid, the size and density of which depend on the parametrization of the surface (compare with pixel-grid of a monitor screen). The method as described so far defines an exact solution for the incidence relation between the curve and the DOFs of the surface. However, the number of dependent control points is finite. Their distribution fully maps the grid structure of a tensor-product surface. The surfaces are piecewise polynomial and continuity of low order derivatives is guaranteed, however higher order derivatives are discontinuous across segment boundaries. Although the solution is perfect in an algebraic sense, it fails to deliver an optically "pleasing" surface. We cannot expect to find a continuous mapping of an arbitrary curve on a discrete grid of control points. The aliasing becomes stronger for low degree B-Spline surfaces (degree ≤ 3) consisting of a high number of patches (compare to the example in Fig. 3). The dependent control points are limited to a relatively narrow "strip" near the curve and the low order of continuity among the patches causes high-frequency "bumps".

Thus, the aliasing problem always occurs when using piecewise polynomial surfaces, whenever the curve does not match the rectangular arrangement of DOFs. The problem seems to be known in the field of data interpolation (cf. [5]). In [12], Hayse introduced *curved knot lines* which cope better with an arbitrary curve. The domain of the surface is defined as a curvilinear mesh of knot lines. The parametrization of the surface can then be better adjusted to match a given curve. Although very powerful and conceptually simple, in practice, elementary algorithms for traditional B-Splines (for example knot insertion and removal, degree raising and lowering) become very complicated with Hayes splines, which might be the reason for low acceptance of this type of surfaces. Nevertheless, it can be assumed that malformed surface will also not be accepted by designers.

3.1. Anti-Aliasing

Several "anti-aliasing" approaches have been proposed. One such approach is to define new constraints, working against the aliasing, in connection with the primary incidence constraint. This could become a very tedious procedure. In [24], using a global constraint on the "smoothness" of the surface is proposed. This kind of constraint usually forces the surface to have minimal bending, tension or similar properties (see e.g. [13] for detailed explanation) and is computationally very difficult. Besides the computational difficulties, if imposed without other

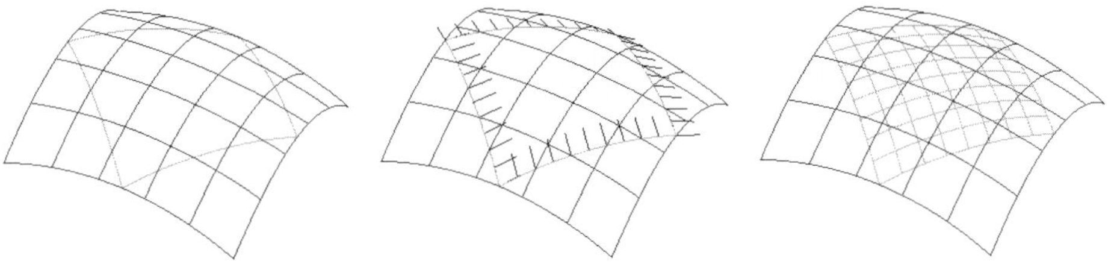

Figure 7. Left: Boundary curves of the new feature, middle: derivatives along the boundary curves assuring C^1 continuity to the original surface, right: the resulting surface $H(s, t)$

$$H(s,t) = \underbrace{S(u(s,t), v(s,t))}_{G} \tag{6}$$

The above expression is a polynomial surface–surface composition. However, if S is a composite B-Spline surface, H cannot in general be written as a tensor product surface anymore [6].

In the following, we derive how a suitable approximation of the surface H can be efficiently obtained. Moreover, we show how a prescribed continuity along the boundaries of H can be achieved.

4.1. The Interpolation Algorithm

We can "scan" arbitrarily many curves representing G, such that s or $t = const.$ and compute their exact representation on S, obtaining a network of 3D curves. Gordon developed a method to interpolate a tensor product surface through an orthogonal network of 3D curves ([11] or [20]). A set of "parallel" curves from surface G are scanned, at suitable values s_i and t_j: $G(s_i, t) = f_i(t)$ and $G(s, t_j) = g_j(s)$, including the given curve in either set. In addition, the vector field curves $\partial f_i(t)/\partial s = d_i(t)$ and $\partial g_i(s)/\partial t = e_j(s)$ are computed. The curves intersect at points $G(s_i, t_j) = h_{ij}$. An algorithm to interpolate the surface H now looks as follows:

1. After inserting $f_i(t)$ and $g_j(s)$ into S we obtain a network of 3D curves incident on S and meeting at points $S(h_{ij})$, as shown on the left of Fig. 7 (in this case, only the four boundary curves are used). The derivative curves d_i and e_j transformed to 3D are vector field curves representing directional derivatives of S with respect to s and t: $\partial S(f_i(t))/\partial s$ and $\partial S(g_j(s))/\partial t$, see Fig. 7, middle.

2. (a) We now have enough information to carry out a cubic interpolation among the curves $f_i \rightarrow f_{i+1}$, and $g_j \rightarrow g_{j+1}$, using the edge derivative conditions $d_i \rightarrow d_{i+1}$, and $e_j \rightarrow e_{j+1}$. Using surface skinning, the surfaces H_1 and H_2 are computed, such that the following relations apply:

$$S(f_i(t)) = H_1(s_i, t)$$
$$S(g_j(s)) = H_2(s, t_j)$$
$$\frac{\partial S(f_i(t))}{\partial s} = \frac{\partial H_1(s_i, t)}{\partial s}$$
$$\frac{\partial S(g_j(s))}{\partial t} = \frac{\partial H_2(s, t_j)}{\partial t}$$

(b) The surface H_3 is obtained as a result of tensor product interpolation of the values $S(h_{ij})$, the derivatives at corner vertices and at the intersection points of the scanned curves:

$$\frac{\partial S(h_{ij})}{\partial s} = \frac{\partial H_3(s_i, t_j)}{\partial s}$$
$$\frac{\partial S(h_{ij})}{\partial t} = \frac{\partial H_3(s_i, t_j)}{\partial t}$$
$$\frac{\partial S^2(h_{ij})}{\partial s\, \partial t} = \frac{\partial H_3^2(s_i, t_j)}{\partial s\, \partial t}$$

3. According to [11], the surface $H(s, t) = H_1 + H_2 - H_3$ interpolates the given network of curves and points at which they intersect (Fig. 7, right). Hence, the surface H is exactly identical to the original surface S along the scanned curves and points, and it approximates the original surface in between. Moreover, due to the derivative information inherited from S, there is at least a C^1 continuous connection of H to S at prescribed curves.

4. The surface H approximates the overall shape of S at the "sub-surface" G and interpolates a set of curves and first derivatives along the curves as scanned from S. The quality of the approximation is determined by measuring the maximum of $\epsilon(s, t) = |S(G) - H|$. Whenever ϵ is larger than a prescribed value, the curve network is refined and the whole process is repeated. In the examples throughout this paper, the refinement is done by recursively inserting a new curve in the middle of each interval of the surface G is each parametric direction. The curve is then composed with the original surface S and added to the interpolation equations for H.

With $\epsilon < 10^{-8}$, in the example from Fig. 7, the approximation of surface H succeeds immediately. The example from Fig. 8 succeeds after three steps (see Fig. 9). The computation takes 0.0039 seconds on a SGI O^2 workstation for the first example (Fig. 7) and 1.9 seconds for the surface in Fig. 8. The degree and knot density of the resulting surface depend on:

- the degree and parameterization chosen for the initial surface G (the shape of the region sketched by the designer)
- the degree and knot density of the original surface.

The degree of a curve resulting from curve-surface composition is given by $d = k(m + n)$, where k is the degree of the domain curve and m and n are the degrees in both parametric directions of a tensor-product surface. In both ex-

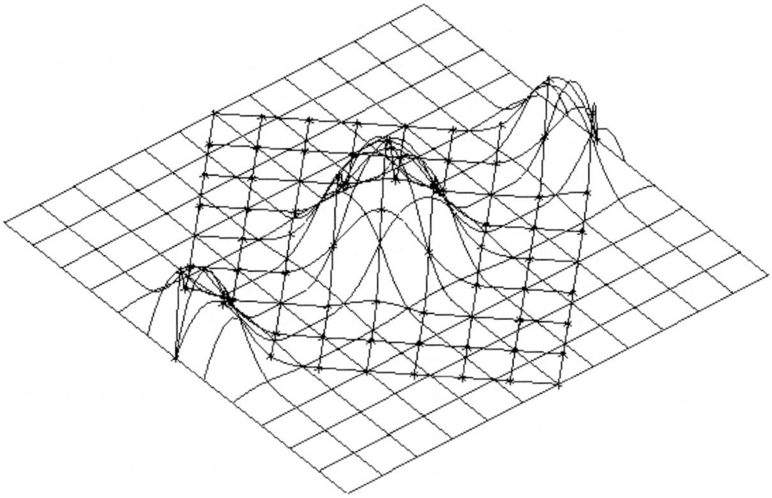

Figure 8. A more sculpted surface, bi-quadratic, 12 × 12 control points. Interpolation of shown curves and points leads to the result shown in Fig. 9

amples, the curves are represented as lines (degree one curves) in the domain of a bi-quadratic surface with 4 × 4 control points (example in Fig. 7) and 22 × 22 control points (Fig. 8). This results in bi-quadratic surfaces with 6 × 6 control points for the first example, and 81 × 81 for the second example.

The interpolation equations are set up using blossom-based methods from ([18]) and solved efficiently with the aid of algorithms for solving sparse and banded linear systems.

5. A Design Example

Figures 10 and 11 demonstrate a design application of the presented method. Here, the designer wants to add a "crater" shaped feature to the surface shown in Fig. 10:

1. Two closed curves are sketched on the surface. The system projects the curves into the domain of the surface and computes their exact representation on the surface. They represent the boundaries of the new feature. The designer can choose a continuity of the crater feature along the boundary curves. Here C^0 and C^1 continuity along both boundaries are specified.
2. The system computes a replacement surface from surface curves as described in the previous section. Two tangency and two incidence constraints along the boundary curves are generated between the new and the original surface. The area covered by the new surface is trimmed away from the original surface, see Fig. 10.

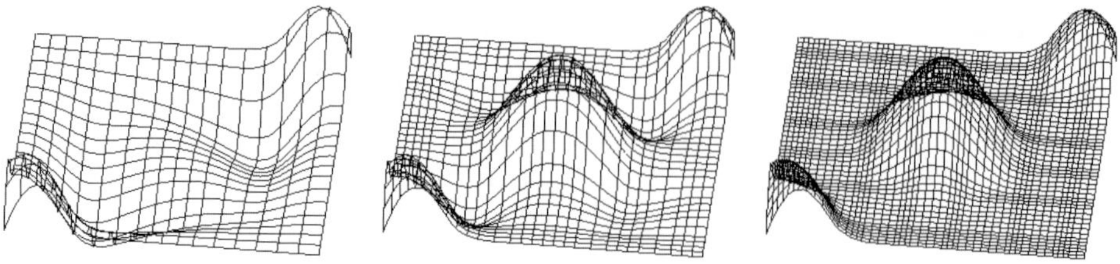

Figure 9. The left-most figure shows the surface H after a first interpolation step (only the four boundary curves and derivatives are interpolated). The approximation error ϵ falls below the prescribed limit (10^{-8} in this example) after twice inserting a curve and derivatives in the middle of each interval (right)

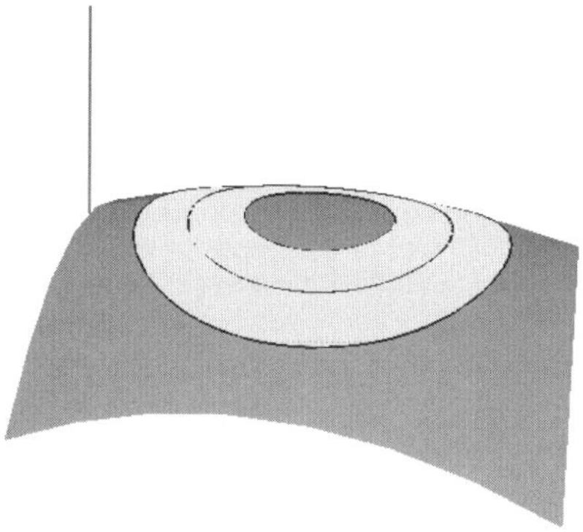

Figure 10. The replacement surface and the selected iso-curve from the crater example

3. The manipulation tool of the designer will be any iso-parametric curve in either direction on the crater surface, which can now be selected by choosing a direction and picking a point anywhere on the surface.

The interactive system offers a manipulation handle for translating, rotating and shaping the selected curve (Fig. 11). The surface reacts as expected: the incidence and tangency constraints along the boundary and feature curves assure the proper connection of the new feature to the original surface. Since all constrained curves are iso-parametric lines in the new surface, no aliasing effects occur.

6. Conclusion and Future Work

The main contribution of this article is a method for adding editable features to a free-form surface model, aligned along arbitrary user-defined curves on the surface. The described algorithm overcomes the difficulties of variational methods applied previously for this purpose ([24], [3]). It efficiently computes a new properly parametrized surface, which replaces the old surface inside a user-defined region, such that the edited curve become an iso-parametric line in the domain of the new surface. The method is very efficient and numerically robust, and it

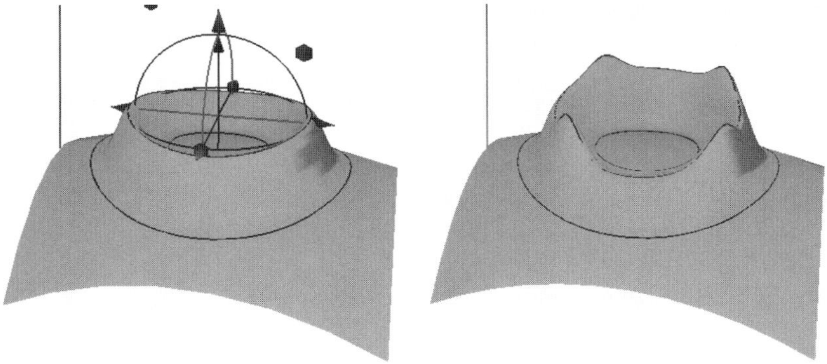

Figure 11. The "crater" design example. The surface on the right shows a local modification of the selected iso-curve

considerably reduces the complexity of the interpolation equations. We re-use the blossom-based methods from our previous work [18], for scanning curves and derivatives from the original surface. Furthermore, the variational approach is revised and it is shown how it can be replaced by a more direct, much more efficient and robust method, directly utilizing the blossom-based composition algorithm.

This work is a step towards the integration of constraint-based modeling and free-from surface sculpting. Our goal is a constraint-based modeling system providing more support in early design phases. In such a system, the designer is not limited to a history of modeling operations. New elements and relationships among them are created; the designer specifies which properties the model should have, instead of defining a sequence of geometric construction steps. For a complete discussion of declarative constraint-based modeling, refer to [14], [7], [2], for example.

The methods introduced here match the declarative modeling concept well; consider the "crater" example from Section 5. The work of the designer is highly interactive and graphics-based. Once the new feature is defined, it is no longer important how it was created; the coherence of the model is maintained by the curve-surface incidence constraints. The methods presented here were already integrated in our prototype system, described in [7] and [2].

Future research will concentrate on generalization and further extensions of the described method. Specifically, the dependency between the added feature and the original surface has to be made bi-directional. In the current application, only the new surface feature can be manipulated, while the incidence and tangency along its boundary curves are maintained. This is accomplished by fixing the position and derivatives of the boundary curves. In order to avoid this, a method applied in "surface pasting" [1] could be used. Translated into the notation of this paper: after each modification of the feature H_0 (resulting in H_0') the actual surface is expressed as a linear combination relative to the shape of the original surface: $H_0' = H_0 + \Delta H(S)$. ΔH denotes a difference surface relative to the original surface

S, expressed in term of normals of S. Thus, if S is changed to S', the feature H_1' is restored as: $H_1' = H_1 + \Delta H(S')$.

Next, the described method will be extended to arbitrary surface models, in which trimmed surface patches can occur. The interpolation algorithm as described in Section 4.1 only works if the surface G in the domain of a single composite or B-Spline surface can be found. General surface models are not limited to composite surfaces (surfaces with common parametrization). For such cases, the interpolation algorithm must be modified.

Acknowledgements

This work was supported in part by a grant from the Ministry of Science and Culture of Thuringia (TMWFK) Germany. Figures 6, 7 and 8 were created using the IRIT solid modeler [8].

References

[1] Barghiel, C., Bartels, R., Forsey, D.: Pasting spline surfaces. pp. 31–40. Vanderbilt University Press, 1999.
[2] Brüderlin, B., Döring, U., Klein, R., Michalik, P.: Declarative geometric modeling with constraints. In: Conference Proceedings CAD 2000 (Iwainsky, A., ed.), Berlin, March 2000. GFAI.
[3] Celniker, G., Welch, W.: Linear constraints for deformable B-spline surfaces. Comput. Graphics *25*, 171–174 (1992).
[4] Coquillart, S.: Extended free-form deformation: a sculpturing tool for 3D geometric modeling. Comput. Graphics *24*, 187–196 (1990).
[5] Cox, M.: Algorithms for spline curves and surfaces. In: Fundamental developments of computer-aided geometric modeling (Piegl, L. A., ed.), pp. 51–75. New York: Academic Press, 1993.
[6] DeRose, T., Goldman, R., Hagen, H., Mann, S.: Functional composition algorithm via blossoming. ACM Trans. Graphics *12*, (2) (1993).
[7] Doering, U., Michalik, P., Brüderlin, B.: A constraint-based shape modeling system. Geom. Constraint Solv. Appl. (1998).
[8] Elber, G.: Users' manual – IRIT, a solid modeling program. Technion Institute of Technology, Haifa, Israel, 1990–1996.
[9] Elber, G.: Free form surface analysis using a hybrid of symbolic and numerical computations. PhD thesis, University of Utah, 1992.
[10] Elber, G., Cohen, E.: Filleting and rounding using trimmed tensor product surfaces. In: Proceedings The Fourth ACM/IEEE symposium on Solid Modeling and Applications, pp. 201–216, May 1997.
[11] Gordon, W. J.: Sculptured surface definition via blending-function methods. In: Fundamental developments of computer-aided geometric modeling (Piegl, L. A., ed.), pp. 117–134. New York: Academic Press, 1993.
[12] Hayes, J.: Nag algorithms for the approximation of functions and data. In: Algorithms for approximation (Mason, J., Cox, M., eds.), pp. 653–668. Oxford: Clarendon Press, 1998.
[13] Hoschek, J., Lasser, D.: Fundamentals of computer aided geometric design. AK Peters, 1989.
[14] Hsu, C., Alt, G., Huang, Z., Beier, E., Brüderlin, B.: A constraint-based manipulator toolset for editing 3D objects. In: Solid modeling 1997, Atlanta, Georgia, ACM Press, 1997.
[15] Kielbasinsky, A., Schwetlick, H.: Numerische lineare Algebra, eine computerorientierte Einführung. Mathematik für Naturwissenshaft und Technik. Berlin: Deutscher Verlag der Wissenschaften, 1988.
[16] LAPACK User's guide release, 3rd ed, 1999.
[17] Lazarus, F., Coquillart, S., Jancéne, P.: Axial deformations: An intuitive deformation technique. Comput. Aided Des. *26*, 607–613 (1994).
[18] Michalik, P., Brüderlin, B.: Computing curve-surface incidence constraints efficiently. In: Proceedings Swiss Conference on CAD/CAM, February 1999.

[19] Mørken, K.: Some identities for products and degree raising of splines. Construct. Approx. *7*, 195–208 (1991).
[20] Piegl, L., Tiller, W.: The Nurbs Book. Berlin Heidelberg New York Tokyo: Springer, 1995.
[21] Ramshaw, L.: Blossoming: A connect-the-dots approach to splines. Technical Report 19, Digital System Research Center, Palo Alto CA, June 1987.
[22] Sederberg, T., Parry, S.: Free-form deformation of solid geometric models. In: Proceedings SIGGRAPH '86, pp. 151–160, 1986.
[23] Singh, K., Fiume, E.: Wires: A geometric deformation technique. In: Proceedings SIGGRAPH '98, 1998.
[24] Welch, W., Witkin, A.: Variational surface modeling. Comput. Graphics *26*, 157–165 (1992).

P. Michalik
B. Bruderlin
Technical University of Ilmenau
Computer Graphics Program
Postfach 100565
D-98684 Ilmenau
Germany
e-mails: paul@prakinf.tu-ilmenau.de
bdb@prakinf.tu-ilmenau.de

Computing [Suppl] 14, 267–280 (2001)

Computing

A Geometrically Motivated Affine Invariant Norm

V. Milbrandt, Norderstedt

Abstract

Based upon the Loewner ellipsoid an affine invariant norm will be presented. This norm will be compared with the norm established by Nielson [10] using results of scattered data interpolation.

AMS Subject Classifications: *41A05, 41A15, 65D05.

Key Words: Affine invariant norm, Loewner ellipsoid, thin plate splines.

1. Motivation for an Affine Invariant Norm

The main purpose for using an affine invariant norm is to obtain methods and techniques which are not affected by affine transformations of the input data. This means, for example, that artificial choices of the origin or the units of measurement do not have any effects on the final results of the methods.

Many widely used methods of CAGD are not invariant with respect to affine mappings. This lack of invariance can be remedied by some modifications. Nielson [10] proposed to replace the standard (Euclidean) norm by a norm which is affine invariant. As an application he gave a modification of thin plate splines interpolation. Nielson and Foley [11] discussed subsequently some further applications.

An affine invariant norm always depends upon a set $\mathscr{X}$ of n given data points $\vec{X}_i \in \mathbb{R}^d (i = 1, \dots, n)$. This dependence is indicated by the subscript in the notation $\|\cdot\|_{\mathscr{X}}$ of the following definition:

Definition 1. *A norm is called* ***affine invariant*** if and only if for any two points $\vec{P}$ *and* $\vec{Q}$ *in the domain of the norm* $\|\cdot\|_{\mathscr{X}}$ *and for any affine transformation* φ *the equation*

$$\|\vec{P} - \vec{Q}\|_{\mathscr{X}} = \|\varphi(\vec{P}) - \varphi(\vec{Q})\|_{\varphi(\mathscr{X})} \tag{1}$$

is satisfied.

2. Nielson's Norm

Nielson introduced his norm in the plane and gave only short remarks on the generalisation to higher dimensions [10, 11]. In [12] a direct formulation for three dimensions can be found. A definition for arbitrary dimensions will just be given here:

Definition 2. *Let n points* $\vec{X}_i = (x_{i1}, \ldots, x_{id})^T$ $(i = 1, \ldots, n)$ *be given.* ***Nielson's affine invariant norm*** *(NAIN) of a point* $\vec{y} \in \mathbb{R}^d$ *will be defined by*

$$\|\vec{y}\|_N := +\sqrt{\vec{y}^T \mathbf{A} \vec{y}} \tag{2}$$

wherein the matrix $\mathbf{A}$ *depends on the points* $\vec{X}_i$ *and is determined by: Calculate the centre of gravity* $\vec{C} = (c_1, \ldots, c_n)^T = \frac{1}{n}\sum_{i=1}^{n} \vec{X}_i$ *and build the* $(n \times d)$*-matrix*

$$\bar{\mathbf{V}} := (\vec{V}_1 - \vec{C}, \ldots, \vec{V}_n - \vec{C})^T = \begin{pmatrix} x_{11} - c_1 & x_{12} - c_2 & \cdots & x_{1d} - c_d \\ \vdots & \vdots & & \vdots \\ x_{n1} - c_1 & x_{n2} - c_2 & \cdots & x_{nd} - c_d \end{pmatrix} \tag{3}$$

consisting of the differences of the coordinates of the points and the centre.

The defining matrix $\mathbf{A}$ *of the norm then results as*

$$\mathbf{B} := \begin{pmatrix} \sum_{x_1 x_1} & \cdots & \sum_{x_1 x_d} \\ \vdots & & \vdots \\ \sum_{x_1 x_d} & \cdots & \sum_{x_d x_d} \end{pmatrix} := \frac{1}{n}\bar{\mathbf{V}}^T \bar{\mathbf{V}}, \quad \mathbf{A} := \mathbf{B}^{-1}. \tag{4}$$

Remark.

1. The rows of the matrix $\bar{\mathbf{V}}$ are the difference vectors of the given points $\vec{X}_i$ and the centre $\vec{C}$.
2. The entries of the matrix $\mathbf{B}$ can also be calculated as

$$\sum_{x_i x_j} = \frac{1}{n}\sum_{k=1}^{n} (x_{ki} - c_i)(x_{kj} - c_j) \quad (1 \leq i, j \leq d) \tag{5}$$

   (analogous to Nielson's calculations in the planar case).
3. The NAIN will always change if a supplementary point is added to the set $\mathcal{X}$ of base points.
4. In statistics the method used by Nielson is called *principal component analysis.*

A geometric motivation for this norm has been given recently in [1]. There the gauge ellipsoid has been characterised as a kind of best approximating ellipsoid to the given points. The volume of this ellipsoid was previously fixed as a "quadratic mean" of volumes of parallelepipeds spanned by the data points and their centre of gravity.

The ellipsoid used for the introduction of the LAIN is also known in the geometry of masses as *Poinsot's central ellipsoid* [5].

3. The Definition of Another Norm

The geometric motivation of Nielson's norm in [1] is not obvious, some calculations had to be done to become aware of it. A further drawback of the metric (2) is that subsequent to addition of only one supplementary point to the set of base points $\mathcal{X}$ the NAIN will always change.

These facts lead to the search for another geometrically founded affine invariant norm. For this purpose the following theorem will be cited:

Theorem 3 (K. Loewner, 1893–1968). *Let $\mathcal{A}$ be a bounded set (with non-empty interior) in $\mathbb{R}^d$. Then there exists one and only one ellipsoid E of minimal volume containing $\mathcal{A}$, the so-called* ***Loewner ellipsoid***.

A proof can be found in [8, p. 143f.].

For the construction of a norm select the convex hull of the point set $\mathcal{X}$ as the bounded set $\mathcal{A}$. Assume that the points of the set $\mathcal{X}$ do not all lie in a hyperplane. This assumption is required for $\mathcal{A}$ having interior points. Let E be the uniquely defined Loewner ellipsoid of $\mathcal{A}$. As E is compact, at least $d+1$ points $\vec{X}_i \in \mathcal{X}$ are on the boundary of E (in $\mathbb{R}^d$) and all others are in the interior or also on the boundary [4].

Definition 4. *Let E be the Loewner ellipsoid as defined above. The ellipsoid E can be characterised by the matrix* $\mathbf{A}$ *and the centre* $\vec{C}$ *with*

$$E = \left\{ \vec{V} \in \mathbb{R}^d : (\vec{V} - \vec{C})^T \mathbf{A} (\vec{V} - \vec{C}) \leq 1 \right\}. \tag{6}$$

Thus an affine invariant norm

$$\|\vec{y}\|_L := +\sqrt{\vec{y}^T \mathbf{A} \vec{y}} \quad \forall \vec{y} \in \mathbb{R}^d \tag{7}$$

is induced by E, which depends on the convex hull of $\mathcal{X}$. The natural origin of the norm is $\vec{C}$. In the following this norm will be called ***Loewnerean (affine invariant) norm*** *(LAIN).*

Remark. The Loewner ellipsoid E is invariant with respect to arbitrary affine transformations $\vec{x} \mapsto \mathbf{T}\vec{x} + \vec{d}$. This can easily be seen by the facts that the (at least $d+1$) critical points on the boundary E remain on the boundary and that interior points remain in the interior of the image of E. If the original ellipsoid E is characterised by $(\mathbf{A}, \vec{C})$ then the affine transformed Loewner ellipsoid will be described by the centre $\tilde{\vec{C}} = \mathbf{T}\vec{C} + \vec{d}$ and the matrix $\tilde{\mathbf{A}} = \mathbf{T}^{-T}\mathbf{A}\mathbf{T}^{-1}$.

4. Determination of the Loewnerean Norm

In general, one has to solve a nonlinear optimisation problem of Fritz John type, i.e. an extremum problem with inequalities as side conditions, for the determination of the Loewner ellipsoid [4].

The positive definite symmetrical matrix $\mathbf{A}$ and the centre $\vec{C}$ of the ellipsoid E have to be calculated. In $\mathbb{R}^d$ this problem has $d(d+3)/2$ unknown parameters. With the volume $\omega_d = \pi^{d/2}/\Gamma(1+d/2)$ of the Euclidean unit sphere S^{d-1} the optimisation problem is:

$$\min_{\mathbf{A},\vec{C}}\left\{\frac{\omega_d}{\sqrt{\det \mathbf{A}}}\right\} \quad \text{or equivalent} \quad \max_{\mathbf{A},\vec{C}}\{\det \mathbf{A}\} \tag{8}$$

with respect to

$$\|\vec{X}_i - \vec{C}\|_L^2 = (\vec{X}_i - \vec{C})^T \mathbf{A}(\vec{X}_i - \vec{C}) \leq 1 \quad \forall i \in \{1,\dots,n\}. \tag{9}$$

Optimisation problems of this kind can be solved numerically by sequential quadratic approximation (SQP), a penalty method, where the nonlinear problem will be replaced locally by quadratic problems to determine the directions of the largest gradient. An implementation of SQP is available in the NetLib library [6]. Ingenious starting values are very important, especially for dimensions $d \geq 3$, as otherwise one may obtain wrong local extrema due to the fact that the objective function (8) is a polynomial of degree d and the constraints (9) are of polynomial degree 3.

5. Another Determination of the Loewner Ellipsoid

Firstly, observe the case of exactly $d+1$ points in $\mathbb{R}^d$. Then all points are characteristic contact points of the set $\mathscr{A}$ and the ellipsoid E. Now recall the following theorem, which can be applied in this situation:

Theorem 5 (Juhnke [4]). *For the minimal ellipsoid* $E_0 = E(\mathbf{C^0}, \vec{x}^0) = \{\vec{t} \in \mathbb{R}^d \mid (\vec{t} - \vec{x}^0)^T \mathbf{C^0}(\vec{t} - \vec{x}^0) \leq 1\}$ *of the d-dimensional simplex* $S := \operatorname{conv}\{\vec{p}^1, \dots, \vec{p}^{d+1}\} \subset \mathbb{R}^d$ *it yields:*

(*a*) *The centre* $\vec{x}^0$ *of the minimal ellipsoid* E_0 *is the centre of gravity* S *and it yields:* $\mathbf{C^0} = (1 + 1/d)\mathbf{P}^{-T}\mathbf{G}\mathbf{P}^{-1}$, *whereof* $\mathbf{P} := (\vec{p}^1 - \vec{p}^{d+1}, \dots, \vec{p}^d - \vec{p}^{d+1})_{d,d}$ *is the matrix of the edge vectors of* S *emanating from one simplex vertex and* $\mathbf{G} = (g_{ik})_{d,d}$ *is given by* $g_{ik} := \begin{cases} 2, & i = k \\ 1, & i \neq k \end{cases}$, $i, k = 1, \dots, d$.

(*b*) *The ratio of the volumina of the simplex* S *and the minimal ellipsoid* E_0 *is only dependent on the dimension. The following equation holds:*

$$\frac{V(S)}{V(E_0)} = \frac{(d+1)^{(d+1)/2}\Gamma(d/2+1)}{d!(d\pi)^{d/2}}$$

(*c*) *The tangential hyperplane of the minimal ellipsoid* E_0 *at the vertex* $\vec{p}^j$ *is parallel to the opposite face* aff $\{\vec{p}^1, \ldots, \vec{p}^{j-1}, \vec{p}^{j+1}, \ldots, \vec{p}^{d+1}\}$ *of the simplex* S.

(*d*) *Every affine mapping, which maps* E_0 *onto a sphere, maps* S *onto a regular simplex.*

Secondly, at least $d+1$ characteristic points lie on the Loewner ellipsoid, as the ellipsoid is compact [4]. Thus, as the minimal ellipsoid in the case of exactly $d+1$ points can be determined by theorem 5, the Loewner ellipsoid can be deduced in the general case of n points ($n > d+1$) in the following way:

Let $\mathscr{W}$ be the set of all subsets containing exactly $d+1$ points from $\mathscr{X}$ ($\#\mathscr{X} = n$), for each subset not all points lying in a hyperplane. Calculate the minimal ellipsoid for each these subsets $T \in \mathscr{W}$ by Juhnke's theorem.

Let $\mathscr{W}' \subset \mathscr{W}$ be the subset of $\mathscr{W}$ such that for each element $T \in \mathscr{W}'$ all points of $\mathscr{X}$ are in the interior or on the boundary of the Loewner ellipsoid given by T.

For the volume of the Loewner ellipsoid L one knows by Juhnke: If S is the simplex spanned by the point set $T \in \mathscr{W}'$, and Vol(S) is its volume, the volume of the minimal ellipsoid belonging to T is

$$\mathrm{Vol}(T) := \mathrm{Vol}(L) = \frac{d!(d\pi)^{d/2}}{(d+1)^{(d+1)/2}\Gamma(d/2+1)}\mathrm{Vol}(S) = \frac{d!(d)^{d/2}\omega_d}{(d+1)^{(d+1)/2}}\mathrm{Vol}(S) \tag{10}$$

As $\#\mathscr{W} \le \binom{n}{d+1}$, it is assured that $\mathscr{W}'$ is finite. Thus the minimum of the volumina of all minimal ellipsoids of the elements of $\mathscr{W}'$ exists. The ellipsoid which takes the minimum is obviously also the Loewner ellipsoid L of the given point set $\mathscr{X}$.

Example. In $\mathbb{R}^3$ the Loewner ellipsoid of the five points $\vec{X}_1 = (1, -3, 0)^T$, $\vec{X}_2 = (2, 5, 0)^T, \vec{X}_3 = (-2, -1, 0)^T, \vec{X}_4 = (-2, -1, 2)^T$ and $\vec{X}_5 = (0, 0, 0)^T$ will be determined.

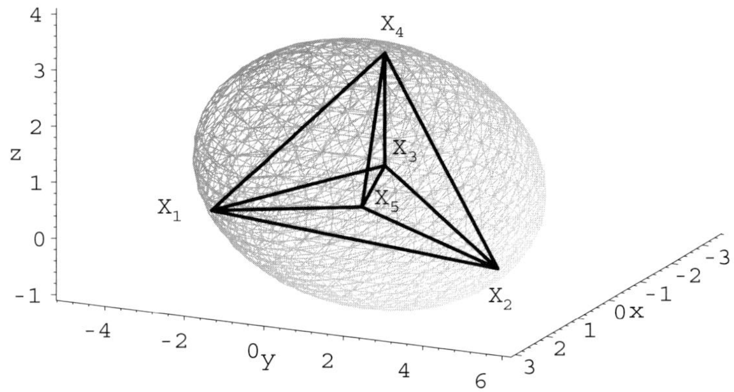

Figure 1. Example: Loewner ellipsoid

Firstly, the set of all 4-elementary subsets, which are not situated in a plane, has to be determined. This leads to $\mathscr{W} = \{\{\vec{X}_1, \vec{X}_2, \vec{X}_3, \vec{X}_4\}, \{\vec{X}_2, \vec{X}_3, \vec{X}_4, \vec{X}_5\}, \{\vec{X}_1, \vec{X}_3, \vec{X}_4, \vec{X}_5\}, \{\vec{X}_1, \vec{X}_2, \vec{X}_4, \vec{X}_5\}\}$, as $\vec{X}_1, \vec{X}_2, \vec{X}_3$ and $\vec{X}_5$ lie in the common plane $z = 0$.

Now all elements of $\mathscr{W}$, where the minimal ellipsoid does not enclose all five given points $\vec{X}_i$, have to be excluded. In this case one gets $\mathscr{W}' = \{\{\vec{X}_1, \vec{X}_2, \vec{X}_3, \vec{X}_4\}\}$, i.e. there is only one possible candidate for the Loewner ellipsoid, as $\vec{X}_5$ is in the convex hull of $\vec{X}_1, \vec{X}_2, \vec{X}_3$ and $\vec{X}_4$. From the volume of the pyramid (3-dimensional simplex) $\mathrm{Vol}(\{\vec{X}_1, \vec{X}_2, \vec{X}_3, \vec{X}_4\}) = 13 \cdot 2/3 = 26/3$ one deduces by Juhnke's theorem 5, part b), the volume of the Loewner ellipsoid of the sole element of $\mathscr{W}'$ as

$$\mathrm{Vol}(L) = \frac{3!\ 3^{3/2}\omega_3}{4^{4/2}} 26/3 = \frac{39\sqrt{3}}{4}\omega_3 = 13\sqrt{3}\pi. \tag{11}$$

On the other hand – with the explicit formulae of Juhnke's theorem, part a) – the following equations result for the only possible Loewner ellipsoid:

$$L := \{\vec{x} \in \mathbb{R}^3 : (\vec{x} - \vec{S})^T \mathbf{C}(\vec{x} - \vec{S})\}$$

- Central point = centre of gravity = $\vec{S} = (\vec{X}_1 + \vec{X}_2 + \vec{X}_3 + \vec{X}_4)/4 = (-1/4, 0, 1/2)^T$.
- Matrix $\mathbf{C} = \frac{1}{39}\begin{pmatrix} 8 & -2 & 8 \\ -2 & 2 & -1 \\ 8 & -1 & 26 \end{pmatrix}$.

Thus, the volume of the Loewner ellipsoid results as

$$\mathrm{Vol}(L) = \frac{1}{\sqrt{\det \mathbf{C}}}\omega_3 = \frac{39\sqrt{3}}{4}\omega_3 = 13\sqrt{3}\pi. \tag{12}$$

The two calculated volumes (11), (12) coincide, as expected.

Remark. 1. With this combinatorical deduction algorithm the numerical optimisation will not be needed any more. But this advantage contrast with the drawback that the number of elements of $\mathscr{W}$ depend on the number of given points in $\mathscr{X}$ by order $d + 1$. Thus the calculation expenses increase dramatically with the number of points – and therefore the optimisation should be preferred again.

2. Theorem 5, part d), provides a simple proof of the properties of the Loewner ellipsoid for the special case of a simplex by affine relation to the circumsphere of a regular simplex. Moreover, part d) is sufficient to construct geometrically the Loewner ellipsoid of simplex.

6. Application to Scattered Data Interpolation

Nielson modified the thin plate splines interpolant by replacing the standard Euclidean norm by his affine invariant one. Over a parameter plane the thin plate splines (TPS) function has the representation

$$S(\vec{X}) = \sum_{j=1}^{n} b_j \|\vec{X} - \vec{X}_j\|_N^2 \log \|\vec{X} - \vec{X}_j\|_N + c_1 x + c_2 y + c_3 \tag{13}$$

where $\vec{X}_j = (x_j, y_j)^T, \vec{X} = (x, y)^T$ and $\|\cdot\|_N$ is Nielson's norm. The coefficients c_1, c_2, c_3 and $b_j (j = 1, \dots, n)$ are determined by the n interpolation conditions $S(\vec{X}_i) = f_i (i = 1, \dots, n)$ and three equations for the balance of forces

$$\sum_{j=0}^{n} b_j = 0, \quad \sum_{j=0}^{n} b_j x_j = 0, \quad \sum_{j=0}^{n} b_j y_j = 0. \tag{14}$$

The substitution of Nielson's norm $\|\cdot\|_N$ by the Loewnerean norm $\|\cdot\|_L$ yields another affine invariant interpolant of the given set $\mathscr{X}$. This follows, as both norms are based on an ellipsoid, i.e. a positive definite quadratic form, which is related in an affinely invariant way to the set of input data points. Firstly, the gauge ellipsoid has to be determined. Then one could perform an affine mapping α of the input data which transfers the ellipsoid into the Euclidean unit sphere. Then one use the standard (Euclidean) algorithm and afterwards one applies the inverse affine mapping α^{-1} on the output. In this way, by transformation, every Euclidean algorithm with linear polynomial part can be made affinely invariant.

6.1. Numerical Results

Several tests were carried through with the modified TPS interpolation method (as well as for similar modifications of other interpolants). For comparison the numerical tests have been run for the (unmodified) thin plate splines and for the modifications with Nielson's and the Loewnerean norm.

Some of the tests involved the application of these methods to data sets where the dependent values have been taken from known functions. In these cases errors (maximal, mean and root mean square (RMS)) could be calculated between the given original function and the constructed interpolant.

6.1.1. Example

The following example is typical for the obtained numerical results. The approximated function is Franke's well-known test function f_1 [3]:

$$\begin{aligned} f_1(x,y) = &\frac{3}{4}\exp\left(-\frac{(9x-2)^2+(9y-2)^2}{4}\right) + \frac{3}{4}\exp\left(-\frac{(9x+1)^2}{49} - \frac{9y+1}{10}\right) \\ &+ \frac{1}{2}\exp\left(-\frac{(9x-7)^2+(9y-3)^2}{4}\right) - \frac{1}{5}\exp(-(9x-4)^2-(9y-7)^2) \end{aligned} \tag{15}$$

and the data sets used are the three sets of Franke and three further data sets containing the same number of points (25, 33, 100 points) which were placed

Table 1. Errors for randomly distributed points in $[0,1]^2$

Number of points	Max. errors			Mean errors			RMS errors		
	TPS	NAIN	LAIN	TPS	NAIN	LAIN	TPS	NAIN	LAIN
25	*0.4462*	0.4439	0.4271	*0.04340*	0.04319	0.04318	*0.07089*	0.07049	0.06965
33	0.2690	*0.2791*	0.2689	0.03105	0.03131	*0.03161*	0.05504	*0.05801*	0.05544
100	0.0354	*0.0427*	0.0412	0.00427	*0.00432*	0.00430	0.00685	*0.00716*	0.00707

Table 2. Errors for the data sets of Franke [3]

Number of points	Max. errors			Mean errors			RMS errors		
	TPS	NAIN	LAIN	TPS	NAIN	LAIN	TPS	NAIN	LAIN
25	0.1214	0.1206	*0.1232*	0.02547	0.02469	0.02538	*0.03502*	0.03393	0.03464
33	0.1569	*0.1601*	0.1569	*0.02967*	0.02886	*0.02967*	0.04264	0.04219	*0.04264*
100	*0.0531*	0.0529	0.0515	*0.00530*	0.00528	0.00519	*0.00957*	0.00952	0.00931

randomly in the area $[0, 1]^2$. These additional data sets can be looked up in [9]. A plot of the function f_1 is shown in Fig. 8.

6.1.2. Errors for Franke's Test Function

In the tables the best (smallest) errors are marked underlined, the worst (largest) *emphasised*. The results from Table 1 for the random data set with 33 points are illustrated in Figs. 2–5.

The scaling of the z-axis has to be observed as it severely differs among the figures. In Fig. 4 (NAIN) the range is $-0.28\ldots+0.18$ whereas in Fig. 5 (LAIN) the range is $-0.27\ldots+0.10$, i.e. the scale is about 20% smaller.

6.2. Results

Examination of the numerical results discloses the dependence of the errors from the given data. The better interpolants of the affine invariant modified thin plate splines were achieved in Table 1 (random data) by the LAIN, in Table 2 (Franke's data) by the NAIN. However, the errors for the original and the two modified affine invariant methods are not very different. As Nielson already remarked, this is due to the fact that the gauge ellipses of the affine invariant norms are very close to a circle for large data sets situated in the area of the square $[0, 1]^2$. But in case that the data sets are scaled in only one direction (for example by factor 10 in x-axis), solely the errors of the original method will increase dramatically (see Table 3, Figs. 6 and 7), whereas the affine invariant modified interpolants will both (with NAIN and LAIN) remain unchanged and will have only the same small interpolation errors as for unscaled data. Thus, the basic ability of thin plate splines to approximate standard functions has been preserved or even improved.

For the special case of $d+1$ points in $\mathbb{R}^d$ the NAIN and LAIN are distinguished between each other by nothing but a factor. In this case the centre of gravity is the

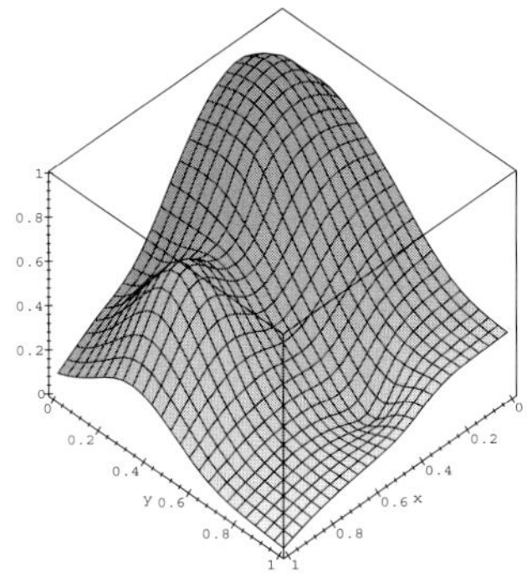

Figure 2. Graph by Nielson's norm

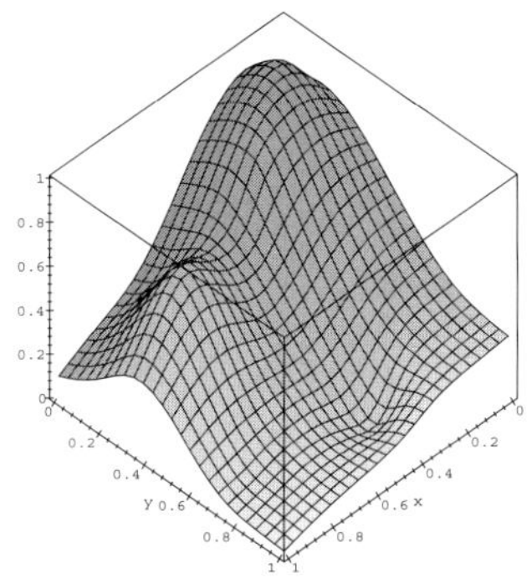

Figure 3. Graph by Loewnerean norm

origin of both norms. But in general, the unit (hyper-) spheres severely differ, neither their centre nor the directions of the major axes of their unit (hyper-) spheres correspond to each other, see Fig. 9 for an example.

Remark. Depending on which norm one introduces, one gets a different distribution of the data points in the domain of the constructed norm. It seems, that the data point distribution affects the quality of the radial basis function interpolants. An interesting topic for further research would be the study of affinely invariant metrics, which optimise the distribution in a certain predetermined way.

Finally, the norms have been applied to thin plate splines for the generation of derivative data. Only by exchanging the Euclidean norm for Nielson's or the Loewnerean norm an improved numerical stability could be achieved. In Figs. 10 and 11 the so-called 9-parameter interpolant of a sphere are shown to demonstrate this fact. In this example, starting by a triangulation of points, derivative data were generated in the vertices by TPS interpolation and in subsequence the 9-parameter interpolant determined by using the previously created derivatives.

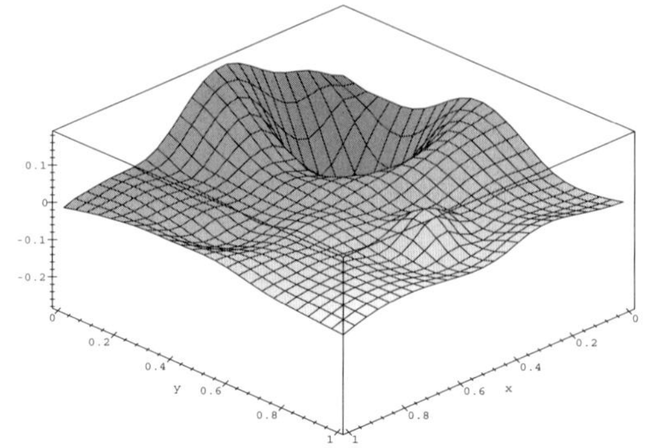

Figure 4. Errors by Nielson's norm

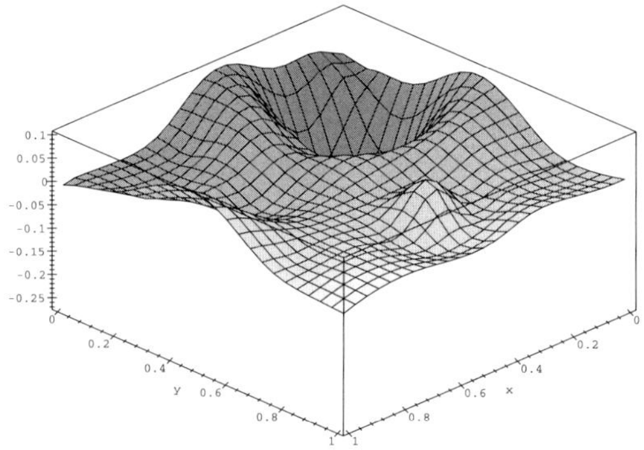

Figure 5. Errors by Loewnerean norm

Table 3. Errors for scaled data sets (unmodified TPS)

Number of points	Random data sets			Franke's data sets		
	Max.	Mean	RMS	Max.	Mean	RMS
25	0.7274	0.11110	0.16090	0.6526	0.10901	0.14342
33	0.6111	0.10598	0.15863	0.4689	0.11174	0.14988
100	0.4168	0.05741	0.08537	0.3716	0.05234	0.07912

7. Application to Knot Selection

In many areas of CAGD parametric curves are an important feature. As most applications use piecewise defined curves a knot selection is involved. It is well known that the parametrisation of the knots has severe effects on the resulting curve [7]. In this section the natural cubic interpolating spline for chord-length parametrisation using the Loewnerean affine invariant norm will be calculated. The knot selection is considered as part of the overall curve interpolation algorithm. Given points $P_i (i = 0, \ldots, n)$ in the plane $\mathbb{R}^2$ a knot sequence $(t_i)_{i=0}^n$ will be defined by $t_0 = 0, t_{i+1} = t_i + \|P_{i+1} - P_i\|_L (i = 1, \ldots, n)$. For comparison with

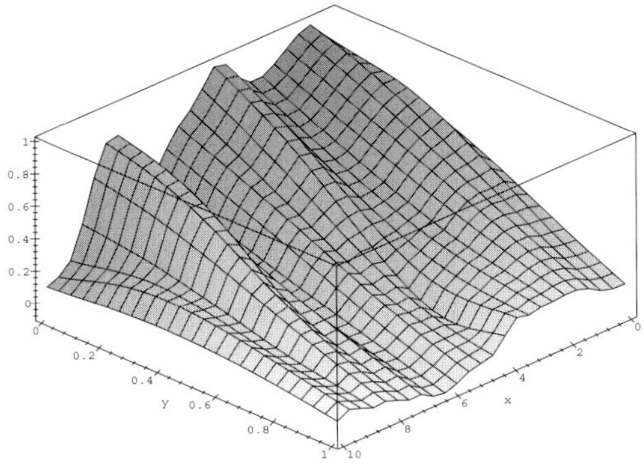

Figure 6. TPS with scaled data (33 random points)

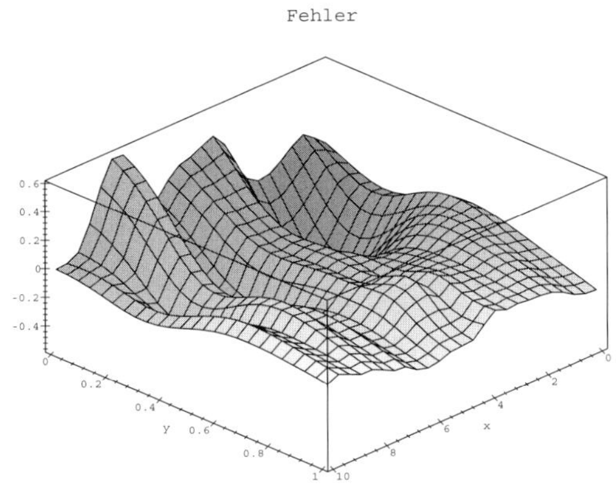

Figure 7. TPS errors for scaled data (33 random points)

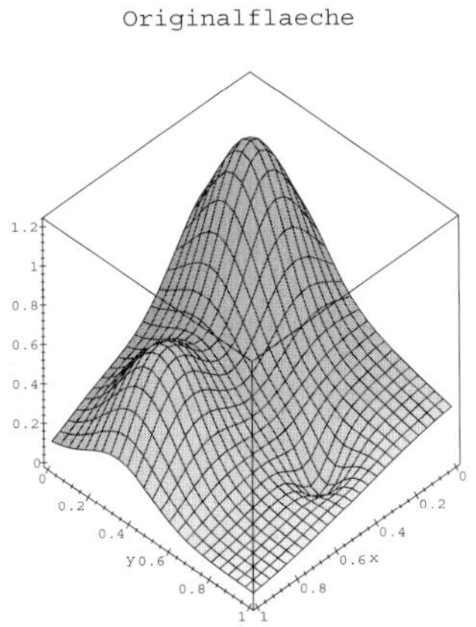

Figure 8. Franke's function f_1

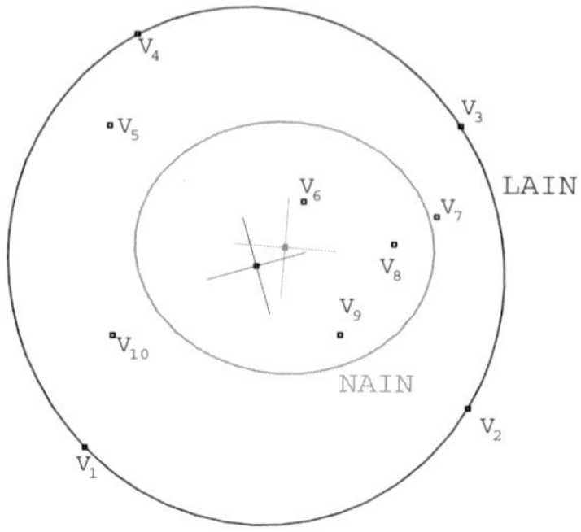

Figure 9. NAIN and LAIN

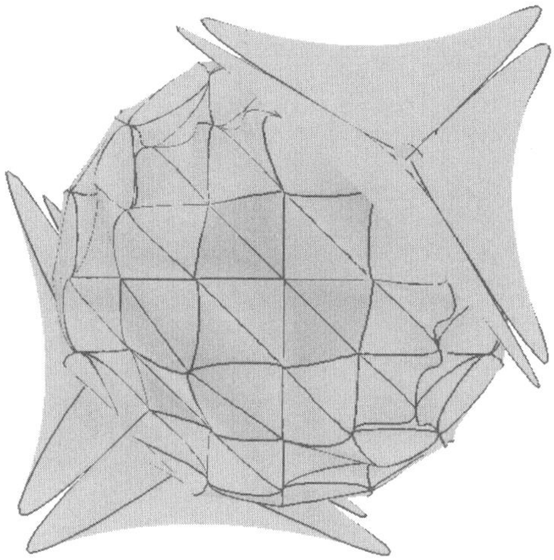

Figure 10. Euclidean norm

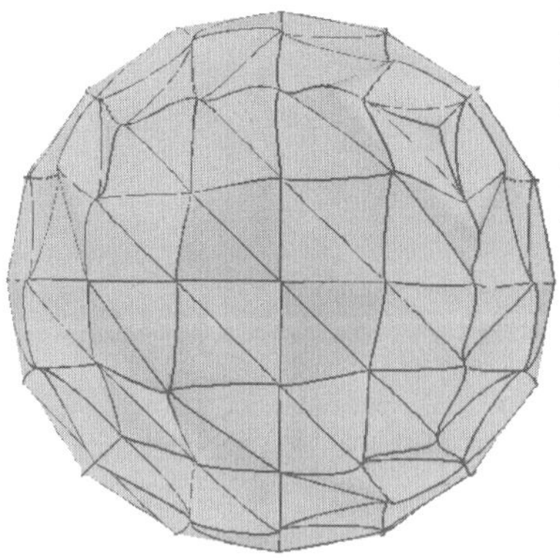

Figure 11. Loewnerean norm

the results of Nielson and Foley [11, §3], their example has been calculated again – this time using the LAIN. By using an affine invariant norm – instead of the standard Euclidean metric – one remedies the lack of affine invariance for the chord-length interpolation method. This modified interpolant works well in most cases. But it leads to the possibility of dissatisfying results in some cases, comparable to those of Nielson and Foley. In Fig. 12 the chord-length knot spacing interpolant is shown for both NAIN and LAIN. The curve slightly improved approximating the polygon is the interpolant using the Loewnerean norm. The shape of the interpolant near the implied indentation is the problem. The results might be further improved by a combination of the LAIN with the method developed by Foley and Nielson in [2].

8. Conclusion

Both affine invariant norms have advantages. Especially in higher dimensions, Nielson's norm can easier be calculated whereas the Loewnerean norm has an obvious geometric meaning – the gauge ellipsoid is the Loewner ellipsoid – and is less intruded by small inaccuracies of the given points (e.g. from measurement errors).

A further advantage of the Loewnerean norm is that it will not change if a point $\vec{X}_{n+1}$ is added to the base data set $\mathscr{X}$ as far as this point lies in the interior or on the boundary of the Loewner ellipsoid E, i.e. for $\|\vec{X}_{n+1}\|_L \leq 1$. For the approximation quality Tables 1 and 2 indicate that one cannot say a priori which one of the both

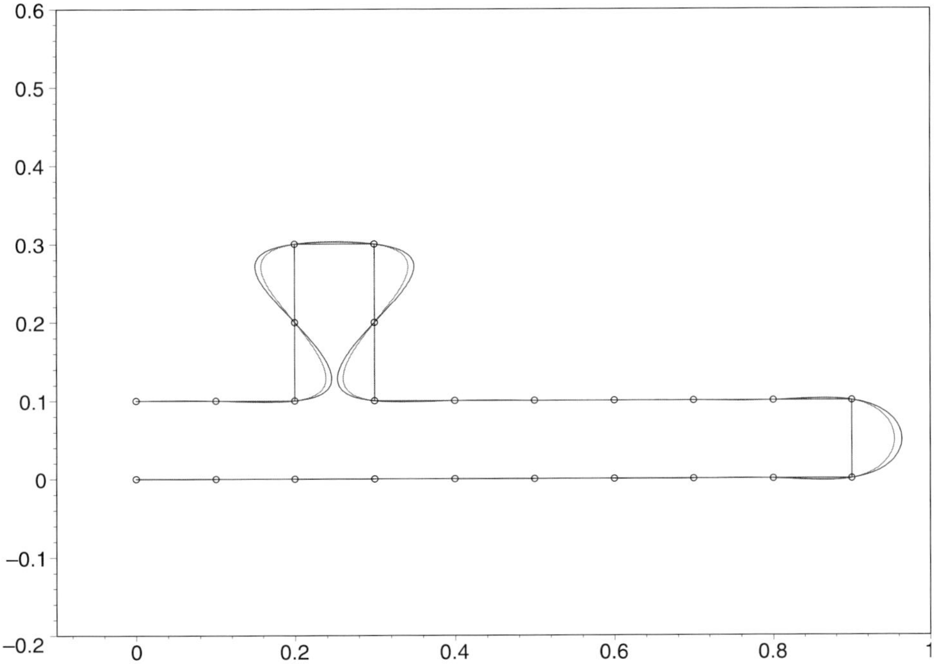

Figure 12. Chord-length knot spacing using affine invariant norms

norms results in the "better" interpolant. But the results for both data sets with 100 points and further examples indicate that in many cases for larger data sets the results of the LAIN interpolant are slightly superior.

Additionally, in $\mathbb{R}^d$ at least $d+1$ points of the given set $\mathscr{X}$ are situated on the Loewner ellipsoid E, thus their norm will be 1. All other points lie in the interior or on the boundary of E, their norms are smaller or equal to 1. Consequently, an upper bound of the norm is known in advance for all given points. This leads to an improved numerical stability of those applications where the LAIN is used (compare Figs. 10 and 11).

Acknowledgement

The author thanks Prof. Dr. W. Degen for his useful suggestion to inspect the Loewner ellipsoid as a starting point.

References

[1] Degen, W. L. F., Milbrandt, V.: The geometric meaning of Nielson's affine invariant norm. Comput. Aided Geom. Des. *15*, 19–25 (1997).
[2] Foley, T. A., Nielson, G. M.: Knot selection for parametric spline interpolation. In: Mathematical methods in computer aided geometric design (Lyche, T., Schumaker, L. L., eds.), pp. 261–271. Boston: Academic Press, 1989.
[3] Franke, R.: A critical comparison of some methods for interpolation of scattered data. Technical Report #NPS-53-79-003, Naval Postgraduate School, 1979.
[4] Juhnke, F.: Volumenminimale Ellipsoidüberdeckungen. Beitr. Alg. Geom. *30*, 143–153 (1990).
[5] Jung, G.: Geometrie der Massen. In: Encyklopädie der mathematischen Wissenschaften, Vol. IV, 1 (Mechanik), pp. 279–344. Leipzig: Teubner, 1903.
[6] Lawrence, C., Zhou, J. L., Tits, A. L.: User's Guide for CFSQP Version 2.5: A C Code for Solving (Large Scale) Constrained Nonlinear (Minimax) Optimization Problems, Generating Iterates Satisfying All Inequality Constraints. University of Maryland, April 1997. Homepage: http://www.isr.umd.edu/Labs/CACSE/FSQP/fsqp.html.
[7] Hoschek, J., Lasser, D.: Grundlagen der geometrischen Datenverarbeitung, 2nd ed. Stuttgart: Teubner, 1992.
[8] Laugwitz, D.: Differentialgeometrie. Stuttgart: Teubner, 1960.
[9] Milbrandt, V.: Affin-invariante Interpolation auf Dreiecksflächen. PhD thesis, Universität Stuttgart. Aachen: Shaker-Verlag, 1999.
[10] Nielson, G. M.: Coordinate free scattered data interpolation. In: Topics in multivariate approximation (Chui, C. K., Schumaker, L. L., Utreras, F. I., eds.), pp. 175–184. Boston: Academic Press, 1987.
[11] Nielson, G. M., Foley, T. A.: A survey of applications of an affine invariant norm. In: Mathematical methods in computer aided geometric design (Lyche, T., Schumaker, L. L., eds.), pp. 445–467. Boston: Academic Press, 1989.
[12] Nielson, G. M., Hagen, H., Müller, H.: Scientific visualization: overviews, methodologies, and techniques, chapter 20. Tools for Triangulation and Tetrahedrization. IEEE Computer Society, 1997.

V. Milbrandt
Frans-Hals-Ring 51
D-22846 Norderstedt
Germany
e-mail: milbrandt@gmx.de

Computing [Suppl] 14, 281–292 (2001)

Computing

Exploiting Wavelet Coefficients for Modifying Functions

A. Nawotki, Kaiserslautern

Abstract

Various methods have been developed to modify and model functions. Even so, we found it worth while to consider a further one, which is based on wavelets. This enables us to separate several aspects of a function and modify one selected exclusively. The quality of this approach is dependent on the choice of the wavelet decomposition. We demonstrate for Haar-wavelets how to estimate changes a priori and how to avoid modifications locally. A more general result is shown for all wavelet decompositions with finite filters. This knowledge can for example be used for a selective encrypting, where only a part of the data must be hidden. We implemented it using a wavelet decomposition, and found the described tools quite handy.

AMS Subject Classifications: I31101, I15023.

Key Words: Wavelets, selective encryption, error estimate, Haar-Wavelets, finite filters.

1. Introduction

Conventional cryptographic algorithms usually encode all the information stored in the data, although very often only a few details are secret and for special applications some information must not be altered. If the information is sorted according to a security classification, than it would be possible to encrypt only those parts, which must not be transmitted to the actual recipient. The 'selectively encrypted' data is still useful for many applications, but it does not include the real secrets. How is it possible to produce such an intermediate version of the data with selectively reduced information?

We need a hierarchy of the information, and we decided to use wavelets for this purpose, because a wavelet decomposition divides a function up in hierarchically ordered levels. We have the desired separation, if it is possible to distribute different aspects of the data to distinct levels. The first step towards this aim is to investigate, which and how coefficients influence the original function, and how the decomposition-level and the included information are related.

The quality of the separation and the correlation between the function and the wavelet coefficients are dependent on the choice of the wavelet decomposition. We start with Haar-wavelets (Section 3), and show a general result for all wavelets with finite filters (Section 4). The developed tools turn out to be very useful: We apply them for a selective encrypting and modify a reflector surface of a headlight in the desired manner (Section 5).

First of all, we start with a short sketch of the wavelet decomposition.

2. The Wavelet Decomposition

A wavelet decomposition may be used to quickly compute different representations of a function. They are attained by the construction of a hierarchy of base functions such that the base functions of the coarsest level describe the overall shape and the finer the hierarchy-level is, the more details are included in it. The theoretical background is described comprehensively for example in [2], [4], and [6]. It is practical for this purpose to work with functions of the following type:

Definition 1. *Let $\varphi(x) \in L^2(\mathbb{R})$ and $\|\varphi(x)\| = 1$. φ is called refinable, if constants $h_k \in \mathbb{R}$ exist such that*

$$\varphi(x) = \sqrt{2}\sum_k h_k \varphi(2x - k). \tag{1}$$

The hat-function

$$\chi_{[0,1)} = \begin{cases} 1, & \text{if } x \in [0,1) \\ 0, & \text{elsewhere} \end{cases}$$

is one of the simplest examples for this class of functions, because it holds $\chi_{[0,1)}(x) = \chi_{[0,1)}(2x) + \chi_{[0,1)}(2x - 1)$. For arbitrary intervals we define the scaled and translated hat-function by

$$\chi_{m,k} := \begin{cases} 2^{-\frac{m}{2}}, & \text{if } x \in [2^m k, 2^m(k+1)) \\ 0, & \text{elsewhere.} \end{cases}$$

Figure 1 shows two neighbored scalings of it, and demonstrates the refinability.

The closure of all linear combinations of integer-translates of a refinable function defines a function-space

$$V_m := \overline{\text{span}\{2^{-\frac{m}{2}}\varphi(2^{-m}x - k) | k \in \mathbb{Z}\}} =: \overline{\text{span}\{\varphi_{m,k}(x) | k \in \mathbb{Z}\}}.$$

The refinability of $\varphi(x)$ implies that the chain of spaces $V_{m+1}, V_m, \ldots$ is nested:

$$\{0\} \subset \ldots \subset V_{m+1} \subset V_m \subset \ldots \subset L^2(\mathbb{R})$$

Thus, we have constructed a hierarchy, where each level describes a different set of functions. With decreasing m the function space V_m is enlarged, and finer and finer details are contained. The resolution index m fixes the amount of details that are available in V_m.

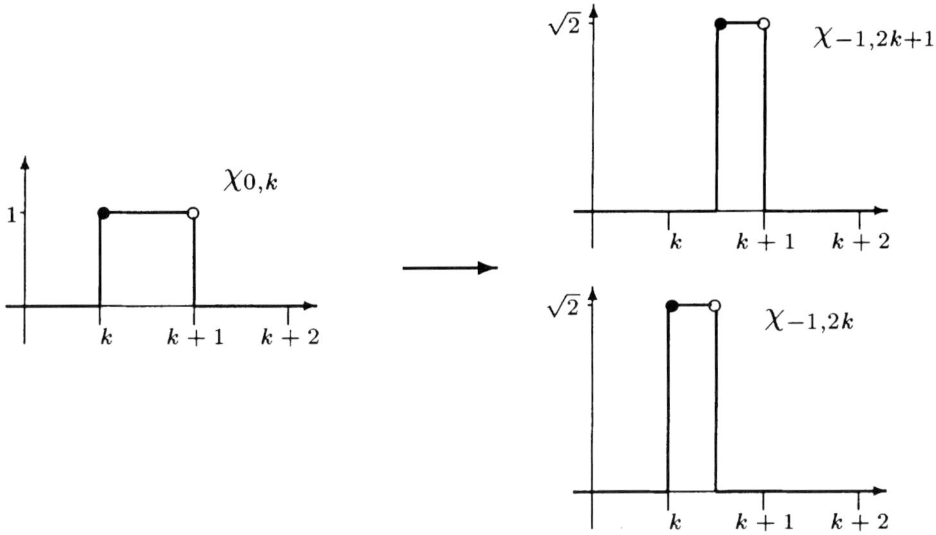

Figure 1. The hat-function in two resolutions: It holds $\chi_{0,k}(x) = \frac{1}{\sqrt{2}}\left(\chi_{-1,2k}(x) + \chi_{-1,2k+1}(x)\right)$

The next step is to define a space W_{m+1} that describes the difference between V_m and V_{m+1}, i.e. W_{m+1} is the orthogonal complement of V_{m+1} in V_m.

We choose for this space base functions of the same structure as the base functions of V_m, i.e. the integer-translates of a refinable function $\psi \in L^2(\mathbb{R})$ with $\|\psi(x)\| = 1$ such that

$$W_{m+1} := \overline{\operatorname{span}\{2^{-\frac{m+1}{2}}\psi(2^{-m-1}x - k) | k \in \mathbb{Z}\}} =: \overline{\operatorname{span}\{\psi_{m+1,k} | k \in \mathbb{Z}\}}.$$

A function ψ fulfilling these conditions is called *wavelet*. If possible, mutually orthogonal translates $\psi_{m+1,k}$ should be selected.

The suitable choice for the hat-function $\chi_{m+1,k}$ is the Haar-wavelet

$$\psi_{m+1,k}(x) := \begin{cases} 2^{-\frac{m+1}{2}}, & \text{if } x \in [2^{m+1}k, 2^m(2k+1)) \\ -2^{-\frac{m+1}{2}}, & \text{if } x \in [2^m(2k+1), 2^{m+1}(k+1)) \\ 0, & \text{elsewhere,} \end{cases}$$

because it holds $\chi_{m,2k}(x) = \chi_{m+1,k}(x) + \psi_{m+1,k}(x)$. Additionally, the intersection of the support of two Haar-wavelets $\operatorname{supp}(\psi_{m+1,k}) \cap \operatorname{supp}(\psi_{m+1,l})$ has measure zero for $k \neq l$, what yields orthogonality.

Now we use the spaces V_{m+1} and W_{m+1} for the computation of different representations of a function. First of all, the base functions of both spaces are included in V_m and, therefore, they can be written down with the base functions of V_m. For V_{m+1} this agrees with the scaling equation (1) written for arbitrary intervals

$$\varphi_{m+1,k}(x) = \sum_{l\in\mathbb{Z}} h_{l-2k}\varphi_{m,l}(x). \tag{2}$$

With $W_{m+1} \subset V_m$ we may write analogously

$$\psi_{m+1,k}(x) = \sum_{l\in\mathbb{Z}} g_{l-2k}\varphi_{m,l}(x) \tag{3}$$

for some $g_l \in \mathbb{R}$. $h = (\ldots, h_l, h_{l+1}, \ldots)$ and $g = (\ldots, g_l, g_{l+1}, \ldots)$ are called scaling and wavelet filter respectively. For Haar-wavelets hold $\varphi_{m+1,k}(x) = \frac{1}{\sqrt{2}}\varphi_{m,2k}(x) + \frac{1}{\sqrt{2}}\varphi_{m,2k+1}(x)$ and $\psi_{m+1,k}(x) = \frac{1}{\sqrt{2}}\varphi_{m,2k}(x) - \frac{1}{\sqrt{2}}\varphi_{m,2k+1}(x)$. Let $c_k^n := \leqslant f, \varphi_{n,k} >_{L^2}$ and $d_k^n := < \bar{f}, \psi_{n,k} >_{L^2}$ respectively be the base coefficients of $f \in V_n$ and $f \in W_n$. Then the relationship between the coefficients of neighbored levels can be expressed in

$$c_k^{m+1} = < f, \varphi_{m+1,k} >_{L^2} \stackrel{(2)}{=} \sum_{l\in\mathbb{Z}} h_{l-2k} < f, \varphi_{m,l} >_{L^2} = \sum_{l\in\mathbb{Z}} h_{l-2k}c_l^m,$$

$$d_k^{m+1} = < \bar{f}, \psi_{m+1,k} >_{L^2} \stackrel{(3)}{=} \sum_{l\in\mathbb{Z}} g_{l-2k} < \bar{f}, \varphi_{m,l} >_{L^2} = \sum_{l\in\mathbb{Z}} g_{l-2k}c_l^m.$$

For Haar-wavelets follow immediately

$$c_k^{m+1} = \frac{1}{\sqrt{2}}(c_{2k}^m + c_{2k+1}^m), \tag{4}$$

$$d_k^{m+1} = \frac{1}{\sqrt{2}}(c_{2k}^m - c_{2k+1}^m). \tag{5}$$

This step can be repeated arbitrary often and results in sets of coefficients $\{c^M, d^m, m = 1, \ldots, M\}, M \in \mathbb{N}$, which describe the function exactly. These sets have the same size as the original coefficient set. (Note that an upper bound for M exists, if the starting set is finite.)

On the other hand we gain a hierarchical order: c^M is the coarsest representation of f, and the size of the details raises with the enlargement of the superscript of the wavelet coefficients.

As well, the original sequence c^0 can be recovered from $\{c^M, d^m, m = 1, \ldots, M\}$. This reconstruction step is sketched in Fig. 2. For example, adding and subtracting Eqs. (4) and (5) leads to the corresponding formulas for the Haar-wavelets:

$$\begin{aligned} c_{2k}^m &= \frac{1}{\sqrt{2}}(c_k^{m+1} + d_k^{m+1}) \\ c_{2k+1}^m &= \frac{1}{\sqrt{2}}(c_k^{m+1} - d_k^{m+1}) \end{aligned} \tag{6}$$

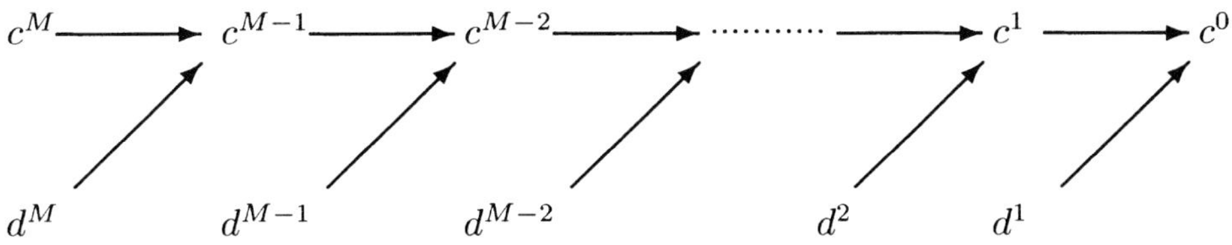

Figure 2. Reconstruction step for a wavelet decomposition

3. The Influence of Haar-Wavelet Coefficients on a Function

Now we are prepared to investigate, if it is possible to influence a function selectively by a change of its wavelet coefficients. For this purpose we need an error estimation. The key for this matter is our new reconstruction formula. It is based on the equations in (6), but it requires no recursion. The first step is to merge the odd and even coefficients into

$$c_k^m = \frac{1}{\sqrt{2}}\left(c_{\lfloor\frac{k}{2}\rfloor}^{m+1} + (-1)^k d_{\lfloor\frac{k}{2}\rfloor}^{m+1}\right),$$

where $\lfloor . \rfloor$ denotes the lower Gaussian bracket ($r \in \mathbb{R}, n \in \mathbb{Z} : \lfloor r \rfloor = n \Leftrightarrow n \leq r < n+1$). Now we can state the following.

Theorem 1.[1] *Let $\{c^M, d^m, m = 1, \ldots, M\}$ be a Haar-wavelet decomposition of depth M. For all $j, l \in \mathbb{N}, j + l \leq M$ holds*

$$c_k^j = \left(\frac{1}{\sqrt{2}}\right)^l c_{\lfloor\frac{k}{2^l}\rfloor}^{j+l} + \sum_{i=1}^{l} (-1)^{\lfloor\frac{k}{2^{i-1}}\rfloor} \left(\frac{1}{\sqrt{2}}\right)^i d_{\lfloor\frac{k}{2^i}\rfloor}^{j+i}. \tag{7}$$

The proof of this assumption is a complete induction over l.

Probably the most interesting situation is $j = 0$ and $l = m$: Every start coefficient c_k^0 is divided into M wavelet coefficients $d_{\lfloor\frac{k}{2^i}\rfloor}^i, i = 1, \ldots, M$, including the details, and one scaling coefficient $c_{\lfloor\frac{k}{2^M}\rfloor}^M$ for the overall shape.

With this formula we can predict, how the change of f depends on the size of the modification. It turns out that this relies on the size of the modification and on the depth where the modification takes place.

Conclusions 1. *Let $\{c^M, d^m, m = 1, \ldots, M\}$ be a Haar-wavelet decomposition of depth M and F_C^j the absolute value of the maximum change of the starting function, if a constant $C \in \mathbb{R}$ is added to (at least) one wavelet coefficient in depth j. It holds*

1. $\frac{F_C^{j+1}}{F_C^j} = \frac{1}{\sqrt{2}}$ *for all $C \in \mathbb{R}\backslash\{0\}$ and $j = 1, \ldots, M-1$,*
2. $F_C^j = |C| F_1^j$ *for all $C \in \mathbb{R}$ and $j = 1, \ldots, M$, and*
3. $F_C^j = |C| (\frac{1}{\sqrt{2}})^{j-1} F_1^1$ *for all $C \in \mathbb{R}$ and $j = 1, \ldots, M$.*

[1]This Theorem and Conclusion 2 were already discussed in [5], but we include them here for completeness.

Proof: Inspection of formula (7).

Another interesting question is, how a part of the function can be kept constant. Formula (7) states how large the region of influence of one single coefficient is.

Conclusions 2. *Let* $\{c^M, d^m, m = 1, \ldots, M\}$ *be a Haar-wavelet decomposition of depth* M. *A change of the wavelet coefficient* d^j_k *effects at most* 2^j *start coefficients, namely* c^0_i, $i = 2^j k, \ldots, 2^j(k+1) - 1$.

Vice versa, c^0_i *is not altered, if* $d^j_{\lfloor \frac{i}{2^j} \rfloor}$, *is constant for all* $j = 1, \ldots, M$.

Proof: Formula (7), again.

Now it is obvious, how an one-dimensional function ($f \in \mathbb{R}$) can be fixed in a region: First, the coefficients c^0_i describing the concerned area are determined. Then we must guarantee nothing but that the corresponding coefficients $d^j_{\lfloor \frac{i}{2^j} \rfloor}$ are constant for all levels j.

This principle can be transferred to two-dimensions. On the occasion we must take into account that different decomposition strategies are possible. The non-standard decomposition decomposes alternately the two space directions, while the so-called standard method works off the directions successively.

Suppose c^{00}_{ij} must stay unchanged, and that we decompose m times in x-direction and after that subdivide n times in y-direction. In that case we must not modify the coefficients $d^{l0}_{\lfloor \frac{i}{2^l} \rfloor j}$ for $l = 1, \ldots, m$ and $d^{mk}_{\lfloor \frac{i}{2^m} \rfloor \lfloor \frac{j}{2^k} \rfloor}$ for all $k = 1, \ldots, n$.

For a non-standard decomposition and $m \geq n$ the coefficients $d^{10}_{\lfloor \frac{i}{2} \rfloor j}$, $d^{11}_{\lfloor \frac{i}{2} \rfloor \lfloor \frac{j}{2} \rfloor}, \ldots,$ $d^{nn}_{\lfloor \frac{i}{2^n} \rfloor \lfloor \frac{j}{2^n} \rfloor}$, $d^{n+1,n}_{\lfloor \frac{i}{2^{n+1}} \rfloor \lfloor \frac{j}{2^n} \rfloor}$, $d^{n+2,n}_{\lfloor \frac{i}{2^{n+2}} \rfloor \lfloor \frac{j}{2^n} \rfloor}, \ldots, d^{mn}_{\lfloor \frac{i}{2^m} \rfloor \lfloor \frac{j}{2^n} \rfloor}$ and for $m < n$ the coefficients $d^{10}_{\lfloor \frac{i}{2} \rfloor j}$, $d^{11}_{\lfloor \frac{i}{2} \rfloor \lfloor \frac{j}{2} \rfloor}, \ldots, d^{mm}_{\lfloor \frac{i}{2^m} \rfloor \lfloor \frac{j}{2^m} \rfloor}$, $d^{m,m+1}_{\lfloor \frac{i}{2^m} \rfloor \lfloor \frac{j}{2^{m+1}} \rfloor}$, $d^{m,m+2}_{\lfloor \frac{i}{2^m} \rfloor \lfloor \frac{j}{2^{m+2}} \rfloor}, \ldots, d^{mn}_{\lfloor \frac{i}{2^m} \rfloor \lfloor \frac{j}{2^n} \rfloor}$ must stay as they are.

Example. The following illustrates the propagation of the fixing of single coefficients during the decomposition. Assume that the scaling coefficients $c^{00}_{05} = *$, $c^{00}_{14} = \star, c^{00}_{23} = \circ, c^{00}_{33} = \bullet, c^{00}_{42} = \oslash, c^{00}_{51} = \otimes, c^{00}_{61} = \oplus$ must not be modified. Thus, the starting coefficients look like:

$$\begin{array}{cccccccc}
c^{00}_{00} & c^{00}_{01} & c^{00}_{02} & c^{00}_{03} & c^{00}_{04} & * & c^{00}_{06} & c^{00}_{07} \\
c^{00}_{10} & c^{00}_{11} & c^{00}_{12} & c^{00}_{13} & \star & c^{00}_{15} & c^{00}_{16} & c^{00}_{17} \\
c^{00}_{20} & c^{00}_{21} & c^{00}_{22} & \circ & c^{00}_{24} & c^{00}_{25} & c^{00}_{26} & c^{00}_{27} \\
c^{00}_{30} & c^{00}_{31} & c^{00}_{32} & \bullet & c^{00}_{34} & c^{00}_{35} & c^{00}_{36} & c^{00}_{37} \\
c^{00}_{40} & c^{00}_{41} & \oslash & c^{00}_{43} & c^{00}_{44} & c^{00}_{45} & c^{00}_{46} & c^{00}_{47} \\
c^{00}_{50} & \otimes & c^{00}_{52} & c^{00}_{53} & c^{00}_{54} & c^{00}_{55} & c^{00}_{56} & c^{00}_{57} \\
c^{00}_{60} & \oplus & c^{00}_{62} & c^{00}_{63} & c^{00}_{64} & c^{00}_{65} & c^{00}_{66} & c^{00}_{67}
\end{array}$$

If the first decomposition is done in the direction of the first index, then c^{00}_{ij} is influenced by $d^{10}_{\lfloor \frac{i}{2} \rfloor j}$, and therefore c^{00}_{05} is affected by d^{10}_{05}, c^{00}_{14} by d^{10}_{04}, c^{00}_{23} and c^{00}_{33} by d^{10}_{13}, c^{00}_{42} by d^{10}_{22}, c^{00}_{51} by d^{10}_{21}, and c^{00}_{61} by d^{00}_{31}. This is shown in the next picture:

$$\begin{array}{cccccccc}
d^{10}_{00} & d^{10}_{01} & d^{10}_{02} & d^{10}_{03} & \star & * & d^{10}_{06} & d^{10}_{07} \\
d^{10}_{10} & d^{10}_{11} & d^{10}_{12} & \circ\bullet & d^{10}_{14} & d^{10}_{15} & d^{10}_{16} & d^{10}_{17} \\
d^{10}_{20} & \otimes & \oslash & d^{10}_{23} & d^{10}_{24} & d^{10}_{25} & d^{10}_{26} & d^{10}_{27} \\
d^{10}_{30} & \oplus & d^{10}_{32} & d^{10}_{33} & d^{10}_{34} & d^{10}_{35} & d^{10}_{36} & d^{10}_{37}
\end{array}$$

In the following step the standard and the non-standard decomposition cause different results.

successively

$$\begin{array}{cccccccc}
d^{20}_{00} & d^{20}_{01} & d^{20}_{02} & \circ\bullet & \star & * & d^{20}_{06} & d^{20}_{07} \\
d^{20}_{10} & \oplus\otimes & \oslash & d^{20}_{13} & d^{20}_{14} & d^{20}_{15} & d^{20}_{16} & d^{20}_{17}
\end{array}$$

alternately

$$\begin{array}{cccc}
d^{11}_{00} & d^{11}_{01} & *\star & d^{11}_{03} \\
d^{11}_{10} & \circ\bullet & d^{11}_{12} & d^{11}_{13} \\
\otimes & \oslash & d^{11}_{22} & d^{11}_{23} \\
\oplus & d^{11}_{31} & d^{11}_{32} & d^{11}_{33}
\end{array}$$

The next iteration works analogously.

successively

$$\begin{array}{cccccccc}
d^{30}_{00} & \otimes\oplus & \oslash & \circ\bullet & \star & * & d^{30}_{06} & d^{30}_{07}
\end{array}$$

alternately

$$\begin{array}{cccc}
d^{21}_{00} & \circ\bullet & *\star & d^{21}_{03} \\
\otimes\oplus & \oslash & d^{21}_{12} & d^{21}_{13}
\end{array}$$

On the left-hand side we now cannot continue to subdivide in the first space direction, and thus we must switch to the second direction.

successively

$$\begin{array}{cccc}
\otimes\oplus & \oslash\circ\bullet & \star * & d^{31}_{03}
\end{array}$$

alternately

$$\begin{array}{cc}
\circ\bullet & *\star \\
\oslash\otimes\oplus & d^{22}_{11}
\end{array}$$

The last two repetitions of both methods match, what is compelling in the last stop, but by chance in the second last.

successively/alternately

$$\begin{array}{cc}
\circ\bullet\oslash\otimes\oplus & *\star
\end{array}$$

successively/alternately

$$*\star\circ\bullet\oslash\otimes\oplus$$

A often desired special case is the conservation of the boundary for a two-dimensional function. The boundary is of interest only, if the support of the function is compact. Thus, we may assume $c^0_{kl} \neq 0$ for finitely many indices only.

Corollary 1. *Let $\{c^{mn}, d^{ij}, i = 1, \ldots, m, j = 1, \ldots, n\}$ be a two-dimensional Haar-wavelet decomposition of depth (m, n) and c^{00}_{kl} for $k = 0, l = 0, k = x_{max}$, or $l = y_{max}$ describe the boundary, which must be maintained.*

1. *If a standard decomposition, which operates in x-direction first, is carried through, than the coefficients $d^{i0}_{0l}, d^{i0}_{\lfloor \frac{k}{2^i} \rfloor 0}, d^{i0}_{\lfloor \frac{x_{max}}{2^i} \rfloor l}, d^{i0}_{\lfloor \frac{k}{2^i} \rfloor y_{max}}, d^{mj}_{0 \lfloor \frac{l}{2^j} \rfloor}, d^{mj}_{\lfloor \frac{k}{2^m} \rfloor 0}, d^{mj}_{\lfloor \frac{x_{max}}{2^m} \rfloor \lfloor \frac{l}{2^j} \rfloor}, d^{mj}_{\lfloor \frac{k}{2^m} \rfloor \lfloor \frac{y_{max}}{2^j} \rfloor}$ must be preserved.*

2. *If a non-standard decomposition is applied, than the coefficients* $d^{ij}_{0\lfloor\frac{l}{2^j}\rfloor}$, $d^{ij}_{\lfloor\frac{k}{2^i}\rfloor 0}$, $d^{ij}_{\lfloor\frac{x_{max}}{2^i}\rfloor\lfloor\frac{l}{2^j}\rfloor}$, $d^{ij}_{\lfloor\frac{k}{2^i}\rfloor\lfloor\frac{y_{max}}{2^j}\rfloor}$, *with* $i = j$ *or* $i = j + 1$ *and*

- *for* $m > n + 1, z = 2, \dots, m - n$ *additionally the coefficients* $d^{n+z,n}_{0\lfloor\frac{l}{2^n}\rfloor}$, $d^{n+z,n}_{\lfloor\frac{k}{2^{n+z}}\rfloor 0}$, $d^{n+z,n}_{\lfloor\frac{x_{max}}{2^{n+z}}\rfloor\lfloor\frac{l}{2^n}\rfloor}$, $d^{n+z,n}_{\lfloor\frac{k}{2^{n+z}}\rfloor\lfloor\frac{y_{max}}{2^n}\rfloor}$,
- *for* $m < n, z = 1, \dots, n - m$ *supplementary the coefficients* $d^{m,m+z}_{0\lfloor\frac{l}{2^{n+z}}\rfloor}$, $d^{m,m+z}_{\lfloor\frac{k}{2^m}\rfloor 0}$, $d^{m,m+z}_{\lfloor\frac{x_{max}}{2^m}\rfloor\lfloor\frac{l}{2^{m+z}}\rfloor}$, $d^{m,m+z}_{\lfloor\frac{k}{2^m}\rfloor\lfloor\frac{y_{max}}{2^{m+z}}\rfloor}$

must be maintained.

Everywhere must hold $1 \leq i \leq m$, $0 \leq j \leq n$, $k = 0, \dots, x_{max}$, *and* $l = 0, \dots y_{max}$.

Putting all statements together, we have the possibility to predict the changes that are caused by modifications of Haar-wavelet coefficients and to limit portions of the function as we like. Now we investigate if for other types of wavelet decompositions hold similar statements.

4. A Statement on Wavelets with Finite Filters

The scaling equations of the Haar-wavelets are comparatively simple, because only two coefficients are non-trivial in both filters. Infinite filters are of very small practical use. Thus, let us consider the general case with finite filters of length $s + 1$ and $\acute{s} + 1$ respectively, i.e.

$$c^{m+1}_k = \sum_{j=0}^{s} h_j c^m_{2k+j},$$

$$d^{m+1}_k = \sum_{j=0}^{\acute{s}} g_j c^m_{2k+j}.$$

It follows immediately that up to $(\acute{s} + 1)(s + 1)^{j-1}$ wavelet coefficients d^j of level j are influenced by one start coefficient c^0. This is a very rough estimation, which can be improved.

Theorem 2. *Let* $\{c^M, d^m, m = 1, \dots, M\}$ *be a wavelet decomposition of depth* M *with a scaling filter of length* $s + 1$ *and a wavelet filter of length* $\acute{s} + 1$. *For all* $j \in \mathbb{N}$ *holds*

$$c^{m+j}_k = \sum_{i_1,\dots,i_j=0}^{s} h_{i_1} \cdots h_{i_j} c^m_{2^j k + 2^{j-1} i_j + \cdots + 2^0 i_1}, \tag{8}$$

$$d^{m+j}_k = \sum_{l=0}^{\acute{s}} \sum_{i_1,\dots,i_{j-1}=0}^{s} g_l h_{i_1} \cdots h_{i_{j-1}} c^m_{2^j k + 2^{j-1} l + 2^{j-2} i_{j-1} + \cdots + 2^0 i_1}. \tag{9}$$

This theorem can be proven easily with an induction, but nevertheless its statement is quite useful, because these formulas determine the correlation between the coefficients in different levels.

Conclusions 3. *Let* $\{c^M, d^m, m = 1, \ldots, M\}$ *be a wavelet decomposition of depth* M *with a scaling filter of length* $s+1$ *and a wavelet filter of length* $\acute{s}+1$*. For all* $m+j \leq M$ *holds:*

1. *The coefficient* c_k^{m+j} *is influenced by* $c_{2^jk}^m, \ldots, c_{2^jk+s(2^j-1)}^m$*, i.e. by* $s(2^j-1)+1$ *scaling coefficients, and* d_k^{m+j} *is affected by* $c_{2^jk}^m, \ldots, c_{2^jk+2^{j-1}(s+\acute{s})-s}^m$*, i.e. by* $2^{j-1}(s+\acute{s})-s+1$ *wavelet coefficients.*

2. c_k^m *influences* $c_{\lceil\frac{k-s(2^j-1)}{2^j}\rceil}^{m+j}, \ldots, c_{\lfloor\frac{k}{2^j}\rfloor}^{m+j}$ *and* $d_{\lceil\frac{k-2^{j-1}(s+\acute{s})+s}{2^j}\rceil}^{m+j}, \ldots, d_{\lfloor\frac{k}{2^j}\rfloor}^{m+j}$*, where* $\lfloor . \rfloor$ *and* $\lceil . \rceil$ *denote the lower and upper Gaussian brackets.*[2]

3. *In every level of the wavelet decomposition up to s scaling coefficients* c_k^j *and* $\frac{s+\acute{s}}{2}$ *wavelet coefficients* d_k^j *influence the start coefficients* c_l^0*.*

Proof: Formulas (8) and (9).

The third conclusion limits the first estimation of the influence at the beginning of this section considerably from $(\acute{s}+1)(s+1)^{j-1}$ to $\frac{s+\acute{s}}{2}$.

Of course, the conclusions hold for Haar-wavelets too: The length of the filters is $s = \acute{s} = 1$. Thus, d_k^{m+j} is influenced by $c_{2^jk}^m, \ldots, c_{2^jk+2^j-1}^m$, altogether 2^j coefficients. Vice versa, c_k^m affects $\{d_{\lceil\frac{k-2^j+1}{2^j}\rceil}^{m+j}, \ldots, d_{\lfloor\frac{k}{2^j}\rfloor}^{m+j}\} = \{d_{\lfloor\frac{k}{2^j}\rfloor}^{m+j}\}$, thus $\frac{s+\acute{s}}{2} = 1$ wavelet coefficient in every level. These statements coincide with Conclusion 2.

5. An Application

Now, we apply the deduced correlations between a function and its wavelet coefficients to a selective encrypting algorithm. The goal of this security procedure is different from standard encrypting methods, and thus our method has very few in common with other encryptions: Here, the data is split into two portions, one consisting of the secret information and the other containing that data only, which can be transmitted without restrictions, or which is necessary for the recipient. Only the delicate part is encrypted and added to the untouched rest. Thus we get a semi-modified data set, which can still be utilized for some uses, and contains public information only.

The technical realization is based on the usage of a wavelet decomposition. The decomposition coefficients are ordered in levels, and each level corresponds to a specific detail size. All we have to do is to find the level where the details have the appropriate size to change the secret information while they do not alter the rest. Therefore we need the derived correlations of Sections 3 and 4.

---

[2] $r \in R, m, n \in Z : \lfloor r \rfloor = m \Leftrightarrow m \leq r < m+1, \lceil r \rceil = n \Leftrightarrow n < r \leq n+1$

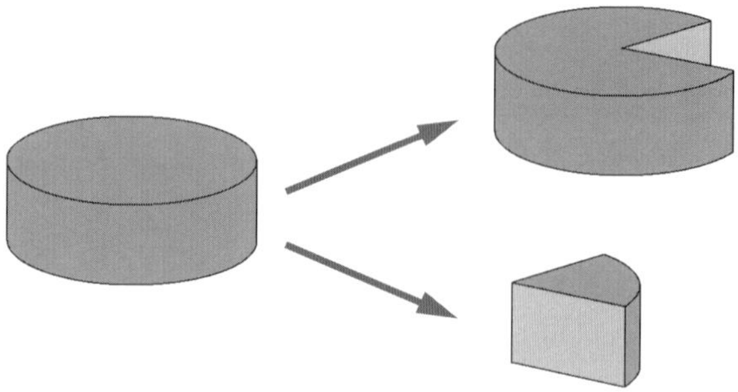

Figure 3. Selective Crypting: The original data is split into a public and a secret part

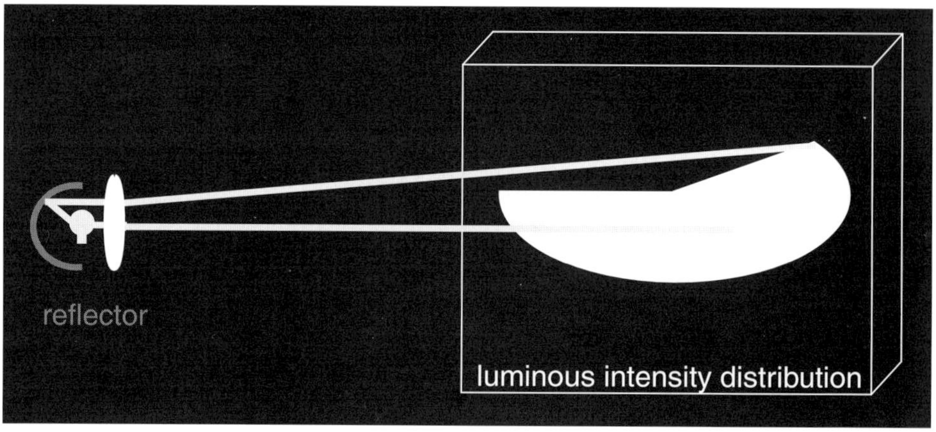

Figure 4. The working-principle of a reflector

As test examples serve reflector surfaces of the car-supplier HELLA KG Hueck & Co. These workpieces must fulfill two demands: The geometrical form must fit into the given volume, and the reflected light-rays emitted from the light source must sum up to a legislative stipulated luminous intensity distribution. That is the pattern which arises on a wall opposite a switched on headlight in a fixed distance. (The principle of a headlight is sketched in Fig. 4.)

Of course, this surface data is given as a continuous function. Thus we cannot use standard image encryption methods, which work on bitmap information. Our goal is to separate the described two aspects by a wavelet decomposition. Unfortunately there does not exist an automatic tool for measuring the quality of a luminous intensity distribution. Thus, a heuristic algorithm searches those coefficients which describe nothing but the functional aspect of the reflector, i.e. the luminous intensity distribution. The result is a semi-destroyed model of the reflector, which can be transmitted without security considerations.

Figure 6 depicts the luminous intensity distribution of Fig. 5 after the encrypting process. The function of the reflector is totally destroyed and the headlight has

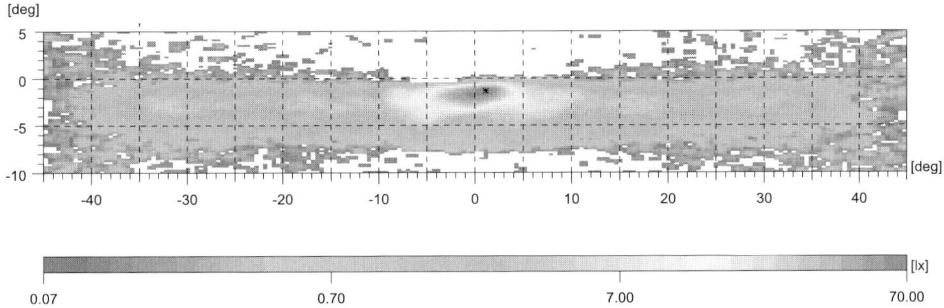

Figure 5. Original luminous intensity distribution

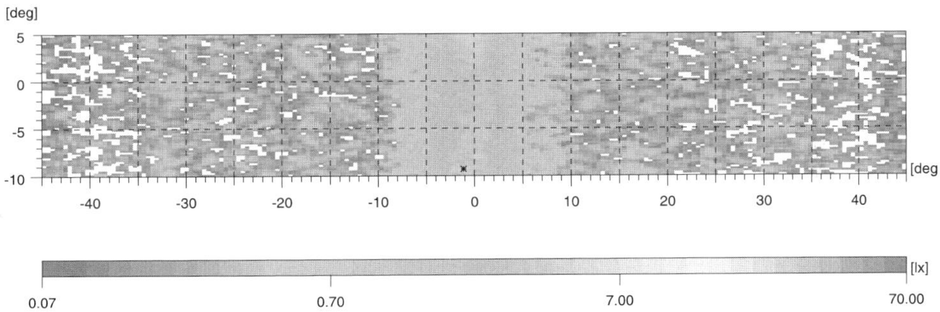

Figure 6. Encrypted distribution (change of geometry $\leq$ 0.91 mm)

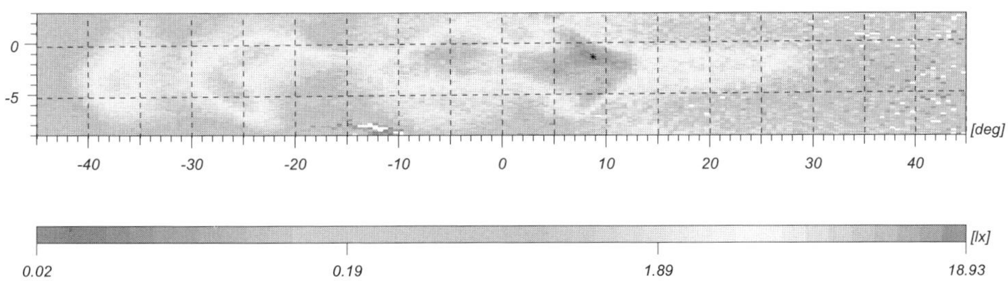

Figure 7. Encrypted distribution (change of geometry $\leq$ 1.79 mm)

been transformed into a spot. In addition the form is almost preserved: The geometries differ at most 0.91 mm! Thus, the change of the form is not visible and the modified reflector still fits into the car. In fact, this change is within the tolerance of mass production.

This example was computed with Haar-wavelets. Our algorithm applies Conclusion 1, which enables us to steer the changes of the surface as we like. Corollary 1 makes it possible to fix the boundary and other important regions of the reflector, for example some sensible connection of headlight and car-body. Thus all user-demands could be fulfilled and the algorithms is in fact used in practice. Other wavelet decompositions can be used for selective encrypting, too. Figure 7 shows a encrypted luminous intensity distribution, which was computed with semi-orthogonal B-spline-wavelets. (For the details of the computations of such B-spline-wavelets, see [3].)

Altogether, we detected some useful relationships between wavelet coefficients and the corresponding function and used them to establish a selective encrypting algorithm.

References

[1] Bartels, R., Beatty, J., Barsky, B.: A introduction to splines for use in computer graphics and geometric modeling. San Francisco: Morgan Kaufmann, 1987.
[2] Chui, C. K.: An introduction to wavelets. New York: Academic Press, 1992.
[3] Finkelstein, A., Salesin, D.: Multiresolution curves, In: Cunningham, S. (ed.): Proceedings of SIGGRAPH pp. 261–268, 1994.
[4] Louis, A. K., Maaß, P., Rieder, A.: Wavelets. Stuttgart: Teubner, 1994.
[5] Nawotki, A.: Selective crypting with haar-wavelets. In: Brunet, P., Hoffmann, C., Roller, D. (eds.) CAD-Tools and Algorithms for Product Design. Berlin Heidelberg NewYork Tokyo: Springer, 1999.
[6] Stollnitz, E. J., DeRose, T. D., Salesin, D. H.: Wavelets for computer graphics. San Francisco: Morgan Kaufmann, 1996.

A. Nawotki
Department of Computer Science
University of Kaiserslautern
P.O. Box 3049
Germany
e-mail: nawotki@informatik.uni-kl.de

Computing [Suppl] 14, 293–308 (2001)

Computing

Parametric Representation of Complex Mechanical Parts Using PDE Surface Generation

M. Robinson, M. I. G. Bloor, and **M. J. Wilson,** Leeds

Abstract

A brief description of the PDE method of surface generation is given, before looking at the way in which this method can be used to generate and parameterise a complex solid; namely an internal combustion engine piston. This paper demonstrates that because of the nature of the PDE method, the surface patches which are generated are smooth, guaranteed to meet perfectly at the boundaries of the patches, and can be constructed with tangent plane continuity at the boundaries where this is required. Furthermore, the method uses relatively few design parameters which allows us to change the shape of the object easily and opens the possibility of linking directly to numerical optimisation techniques.

AMS Subject Classifications: 65D17, 68U07, 68U05, 65D17.

Key Words: Parametric design, blending, PDE method.

1. Introduction

Conventional designs for many complex mechanical parts are based on a combination of the part's engineering requirements, the available methods for manufacturing the part, and the ability to represent the part with either traditional two-dimensional drawings or CAD packages (see [1]). In many cases, the constraints of what it is possible to 'draw' using the CAD package has precluded the use of designs which may otherwise satisfy the engineering requirements.

For example, many complex mechanical parts are built up from an intersecting series of simple geometric solids which form 'primary' surfaces, and secondary blend surfaces which form smooth transitions between the primary surfaces (see for example [2, 4]). It is not clear the extent to which this straightforward geometric design is determined by the engineering requirements, the manufacturing process, or the ability to specify a blend radius simply, on paper or using a CAD package.

There exist a variety of different methods for producing blend surfaces, many of which are summarized in the review article of Vida et al. [3]. Conceptually, perhaps the simplest method is the rolling ball blend, and work on this has long been considered in the literature; see [5, 6]. Often the primary surfaces of mechanical objects can be expressed as quadrics and a number of blending methods have been devised for just this situation, for generating both parametric blends, e.g. [10], or implicit blends, e.g. [7–9, 11].

An alternative approach to generating blends using partial differential equations has been described by [12, 13]. In essence, the problem of generating the blend is treated as a boundary value problem, where the required position and 'direction' of the primary surfaces is known on some trimlines, and the method uses these boundary conditions to generate the secondary surface.

Using this boundary value approach has certain benefits. Firstly, there are circumstances where the boundary itself must take some specified form in order to satisfy the design requirements. Secondly, even where this is not the case, working from the boundaries of the surface patches makes it easier to ensure continuity (to whatever degree is required) between surface patches. Furthermore, as will be illustrated below, it allows for the creation for a parameteric description of the whole object that includes not just the simple primary surfaces but the complex freeform blends. Thus, when the geometry is altered by changes in the values of the design parameters, the blending surfaces adjust themselves to the changes in shape whilst mainting surface continuity.

Mathematically, we can consider this as looking for a function $\mathbf{X}$ on a domain Ω with boundary $\partial\Omega$, on which boundary data is specified. Various elliptical partial differential equations could be used, although generally we have used an equation based on the biharmonic equation $\nabla^4\phi = 0$, namely

$$\left(\frac{\partial^2}{\partial u^2} + a\frac{\partial^2}{\partial v^2}\right)^2 \mathbf{X}(u, v) = 0, \tag{1}$$

where u and v are co-ordinates of a point in Ω and $\mathbf{X}$ is a mapping from that point in Ω to a point in three dimensional space. The reasons for using this equation have been described in [12, 13] but it is worth recalling them briefly here. By choosing a fourth order equation, we are able to specify both position and derivative boundary conditions which ensures tangent continuity along the edges of surface patches. The resulting solutions of this equation are smooth – which is a physical requirement, and the addition of the factor a allows us some control of the smoothing of the surface, which we consider later in this paper.

This equation requires boundary conditions on the function value and its normal parametric derivatives on the trimlines, $\partial\Omega$. By taking the function value directly from the parameterisation of the trimlines on the primary surface, and ensuring that the direction of the normal vector is equal to that on the primary surface, we ensure continuity of position and tangent plane on the trimlines.

The *magnitude* of the derivatives allows control over the speed at which the generated surface approaches the trimlines, thereby affecting the shape. The other parameter which governs the shape of the generated surface is the smoothing parameter a which controls the relative smoothing in the u and v directions. The changes in the u direction occur over a length scale $1/a$ times the length scale in the v direction, so by changing the value of a we can change the properties of the surface. This is demonstrated by the examples given by [13].

In this paper, we consider the use of PDE surfaces particularly in respect of the blends between primary surfaces, since these are critical in reducing the maximum stresses levels in a piston. However the benefits of this technique are not limited to just the generation of blend surfaces, although they have distinct benefits there. Primary surfaces can be generated from specified boundary conditions, and complex parts can be generated from a number of surface patches. Because of the boundary value approach to the problem, it is easy to ensure continuity between surface patches, and to ensure that there are no holes in the generated surface mesh. The free form surfaces which are generated are generally described by a small number of variable parameters (i.e. the smoothing parameter a and the boundary conditions $\mathbf{X}$, $\mathbf{X}_u$ and $\mathbf{X}_v$ where the subscripts u and v represent differentiation with respect to u and v respectively).

This is a crucial aspect of the approach which ensures that a parametric description of a complex shape is achieved with a low number of parameters. This is particularly important if we wish to link the design with some type of optimisation process. For example with the piston head which we shall consider later in this paper, we may wish to minimise the mass of the part (subject to certain constraints, e.g. that the part is strong enough to withstand the stresses). With conventional design methods the number of independent parameters is often so large as to make numerical optimisation techniques prohibitively expensive. Thus to make optimisation feasible, we need to limit the number of parameters. In addition, the generation of PDE surfaces is very efficient which further facilitates any optimisation process.

It is worth noting that it is possible to make the boundaries themselves parameters in any optimisation process, although in this case, to minimise the number of parameters, we have not done this.

2. Solution of PDEs

There are various ways of determining the solution of Eq. (1). In some cases where the boundary conditions can be expressed as relatively simple functions of u and v it is possible to find a closed form solution. In other cases, numerical methods are necessary.

In this paper we will restrict ourselves to considering periodic patches, i.e. where Eq. (1) has to be solved over the region $0 \leq u \leq 1$ and $0 \leq v \leq 2\pi$, in which case the boundary conditions can be expressed as

$$\mathbf{X}(0, v) = \mathbf{f}_0(v) \tag{2}$$

$$\mathbf{X}(1, v) = \mathbf{f}_1(v) \tag{3}$$

$$\mathbf{X}_u(0, v) = \mathbf{s}_0(v) \tag{4}$$

$$\mathbf{X}_u(1, v) = \mathbf{s}_1(v) \tag{5}$$

and the general solution (see [14]) is given by

$$\mathbf{X}(u,v) = \mathbf{A}_0(u) + \sum_{n=1}^{\infty}[\mathbf{A}_n(u)\cos nv + \mathbf{B}_n(u)\sin nv] \tag{6}$$

where

$$\mathbf{A}_0(u) = \mathbf{a}_{00} + \mathbf{a}_{01}u + \mathbf{a}_{02}u^2 + \mathbf{a}_{03}u^3 \tag{7}$$

$$\mathbf{A}_n(u) = \mathbf{a}_{n1}e^{anu} + \mathbf{a}_{n2}ue^{anu} + \mathbf{a}_{n3}e^{-anu} + \mathbf{a}_{n4}ue^{-anu} \tag{8}$$

$$\mathbf{B}_n(u) = \mathbf{b}_{n1}e^{anu} + \mathbf{b}_{n2}ue^{anu} + \mathbf{b}_{n3}e^{-anu} + \mathbf{b}_{n4}ue^{-anu} \tag{9}$$

and $\mathbf{a}_{n1}, \mathbf{a}_{n2}, \mathbf{a}_{n3}, \mathbf{a}_{n4}, \mathbf{b}_{n1}, \mathbf{b}_{n2}, \mathbf{b}_{n3}, \mathbf{b}_{n4}$ are vector constants, determined by the boundary conditions imposed on $u = 0$ and $u = 1$.

Where the boundary conditions can be expressed exactly in terms of a finite Fourier series, the solution given by Eq. (6) will also be finite. However, this is often not possible, in which case the solution will be the infinite series given in Eq. (6).

An efficient method for finding an approximation to $\mathbf{X}$ is given by [14] based on the sum of the first few Fourier modes and a 'remainder term', i.e.

$$\mathbf{X}(u,v) = \mathbf{A}_0(u) + \sum_{n=1}^{N}[\mathbf{A}_n(u)\cos nv + \mathbf{b}_n(u)\sin nv] + \mathbf{R}(u,v) \tag{10}$$

where $\mathbf{R}(u,v)$ is determined such that the boundary conditions are exactly satisfied by the approximation to the solution $\mathbf{X}(u,v)$, in the following way:

The function $\mathbf{R}(u,v)$ is chosen to be

$$\mathbf{R}(u,v) = \mathbf{r}_1(v)e^{\omega u} + \mathbf{r}_2(v)ue^{\omega u} + \mathbf{r}_3(v)e^{-\omega u} + \mathbf{r}_4(v)ue^{-\omega u}. \tag{11}$$

To find the coefficient functions $\mathbf{r}_1(v)$, $\mathbf{r}_2(v)$, $\mathbf{r}_3(v)$, $\mathbf{r}_4(v)$ we define a function $\mathbf{F}(u,v)$ such that

$$\mathbf{F}(u,v) = \mathbf{A}_0(u) + \sum_{n=1}^{N}[\mathbf{A}_n(u)\cos nv + \mathbf{b}_n(u)\sin nv] \tag{12}$$

and then define four functions $\mathbf{df}_0(v)$, $\mathbf{df}_1(v)$, $\mathbf{ds}_0(v)$, $\mathbf{ds}_1(v)$ which give the difference between the boundary conditions required and the ones satisfied by $\mathbf{F}(u,v)$, i.e.

$$\mathbf{df}_0 = \mathbf{f}_0(v) - \mathbf{F}(0,v) \tag{13}$$

$$\mathbf{df}_1 = \mathbf{f}_1(v) - \mathbf{F}(1,v) \tag{14}$$

$$\mathbf{ds}_0 = \mathbf{s}_0(v) - \mathbf{F}_u(0,v) \tag{15}$$

$$\mathbf{ds}_1 = \mathbf{f}_1(v) - \mathbf{F}_u(1,v). \tag{16}$$

The functions $\mathbf{r}_1(v)$, $\mathbf{r}_2(v)$, $\mathbf{r}_3(v)$, $\mathbf{r}_4(v)$, are then determined from

$$\mathbf{df}_0 = \mathbf{R}(0, v) \tag{17}$$

$$\mathbf{df}_1 = \mathbf{R}(1, v) \tag{18}$$

$$\mathbf{ds}_0 = \mathbf{R}_u(0, v) \tag{19}$$

$$\mathbf{ds}_1 = \mathbf{R}_u(1, v). \tag{20}$$

Thus the approximation to the solution of Eq. (1) satisfies the original boundary conditions exactly.

The constant ω offers a further element of control over the surface design, in that it controls the rate at which $\mathbf{R}(u, v)$ decays away from the boundaries. With the two smoothing parameters, a and ω, we are able to influence the smoothing rate for long and short length scale features independently.

The values of the vector constants $\mathbf{a}_{n1}, \mathbf{a}_{n2}, \mathbf{a}_{n3}, \mathbf{a}_{n4}, \mathbf{b}_{n1}, \mathbf{b}_{n2}, \mathbf{b}_{n3}, \mathbf{b}_{n4}$ are determined from a Fourier analysis of the boundary conditions.

This solution method is considerably faster than looking for a very accurate solution to Eq. (1) using numerical methods such as finite-element or finite difference schemes. Although we have not considered here how close the resulting approximation will be to the real solution away from the boundaries, this is not too important. What we can guarantee is that the approximation to the solution will be exact on the boundaries. (In fact, the approximation is good even away from the boundaries, even taking $N = 5$; see [14].)

3. Traditional Design of Piston

By way of example of how this method can be used, we now consider a typical design for the inside of a piston from an internal combustion engine. The traditional design from one leading piston manufacturer is constructed from a series of relatively simple geometric parts; the major components of this are given as follows:

The internal 'bowl' of the piston is constructed by rotating a given profile through 2π about the z-axis. This is intersected by two surfaces, formed by translating a second profile in the direction of the x-axis; we shall refer to these surfaces as the piston 'wall'. The walls are in turn intersected by two cylinders which form the 'boss'.

The result of this is shown in Fig. 1, where the other half of the (symmetrical) part is removed for the sake of clarity.

These basic primary surfaces then have blends added at all sharp edges, and various cuts made into them, which we are not going to consider in detail here. The blends are traditionally formed as simple 'rolling-ball' blends, albeit of variable radii.

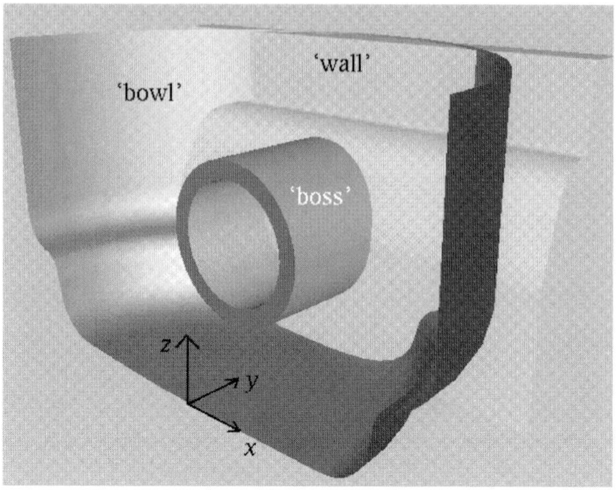

Figure 1. The basic components of the traditional model

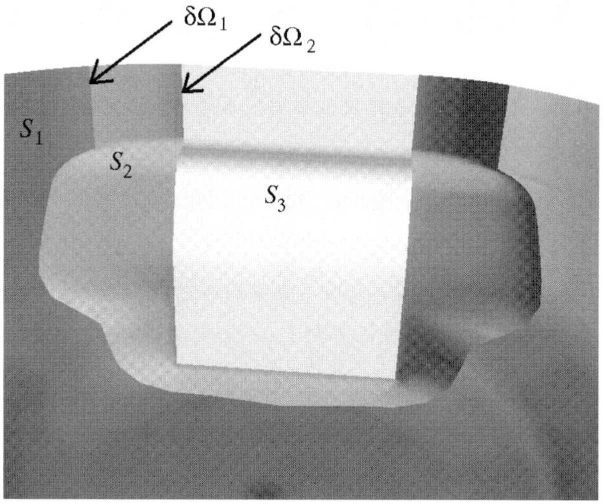

Figure 2. The model without the boss

4. A PDE Model of a Piston

We consider first a representation of the model of the interior of the piston without the bosses, as shown in Fig. 2. We will create three surfaces, labelled S_1, S_2 and S_3.

The piston bowl, which we refer to as S_1, is constructed in the same way as in the original model, by rotating a pre-determined profile about the z-axis. Future work may examine alternative ways of constructing this surface, but for the moment we are content to leave this unchanged.

Surface S_3 is also formed in a similar way to the traditional model, in that it is formed by the projection of a given profile (the same as in the traditional model) in the x-direction. However, we choose to use a smaller part of the original surface, from $y = -d$ to $y = +d$ and with $z \geq e$ where d and e are design parameters.

We now wish to find a surface, S_2 which will be the first PDE surface and which forms a smooth transition between the wall and the bowl, i.e. such that there is tangent plane continuity at the boundaries, marked $\delta\Omega_1$ and $\delta\Omega_2$ in Fig. 2.

The position of the boundary $\delta\Omega_2$ is already given from the edge of surface S_3, and the derivative boundary conditions are easy to determine. We can determine the derivative conditions on the S_3 side of the boundary directly from the numerical representation of the surface: if the surface is represented by grid points $\mathbf{x}_{i,j}$ where (i, j) are the indices of the grid points, increasing in the (u, v) parameter directions by a distance $(\delta u, \delta v)$ per grid point, then on the boundary corresponding with $u = 0$, for example, the derivative boundary conditions are given by

$$\mathbf{X}_u(0, v) = \frac{1}{\delta u}(\mathbf{x}_{2,j} - \mathbf{x}_{1,j}) \tag{21}$$

and the derivatives on the other boundaries can be determined in a similar way.

However, in this case, it is possible to express the direction of the boundary conditions analytically, namely on the vertical sections of $\delta\Omega_2$ the direction of the derivative on the S_3 side of the boundary is given by

$$\hat{\mathbf{X}}_u = \begin{pmatrix} \pm 1 \\ 0 \\ 0 \end{pmatrix} \tag{22}$$

where the choice of positive or negative depends on whether we are on the $u = 0$ or $u = 1$ boundary, and on the horizontal section,

$$\hat{\mathbf{X}}_v = \begin{pmatrix} 0 \\ 0 \\ 1 \end{pmatrix} \tag{23}$$

The easiest way to ensure that we have tangent plane continuity is to use the same direction vectors on either side of the boundary. Note, however, that there is no necessity for them to be of the same magnitude, nor for them to correspond to derivatives with respect to the same parameter. For surface S_2 we are going to take the u parameter measured from boundary $\delta\Omega_1$ towards $\delta\Omega_2$, and the v parameter to be measured along the boundaries $\delta\Omega_1$ and $\delta\Omega_2$ (suitably scaled so that v lies in the range $0 \leq v \leq 2\pi$). Thus on the S_2 side of the boundary $\delta\Omega_2$ we can take

$$\mathbf{s}_2(v) = s_{22}\begin{cases} \hat{\mathbf{X}}_u & \text{on the vertical sections} \\ \hat{\mathbf{X}}_v & \text{on the horizontal section,} \end{cases} \tag{24}$$

where $\mathbf{X}_u$ and $\mathbf{X}_v$ are those given by Eqs. (22) and (23).

The scalar s_{22} is the magnitude of the derivative vector. There is no reason why this cannot vary with v, though for the moment we shall consider the simpler case where s_{22} is taken to be a constant design parameter.

Thus we have the boundary conditions on $\delta\Omega_2$. We now turn our attention to the boundary $\delta\Omega_1$. The position of this is slightly less easy to determine; clearly it must lie on the rotated surface S_1 but the position of this is not fixed.

We might expect the boundary $\delta\Omega_1$ to be formed by the intersection of the original model's 'bowl' and 'wall'. However, since we want the surface S_2 to include the blend between the bowl and the wall, we are going to position the boundary curve slightly away from the intersection of these two surfaces. We translate the original wall surface through a small distance $(-\delta y, \delta z)$ and find the intersection of this new surface with the bowl surface to find the boundary curve $\delta\Omega_1$.

This boundary could be described in a variety of different ways, such as an isoparametric curve in the bowl surface, although in this case it is simpler to find this intersection in terms of the radius and angle of rotation as functions of the vertical height, i.e. $r(z), \phi(z)$. Again, the direction of the derivative boundary condition could be taken directly from the grid representing the surface S_1, but it is simpler in this case to express it as

$$\hat{\mathbf{X}}_u = \begin{pmatrix} -\sin\phi \\ \cos\phi \\ 0 \end{pmatrix} \tag{25}$$

and so for S_2 we choose the derivative boundary condition to be $\mathbf{s}_0 = s_{21}\hat{\mathbf{X}}(u)$ providing $\phi \neq \pi/2$ where s_{21} is the magnitude of the derivative vector for surface 2, boundary 1. Again, there is no necessity for s_{21} to be constant, though it is simpler to demonstrate the effectiveness of this method if we decide to keep it constant.

At $\phi = \pi/2$, corresponding to $x = 0$, the derivative with respect to u on surface S_1 is tangential to the boundary; at this point we need to take $\mathbf{s}_0 = s_{21}\hat{\mathbf{X}}(v)$ which we determine from the discrete representation of surface S_1 in the manner described above.

We now have boundary conditions for $\delta\Omega_1$ and $\delta\Omega_2$. However, this does not completely enclose the surface S_2; to solve Eq. (1) we need to specify boundary conditions on all sides of the surface patch. We could specify a boundary curve with derivative conditions to join together $\delta\Omega_1$ and $\delta\Omega_2$, but there is a simpler method in this case. Since we have a simple solution method for periodic patches, we shall force our non-periodic patch to be periodic in the following manner. We consider a reflection of boundaries $\delta\Omega_1$ and $\delta\Omega_2$ in the plane $z = H$, where H is the height of the part. Both the position and derivative vectors along the reflected portions of the part are easy to find, being simple mirror images of the portions already determined, and we now have a periodic patch between the extended $\delta\Omega_1$ and $\delta\Omega_2$ which we can solve using the spectral approximation outlined above.

This is done at very little extra computational cost and the result is shown in Fig. 3. It is then a simple matter to discard the upper half of this surface in constructing the piston model.

4.1. The Piston Boss

We turn our attention to the piston boss. To construct this, we need to cut a hole in the surface S_3, the parameterisation of which is shown in Fig. 4. In this case, we choose the hole such that it is the entire width of the surface S_3 and extends from the bottom of the surface to a point which is n_b grid points down from the top. We label the newly created boundary $\delta\Omega_3$.

Using the PDE method it would be possible to construct a surface to form the piston's boss directly between the enclosed curves formed by boundaries of this hole and a circular boundary marking the near point of the boss. However, the influence of the discontinuities at the corners of the newly cut hole would be felt over some portion of the boss. To avoid this, we introduce a patch which moves from the four-sided boundary (created from sections of $\delta\Omega_2$ and the line $\delta\Omega_3$) to a continuous boundary $\delta\Omega_4$ which is also on the surface S_3.

In general, if we have a four sided surface patch which can be parameterised in terms of u and v, as shown in Fig. 5, this is represented by a rectangle in (u, v) parameter space. To create a new boundary on this patch, we can choose an ellipse in (u, v) parameter space and reparameterise the surface in terms of (μ, θ),

Figure 3. The extended version of surface S_2

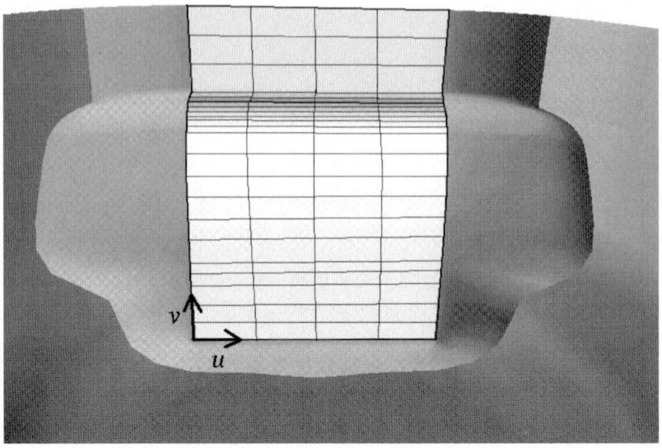

Figure 4. The parameterisation of the surface S_3

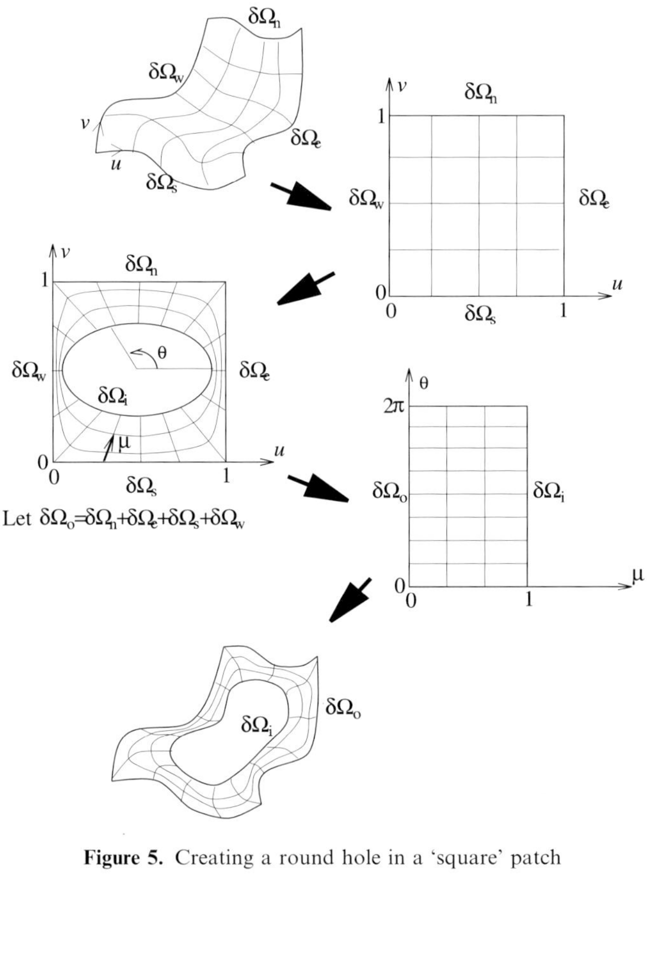

Figure 5. Creating a round hole in a 'square' patch

say, where μ goes from zero on the outer edge of the patch to unity on the inner edge, and θ goes from zero to 2π.

If the ellipse is positioned centrally on the (u, v) parameter space, as in the piston example, we introduce two new design parameters, α and β, which give half the length of the major and minor axes of the ellipse.

It is simple to find the positions of the new boundary; the derivative boundary conditions are only slightly less straightforward. Since we want the new annular surface, S_4 to be close in position to the original surface S_3 we want to choose derivatives as follows:

Consider the unit square in (u, v) space which represents the original quadrilateral patch, as shown in Fig. 5. We use a new polar co-ordinate system (r, θ) with the origin in the centre of the ellipse, at (u_0, v_0) say. The two co-ordinate systems are thus related by the equations

$$u = u_0 + r\cos\theta \quad \text{and} \quad v = v_0 + r\sin\theta. \tag{26}$$

For any given θ, we can easily calculate the value of r corresponding to the two points on the outer quadrilateral boundary and inner ellipsoidal boundary, which we shall call r_0 and r_1 respectively. In reparameterising the patch in terms of μ, θ we choose μ to vary linearly from $r = r_0$ to $r = r_1$, i.e.

$$r = (1 - \mu)r_0 + \mu r_1 \tag{27}$$

and so using the fact that

$$\frac{\partial \mathbf{X}}{\partial \mu} = \frac{\partial u}{\partial \mu}\frac{\partial \mathbf{X}}{\partial u} + \frac{\partial v}{\partial \mu}\frac{\partial \mathbf{X}}{\partial v} \tag{28}$$

we can show that

$$\mathbf{X}_\mu = (r_1 - r_0)[\mathbf{X}_u \cos\theta + \mathbf{X}_v \sin\theta] \tag{29}$$

where $\mathbf{X}_u$ and $\mathbf{X}_v$ are taken from the original, uncut version of surface S_3.

Thus we have full boundary conditions for a periodic PDE patch between the quadrilateral boundary created by sections of $\delta\Omega_2$ and $\delta\Omega_3$ and the newly created boundary $\delta\Omega_4$.

The final PDE surface is constructed between the boundary we have just created, $\delta\Omega_4$ and a curve which represents the near edge of the boss, $\delta\Omega_5$. This is a simple circle described by

$$\mathbf{f}(v) = \begin{pmatrix} r_b \cos v \\ y_b \\ z_b \sin v \end{pmatrix} \tag{30}$$

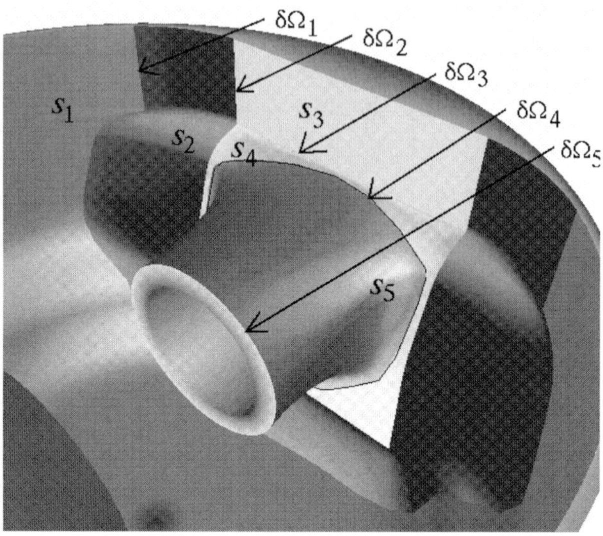

Figure 6. The boundaries of the PDE patches

where r_b is the radius of the boss and y_b and z_b are constants which determine the offset of the centre of the circle from the $y = 0$ and $z = 0$ planes respectively.

The derivative boundary conditions are given by

$$\mathbf{s}(v) = \begin{pmatrix} 0 \\ s_b \\ 0 \end{pmatrix} \tag{31}$$

where s_b is the magnitude of the derivative boundary condition. Again, there is no reason why this could not be a function of v, but in this case we have chosen it to be a constant.

Figure 7 shows the resulting PDE surfaces, along with some simple geometrically defined surfaces to give the exterior of the model, so that we have a solid part. As we have mentioned before, by using a boundary generated model, it is easy to ensure that there are no gaps in the exterior of the surface.

5. Changing the Design Parameters

By constructing the model in the manner described above, we have introduced a number of design parameters, summarised in Table 1. In addition to those listed, there are design parameters associated with the other surface of the piston head which we have not considered here, such as the inner radius of the boss and the outer radius of the piston. In addition, the method uses the given profiles of the original piston bowl and wall.

Figure 7. The PDE model

Table 1. Design parameters for the PDE model

Symbol	Description
d	horizontal cuttoff point for surface S_3
e	vertical cuttoff point for surface S_3
r_b	outer radius of the boss
s_{21}	magnitude of derivatives on $\delta\Omega_1$ for surface S_2
s_{22}	magnitude of derivatives on $\delta\Omega_2$ for surface S_2
s_p	magnitude of derivatives on $\delta\Omega_4$ for surface S_4
s_b	magnitude of derivatives on $\delta\Omega_5$ for surface S_4
y_b	horizontal offset distance of the curve $\delta\Omega_5$
z_b	vertical offset distance of the curve $\delta\Omega_5$
α	half the major axis of the ellipse in (u, v) parameter space for surface S_3
β	half the minor axis of the ellipse in (u, v) parameter space for surface S_3
δx	offset distance to find boundary $\delta\Omega_1$
δy	offset distance to find boundary $\delta\Omega_1$
a_2	smoothing factor for surface S_2
ω_2	smoothing factor for surface S_2
a_4	smoothing factor for surface S_4
ω_4	smoothing factor for surface S_4
a_5	smoothing factor for surface S_5
ω_5	smoothing factor for surface S_5

In fact, we can often choose to link the design parameters together. To illustrate the effect of this, and of varying the design parameters, let us consider some of those which alter the shape of the piston boss.

The parameters which govern the position and size of boundary curve $\delta\Omega_5$ are the outer radius of the boss, r_b; and the position of the centre of the circle, $(0, y_b, z_b)$. If we were to alter the value of the radius r_b, it is extremely likely that we would also wish to alter the parameters which affect the other end of the boss, associated with boundary curve $\delta\Omega_4$. These can be summarised as the horizontal cutoff point for the surface S_3, d, and the values of the design parameters α and β which determine the size of the ellipse in (u, v) parameter space which we cut in surface S_3 to form surface S_4. We have chosen to link these parameters in the following way:

$$r_b = \frac{2}{3}d \tag{32}$$

$$\alpha = 0.95\, u_{cut} \tag{33}$$

$$\beta = 0.95\, v_{cut} \tag{34}$$

where (u_{cut}, v_{cut}) are the values of (u, v) across the quadrilateral patch which was cut in surface S_3 to form S_4.

By choosing this relationship between the design parameters, we guarantee that the boss will be approximately the same cross sectional area along its length, rather than much thinner at one end than at the other (though clearly, we are in effect introducing new design parameters in the form of the fractions in Eqs. (32) to (34)).

In this example, we have not varied any of the other design parameters which affect the shape of the boss, notably the derivative boundary conditions.

Figures 8 to 10 show the effect of now varying d on the shape of the piston boss. By linking together the design parameters, we can see that the radius of the boss r_b varies with d. The crucial thing to note is that in varying this one parameter, we can significantly affect the shape of the piston but the solid is still generated with smooth surfaces, no holes in the surface, and with tangent plane continuity at all the boundaries.

It is worth noting here that it is possible, through altering the parameters, to produce surfaces which interpenetrate. The most straightforward way to detect

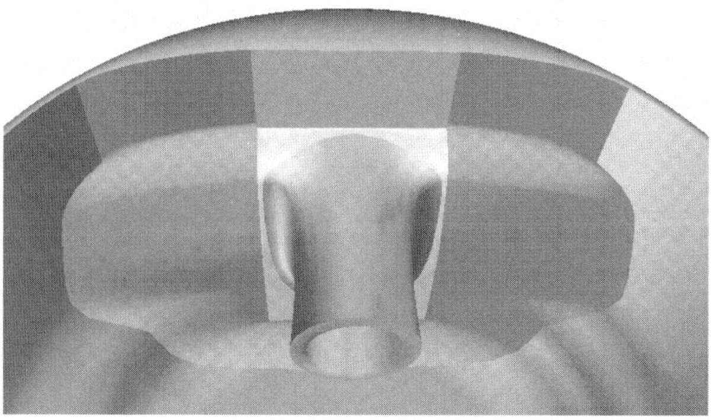

Figure 8. The boss with $d = 12$

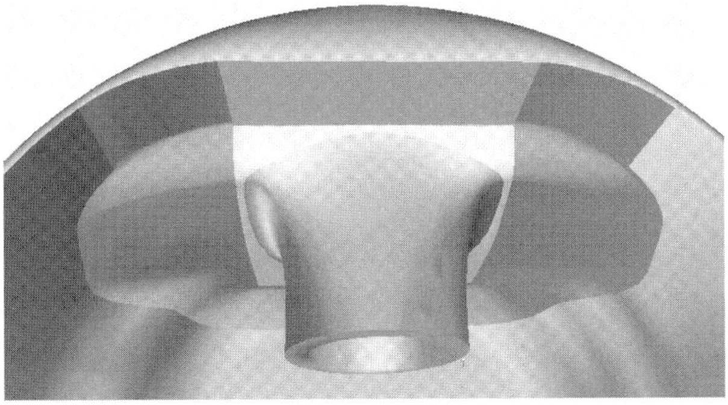

Figure 9. The boss with $d = 17$

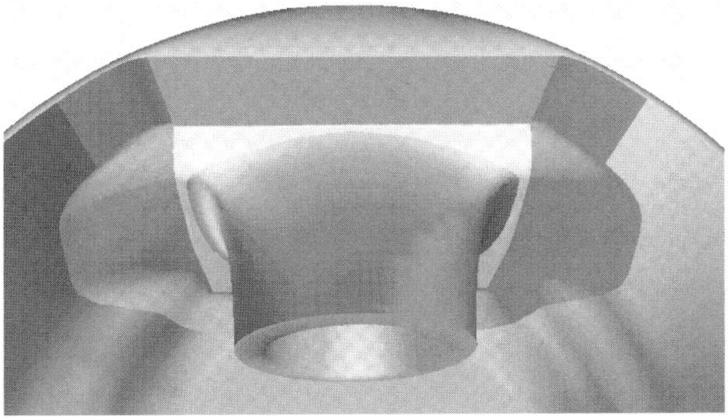

Figure 10. The boss with $d = 22$

this is visually, although reasonable estimates of the permissable upper bound on the derivative magnitudes can be obtained from the geometry of the boundary curves, and it is also possible to devise a numerical check on the surface. If the surfaces do interpenetrate, it is a simple matter to adjust the magnitude of the boundary derivatives and/or the smoothing parameter.

6. Conclusions

In this paper we have shown that the PDE method can be used to generate surface patches as part of a complex mechanical part. The surfaces are generated from the boundary conditions at the edge of the patch, which can be sometimes be expressed in simple analytical form and in other cases can be determined from the numerical representation of other primary surfaces. The choice of PDE which is solved means that we can impose both position and derivative boundary conditions around the edge of the patch, guaranteeing tangent plane continuity where we want it, and the surfaces which are generated are smooth.

The solution of the PDE is generated by creating a spectral approximation to the true solution, and this is extremely fast – practically instantaneous – to do computationally.

Defining the surfaces only in terms of the conditions at the boundary has significant advantages: in addition to the ability to specify the tangent plane at the boundary, we can easily ensure that their are no gaps in the discrete representation of the surface by using the same discretisation along the boundary curves, and crucially the number of parameters is significantly smaller than in many convention surface design methods.

By reducing the number of parameters which describe the model, we open the possibility of linking this design method directly to optimisation.

Acknowledgement

The authors would like to acknowledge the support of EPSRC Grant GR/L05730, and thank Michael Hildyard of AEG Automotive for his interest in the work.

References

[1] Mortenson, M. E.: Geometric modelling. New York: Wiley 1985.
[2] Rockwood, A. P., Owen, J. C.: Blending surfaces in solid modelling. In: Geometric modeling: algorithms and new trends (Farin, G. E., ed.), pp. 367–383. Philadelphia: SIAM 1987.
[3] Vida, J., Martin, R. R., Varady, T.: A survey of blending methods that use parametric surfaces. Comput. Aided Des. *26*, 341–365 (1994).
[4] Woodwark, J. R.: Blends in geometric modelling. In: Mathematics of Surfaces II, pp. 225–297. Oxford: OUP, 1987.
[5] Rossignac, J., Requicha, A. A. G.: Constant-radius blending in solid modelling. Comput. Mech. Eng. *3*, 65–73 (1985).
[6] Kos, G., Martin, R. R., Varady, T.: Methods to recover constant radius rolling ball blends in reverse engineering. Comp. Aided Geom. Des. *17*, 127–160 (2000).
[7] Pratt, M. J.: Cyclides in computer aided geometric design. Comp. Aided Geom. Des. *7*, 221–242 (1990).
[8] Pratt, M. J.: Cyclides in computer aided geometric design II. Comp. Aided Geom. Des. *12*, 131–152 (1995).
[9] Srinivas, Y. L., Dutta, D.: Blending and joining using cyclides. ASME Trans. J. Mech. Des. *116*, 1034–1041 (1994).
[10] Wallner, J., Pottmann, H.: Rational blending surfaces between quadrics. Comp. Aided Geom. Des. *14*, 407–419 (1997).
[11] Allen, S., Dutta, D.: Cyclides in pure blending I. Comp. Aided Geom. Des. *14*, 51–75 (1997).
[12] Bloor, M. I. G., Wilson, M. J.: Generating blend surfaces using partial differential equations. CAD *21*, 165–171 (1989).
[13] Bloor, M. I. G., Wilson, M. J.: Blend design as a boundary-value problem. In: Theory and practise of geometric modelling (Straber, W., Seidel, H.-P., eds.), pp. 221–234. Berlin Heidelberg New York Tokyo: Springer 1989.
[14] Bloor, M. I. G., Wilson, M. J.: Spectral approximations to PDE surfaces. Comput. Aided Des. *28*, 145–152 (1996).

M. Robinson,
M. I. G. Bloor
M. J. Wilson
Department of Applied Mathematics
University of Leeds
Leeds LS2 9JT, UK
e-mail: Mike@amsta.leeds.ac.uk

Computing [Suppl] 14, 309–321 (2001)

Computing

Data-Dependent Triangulation in the Plane with Adaptive Knot Placement

R. Schätzl and **H. Hagen**, Kaiserslautern, **J. F. Barnes**, Nashville, TN, and **B. Hamann** and **K. I. Joy**, Davis, CA

Abstract

In many applications one is concerned with the approximation of functions from a finite set of scattered data sites with associated function values. We describe a scheme for constructing a hierarchy of triangulations that approximates a given data set at varying levels of resolution. Intermediate triangulations can be associated with a particular level of a hierarchy by considering their approximation errors. We present a data-dependent triangulation scheme using a Sobolev norm to measure error instead of the more commonly used root-mean-square (RMS) error. Triangles are split by selecting points in a triangle, or its neighbors, that are in areas of potential discontinuities or areas of high gradients. We call such points "significant points".

AMS Subject Classifications: 65D05, 65D07, 65D15, 65D17, 68U05.

Key Words: Approximation, data-dependent triangulation, knot Selection, multiresolution, Sobolev norm, splines, triangulation.

1. Introduction

We describe a method to create piecewise linear approximations for scattered bivariate data of the form $\{(x_i, y_i, f_i)|i = 1, \ldots, N\}$. Our algorithm creates an initial triangulation of the region defined by the boundary polygon of the convex hull of the given data. Using this triangulation, a refinement process produces a sequence of piecewise linear functions that improve the approximation of the given scattered data in each step. The method can be applied to general multi-valued scattered data, defined as a set

$$\{(x_i, y_i, f_{i,1}, f_{i,2}, \ldots, f_{i,k})|i = 1, \ldots, N\}, \tag{1}$$

where multiple function values $f_{i,j}$ are associated with each site (x_i, y_i).

The input to our method is a set of error tolerances, denoted as $\varepsilon_1, \varepsilon_2, \ldots, \varepsilon_n$, each of which specifies the allowable error per triangulation level. We iteratively refine intermediate triangulations by triangle subdivision until the next error tolerance is met. Each triangulation implies a piecewise linear approximation of the given scattered data. Refinement is performed until we have n triangulations that meet the n prescribed error tolerances. These n triangulation levels define a "hierarchy", which is illustrated in Fig. 1.

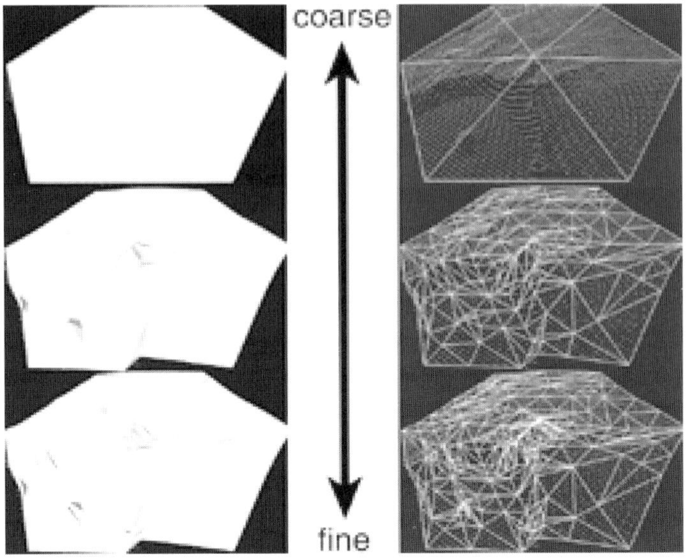

Figure 1. Hierarchy of triangulations (left: flat-shaded triangulated surface, right: triangulation)

Our method does not require connectivity information for the given sites. First, we create a coarse triangulation. This is done by calculating the boundary polygon of the convex hull of the set of all given sites in the plane and triangulating the region defined by the point subset defining the boundary polygon.

We perform triangle subdivision to improve an intermediate linear spline approximation. The triangle with the greatest local error is split into at least two and at most four subtriangles by using at most one split point per edge. This process is then iterated.

We have used different types of error metrics to determine estimates of the local error of a triangulation. The Sobolev norm, which also considers the gradient of the original data, leads to very good results. By considering the gradients, triangles containing "significant" data sites, like discontinuities or high-gradient data, have larger associated errors than triangles in relatively low-gradient areas. We do not need the gradient to be part of the given data set, as it can be approximated in a preprocessing step.

To get an approximation of the gradient, we approximate the surface at each original data site by using the original data site and its ten closest neighbors for a discrete Gaussian least-square fit.

When we perform triangle subdivision to improve an approximation, we consider two different refinement schemes, which we refer to as "Type-A" and "Type-B" refinement. Type-A refinement splits triangles by generating split points along one or all three edges of a triangle. An example of this technique is shown in Fig. 2a for three points on the edges of a triangle. When a triangle is split, so-called "implied splits" must be performed in neighboring triangles ("edge neighbors").

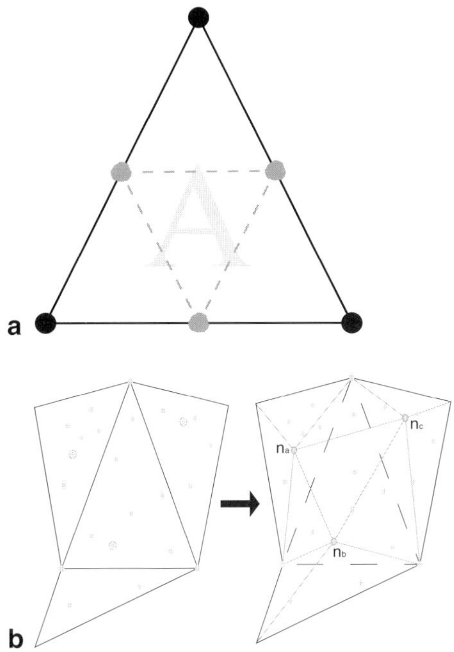

Figure 2. Two types of triangle subdivision. **a** Type-A refinement: The original black triangle is subdivided into four subtriangles. **b** Type-B refinement: The original triangle is subdivided into four triangles using existing data sites

There are some problems with Type-A refinement. These are due to the fact that Type-A refinement introduces split points lying exactly on triangle edges. As a result of this restriction, long edges in a coarse initial triangulation remain visible in all subsequent higher-resolution triangulation levels, leading to artifacts in renderings.

We address this problem by extending the Type-A refinement scheme to choose split points that are *not* necessarily located on the edges of a triangle being refined. We are identifying significant data sites lying inside the triangle or inside one of its neighbors. It is preferable to use original data sites whenever possible. We call this method Type-B refinement. An example of this technique is illustrated in Fig. 2b. Our overall refinement algorithm operates as follows:

- INPUT: N scattered bivariate data points; n error tolerances
- OUTPUT: n triangulations
- ALGORITHM:
  - Compute minimal point set defining the boundary polygon of the convex hull.
  - Compute initial data-dependent triangulation for the region defined by this point set.
  - *Refinement*. Compute n triangulations by performing the following steps:

*While error is greater than the current tolerance ε do

- Refine the triangle with the greatest error.
- Perform all possible refinements and determine how they impact the error.
- Choose a refinement that maximally decreases the error.
- Re-calculate vertex values for those vertices affected by the re-triangulation step.

2. Related Work

A data-dependent triangulation scheme adaptively generates a triangulation by considering approximation error. The techniques described in [13], [14], and [15] deal with the problem of decimating triangular surface meshes and adaptive refinement of tetrahedral volume meshes. These approaches are aimed at concentrating points in regions of high curvatures or high second derivatives. This paradigm can be used to either eliminate points in nearly linearly varying regions (*decimation*) or to insert points in highly curved regions (*refinement*). The data-dependent triangulation scheme we describe is based on the principle of refinement. Our algorithm refines a triangulation by either using existing data sites or inserting new points.

In principle, our technique is related to the idea of constructing a *multiresolution pyramid*, i.e., a data hierarchy of triangulations with increasing precision, see [10]. Figure 2 shows a multiresolution hierarchy of triangles, where the top level is a coarse triangulation, and, as we *descend* the hierarchy, finer triangulations become visible. The pyramid concept has also been extended to the adaptive construction of tetrahedral meshes for scattered scalar-valued data, see [3] and [6]. *Multiresolution methods* have been applied to polygonal (triangular) approximations of surfaces. Such approaches are described in [7], [8], and [18]. Our data-dependent technique can be viewed as a hierarchical method for representing scattered data by multiple levels of triangulations, but our approach is not based on the construction or application of orthogonal basis systems, such as *wavelet* bases.

Scarlatos and Pavlidis discuss a scheme [22] that recognizes the linear "coherence" of discontinuties. In their refinement scheme, they attempt to place a triangle edge along discontinuities in a data set. A primary difference between their work and our scheme is that we allow knots (=mesh vertices) that do not necessarily coincide with the original data sites to be introduced when there is no other option.

An alternative to constructing a triangulation hierarchy is to start with a fine mesh and decimate vertices, edges, or faces. Hoppe [16] discusses a technique for collapsing edges. In [26], an alternative scheme based on collapsing faces is discussed.

Survey papers of scattered data approximation for bivariate and trivariate data are [19], [2] and [11]. In [20], various scattered data interpolation techniques (scalar-valued, trivariate case) are discussed and compared. Our scheme relies on concepts from *geometric modeling* and *computational geometry*. These are discussed in [9] and [21].

3. Adaptive Triangle Refinement

The input to our refinement scheme is a set of points in the plane with height values. The data sites do not have to lie on a regular grid though our examples all have this property.

We start the adaptive refinement process by creating an initial coarse triangulation. We iteratively refine intermediate triangulations until a triangulation is obtained whose associated global approximation error is smaller than a prescribed tolerance. We use a Sobolev norms to measure error, see Eq. (2).

We apply triangle refinement to improve a piecewise linear approximation. In each refinement step, we identify the triangle that deviates the most from the given data and subdivide it. Refining a single triangle consists of these basic steps:

1. Identify appropriate points within the triangle and its edge neighbors. If such points exist, then use them in the refinement step.
2. If no appropriate points are found, then generate new vertices along the edges of the triangle to be split.
3. Approximate the function values for a new vertex and certain existing vertices in the neighborhood where refinement is performed.
4. Construct a new triangulation of the set of original and newly inserted vertices.
5. Compute an error estimate for the new triangulation.

These steps are iterated until a certain error tolerance is met. Our scheme is adaptive in two ways: (i) An intermediate triangulation is refined locally in regions with large errors, and (ii) the locations of vertices are chosen in order to minimize error. We analyze all possible refinements of a triangle and compare them to determine which one leads to the best fit, i.e., leading to maximal error decrease. We apply tests to guarantee that the location of vertices does not lead to overlapping triangles after subdivision.

3.1. Initial Triangulation

The initial triangulation of a given scattered data set defines the domain over which the algorithm is executed. It should consist only of existing data sites of the data set. There are different possibilities to generate valid initial triangulations. We consider the boundary polygon of the convex hull of the set of given data sites as *natural* boundary. We use *Graham's scan* algorithm [12] to compute the points defining this boundary polygon. This results in a set of points which, when triangulated, define the domain for our linear spline hierarchy.

Using the boundary polygon of the convex hull, we compute a data-dependent triangulation of the minimal point set defining the polygon. In general, one has to consider all possible triangulations of this point set and select the one that minimizes a chosen error measure. Computing all possible initial triangulations cannot be done efficiently when the boundary polygon is defined by a relatively large number of points. Therefore, we propose to construct any triangulation of the boundary points and then apply *simulated annealing* in order to obtain a better, possibly optimal, data-dependent initial triangulations, see, e.g., [17] and [23]. This is also preferable in the case of non-convex data.

Another approach proved to produce good results in certain situations: Often, it is possible to obtain a general idea of how data behave in the interior of the domain by analyzing the behavior on the boundary. Following this idea, we compute *all* the points that lie on the boundary and identify the significant points of the boundary polygon. We then use the significant points of the boundary polygon to construct the initial triangulation using the Delaunay triangulation. (Thus, we apply a data-dependent point selection step on the boundary.)

Remark. For many practical applications, it might be sufficient to simply use the four vertices defining the corners of the bounding box containing all original sites. Several real-world data sets are defined on a uniform, rectilinear grid whose convex hull coincides with its bounding box.

Another practical solution is to define the start triangulation manually. This opens the possibility to concentrate on special areas of interest.

3.2. Approximation Error Estimates

In this section, we describe the approximation error estimate that we use in our refinement scheme. We assume that the given scattered data in the plane and the vertices of all intermediate triangulation levels have the same convex hull.

It is the objective of data-dependent triangulation to refine a triangulation in "high-detail" areas by using more and, if necessary, skinnier triangles than in "low-detail" areas. In most data sets, the significant points are points close to discontinuities or points in high-gradient regions. Thus, the error norm should assign more weight to points in those regions.

The error norm S we use is derived from the Sobolev norm [1, 25]. It is defined by Eq. (2), where f denotes the original function to be approximated, L is a linear spline approximation, and $c > 0$ is a constant. The constant c can be chosen arbitrarily. After some experiments, we decided to use the area of the triangle as value for c. By considering not only the difference in function value but also in gradient value, significant areas are more readily identified and captured in the triangulations.

We compute a *local error* for each triangle and a *global error* for each triangulation: If there are m original data sites lying inside the triangle τ (including its boundary), we define the local Sobolev error E_{LSOB} as in Equation 3, where f_i is the value at a given site (x_i, y_i), f_i^x and f_i^y are the two components of the gradient at site (x_i, y_i), $L(x_i, y_i)$ is the value of the linear polynomial over the triangle containing (x_i, y_i), and τ_A is the area of the triangle containing (x_i, y_i). The global error associated with an entire triangulation is defined as the maximum of all E_{LSOB} values.

$$S = \int \|L(x,y) - f(x,y)\| dx\, dy + c \int \left\| \frac{\partial}{\partial x} L(x,y) - \frac{\partial}{\partial x} f(x,y) \right\| + \left\| \frac{\partial}{\partial y} L(x,y) - \frac{\partial}{\partial y} f(x,y) \right\| dx\, dy \tag{2}$$

$$E_{LSOB} = \sqrt{\frac{1}{m} \sum_{i=1}^{m} (L(x_i, y_i) - f_i)^2 + \tau_A \left(\left(\frac{\partial}{\partial x} L(x_i, y_i) - f_i^x \right)^2 + \left(\frac{\partial}{\partial y} L(x_i, y_i) - f_i^y \right)^2 \right)} \tag{3}$$

3.3. Refining a Triangle

When refining a triangle, we are searching for one or three points in the triangle or in its edge neighbors being close to the edges of the triangle to be split. This leads to two different split types, shown in Fig. 3. (To demonstrate the basic idea we chose split points exactly on the edges.) In each case, we determine a new split point within the part of the original data set that lies inside the triangles. If we do not find any appropriate data site, then we take the mid-point of an edge and approximate a function value for this point as described in Section 3.4.

In the first case, see Fig. 3a, we are searching among the existing data sites in triangle A and its edge neighbor B for an appropriate split point. The chosen data site has do be within a certain convex region bounded by the areas of the two triangles. We have to consider the situations shown in Fig. 4: Here, original data sites within the shaded regions cannot be used. The data sites have to lie in a region that is calculated in the following way:

1. Calculate the intersection of the two lines passing through γ and δ and through α and $\beta \rightarrow S_1 =$ intersection point.
2. Calculate the intersection point S_2 in the same way.
3. If S_1 is between α and β, then use the triangle δ, S_1, β; otherwise, use the triangle α, β, δ.

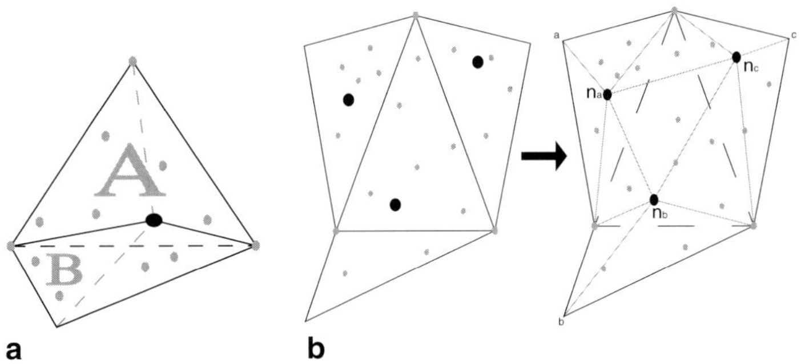

Figure 3. The two different split types. **a** Choosing one split point. **b** Choosing three split points

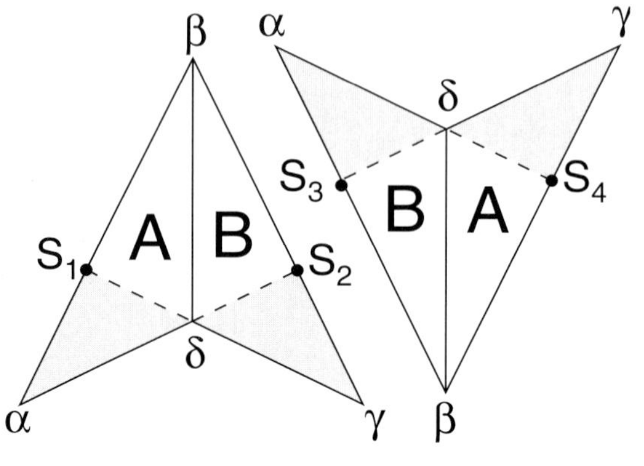

Figure 4. Generating a convex region

4. If S_2 is between β and γ, then use the triangle β, S_2, δ; otherwise, use the triangle β, γ, δ.

5. Similar calculations have to be done to obtain the points S_3 and S_4 in the symmetrical case, shown in Fig. 4 on the right-hand side.

Remark. To avoid very skinny triangles the distance between a chosen data site and the common edge of the two triangles has to be shorter than the distance (perpendicular) to any of the other edges of the triangles.

Every data site satisfying the conditions described above is investigated concerning its "significance". In our current approach, we choose the data site that is approximated worst with respect to the Sobolev norm. If there exist data sites with the same deviation, we choose the one that is closer to the midpoint of the common edge of the two triangles being split. Especially in rather linear regions data sites are chosen that are positioned more in the middle of the triangles to produce more uniform triangles. On the other hand, if there is a significant data site within these two triangles, then it is chosen. In this case, the triangle may become skinnier but more appropriate in the sense of data-dependent triangulation.

If there exists no data site satisfying these conditions, then we generate a new data site that is the midpoint of the common edge. The function value of this new data site is approximated as described in Section 3.4.

The second type of refinement chooses three points lying inside the triangle or inside one of its up to three edge neighbors. This is illustrated in Fig. 3b.

To get a correct triangulation we have to place the new points, called n_a, n_b, and n_c in Fig. 3b, so that none of the new edges intersect each other or the boundary polygon of the union of the triangle to be refined and its edge neighbors.

We determine a data site for each internal edge that has the closest perpendicular distance to the midpoint of that edge. If such a point does not exist or the data site has a smaller distance to any of the midpoints of the other internal edges, then we insert the midpoint of the edge as a new vertex.

3.4. Approximating Function Values

We approximate function values, i.e., the coefficients of our linear spline approximation, at mesh vertices using a local approximation scheme. We use a modified, localized *Shepard's method*, see [24]. We need to determine a local point set to be considered when calculating the function value at a particular vertex. The original scattered data that we use for this local approximation are the points lying within the *tile* around a particular vertex, shown in Fig. 5. The tile of a vertex is constructed by connecting the midpoints of all edges emanating from the vertex and the centroids of all triangles that share the vertex as a common vertex.

We subdivide a tile into triangles and perform an inside/outside test for this set of triangles to determine the original sites that lie inside the tile. We consider this subset of data to estimate a function value f_{app} for the central vertex v. The function value f_{app} is a weighted average defined as

$$f_{app} = \frac{\sum_{i=1}^{M} f_i/d_i^2}{\sum_{i=1}^{M} 1/d_i^2} \quad . \tag{4}$$

Here, M is the number of original sites inside the tile, f_i is the function value associated with a given site (x_i, y_i) inside the tile, and d_i^2 is the squared Euclidean distance between v and (x_i, y_i).

Whenever triangles are refined as a result of inserting additional vertices, we must estimate new function values for all vertices in the triangulation whose associated

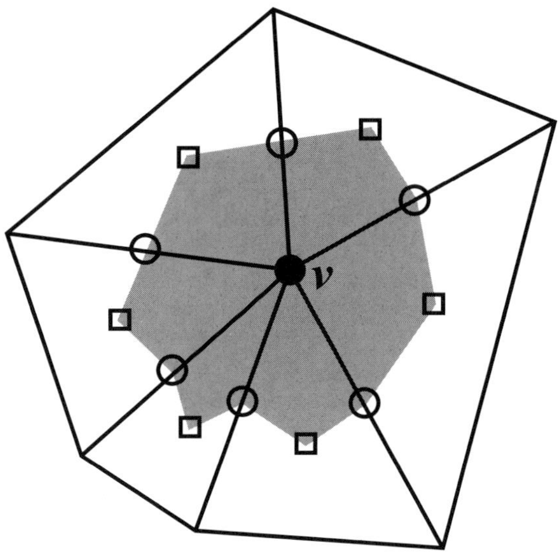

Figure 5. Construction of a tile for mesh vertex v

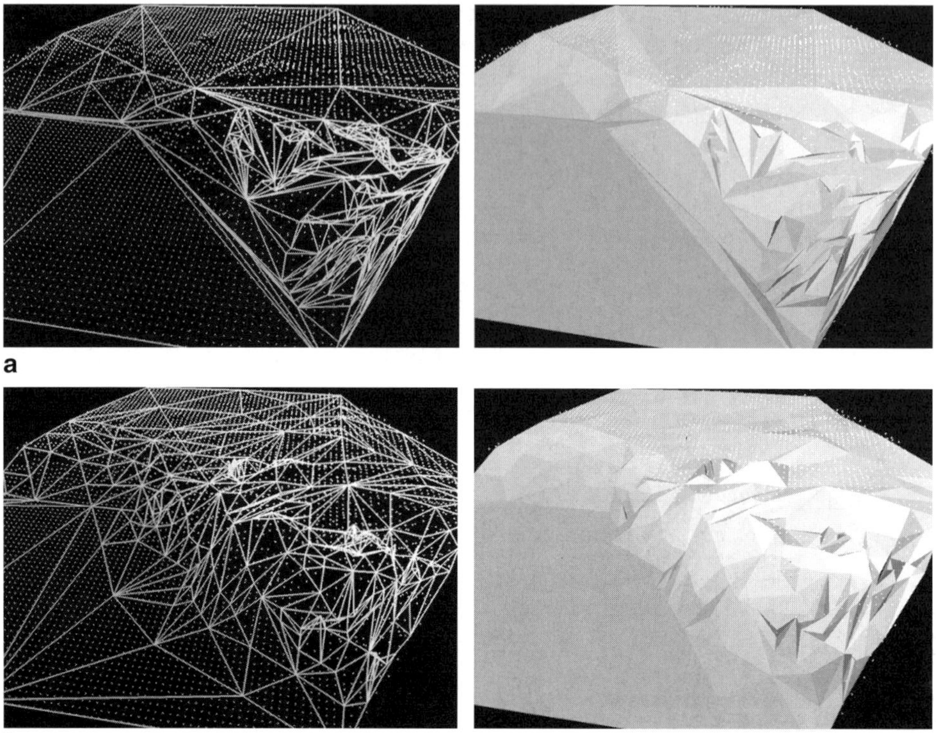

Figure 6. Lake Marquette data set (10000 sample points; 99 refinement steps). a Using RMS norm; 377 triangles. b Using Sobolev norm; 482 triangles

tiles change as a result of the refinement process. This set of vertices is given by the set of points becoming endpoints of new edges in the triangulation.

4. Results

We have applied our method to data sets with and without high-gradient regions and discontinuities. To demonstrate the usefulness of the chosen Sobolev norm we have performed refinement for the same data sets using the RMS error. We have applied our method to the following data sets:

- A discrete Mount St. Helens digital-elevation model (DEM) data set, provided on a uniform rectilinear grid, shown in Fig. 7.
- A Lake Marquette DEM, shown in Fig. 6.

As one can see in both cases, using the RMS error leads to very skinny triangles even in low-gradient regions. Most of the refinement takes place in isolated regions. On the other hand, using our Sobolev norm leads to much improved triangulations. Even smaller features in the data sets are approximated well.

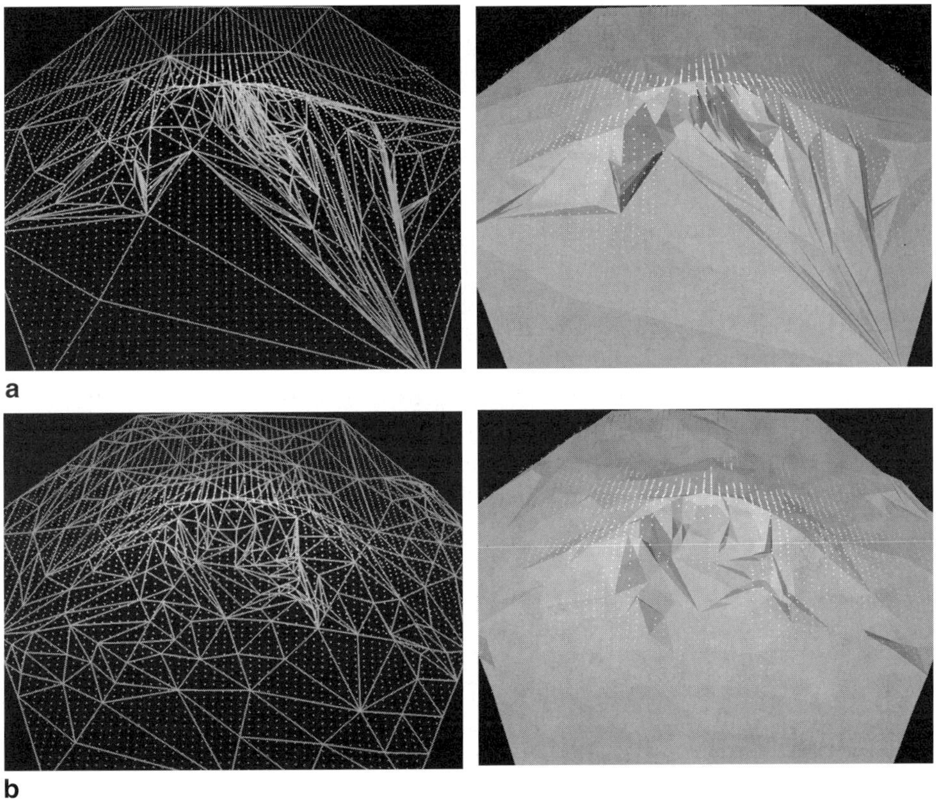

Figure 7. Mount St. Helens DEM (9396 sample points; 99 refinement steps). **a** Using RMS norm; 402 triangles. **b** Using Sobolev norm; 496 triangles

The Mount St. Helens data set demonstrates the usefulness of our approach for approximating data with narrow *cliff regions*. In this image, a drawback of using the Sobolev norm becomes apparent: The Sobolev norm tends to over-smooth the triangulation.

Considering the Lake Marquette data set, one can see how effectively our method handles data sets with high- and low-gradient regions. In the foreground of those pictures, the lake is a low-gradient region, which is approximated by a few large triangles. The fine-structured coastline is approximated by several small triangles. The higher number of triangles in the flat regions results from the use of the gradient in the error norm, as one of the edges in the initial triangulation is right on the border of the coastline.

The computational cost of our algorithm depends on the different algorithmic approaches used. The computation of the initial triangulation has a time complexity of $O\ (n \log n)$, and the gradient approximation can be done in $O\ (n \log n)$ time. The individual refinement step has to check all the original data points lying in the involved triangles, so the time complexity of each refinement step is $O\ (n)$.

How often the iteration step is executed depends on the error value given as input. As a general rule, we can assume that no more iterations should be done than there are original data sites. Thus, the overall complexity is $O(n^2)$.

5. Conclusions and Future Work

We have discussed a new technique for the construction of data-dependent triangulations for bivariate scattered data. Our scheme preserves high-gradient regions or potential discontinuities that might exist in a given data set by using the Sobolev norm. We have tested our method for various examples. We plan on introducing a quality measure that depends on the relative flatness of the region to prevent the generation of too many very skinny triangles. We are currently investigating local re-triangulations (through edge swapping) to eliminate the artifacts that currently result when using the Sobolev norm.

Acknowledgements

This work was supported by the National Science Foundation under contract ACI 9624034 (CAREER Award), through the Large Scientific and Software Data Set Visualization (LSSDSV) program under contract ACI 9982251, and through the National Partnership for Advanced Computational Infrastructure (NPACI); the Office of Naval Research under contract N00014-97-1-0222; the Army Research Office under contract ARO 36598-MA-RIP; the NASA Ames Research Center through an NRA award under contract NAG2-1216; the Lawrence Livermore National Laboratory under ASCI ASAP Level-2 Memorandum Agreement B347878 and under Memorandum Agreement B503159; and the North Atlantic Treaty Organization (NATO) under contract CRG.971628 awarded to the University of California, Davis. We also acknowledge the support of ALSTOM Schilling Robotics, and Silicon Graphics, Inc. We thank the members of the Visualization Thrust at the Center for Image Processing and Integrated Computing (CIPIC) at the University of California, Davis.

References

[1] Adams, R. A.: Sobolev spaces. New York: Academic Press 1975.

[2] Alboul, L., Kloosterman, G., Traas, C. R., van Damme, R. M. J.: Best data-dependent triangulations. Technical Report Memorandum No. 1487, University of Twente, Faculty of Mathematical Sciences, 1999.

[3] Bertolotto, M., De Floriani, L., Marzano, P.: Pyramidal simplicial complexes. In: Third Symposium on Solid Modeling and Applications (Hoffmann, C., Rossignac, J. eds.), pp. 153–162. New York: ACM Press, 1995.

[4] Bonneau, G. P.: Multiresolution analysis on irregular surface meshes. IEEE Trans. Visual. Comput. Graph. *4*, 365–378 (1998).

[5] Bonneau, G. P., Gerussi, A.: Level-of-detail visualization of scalar data sets defined on irregular surface meshes. In: Proceedings of the IEEE Visualization (Ebert, D. S., Hagen, H., Rushmeier, H. E., eds.), pp. 73–77. Los Alamitos: IEEE Computer Society Press, 1998.

[6] Cignoni, P., De Floriani, L., Montani, C., Puppo, E., Scopigno, R.: Multiresolution modeling and visualization of volume data based on simplicial complexes. In: 1994 Symposium on Volume Visualization (Kaufman, A. E., Krüger, W., eds.), pp. 19–26. Los Alamitos: IEEE Computer Society Press, 1994.

[7] DeRose, A. D., Lounsbery, M., Warren, J.: Multiresolution analysis for surfaces of arbitrary topological shape. Technical Report 93-10-05, Department of Computer Science and Engineering, University of Washington, Seattle, WA, 1993.

[8] EcK, M., DeRose, A. D., Duchamp, T., Hoppe, H., Lounsbery, M., Stuetzle, W.: Multiresolution analysis of arbitrary meshes. In: Proceedings of SIGGRAPH 1995 (Cook, R., ed.), pp. 173–182. New York: ACM Press, 1995.

[9] Farin, G.: Curves and surfaces for CAGD, 4th ed. San Diego: Academic Press, 1997.

[10] De Floriani, L.: A pyramidal data structure for triangle-based surface description. IEEE Comput. Graphics Appl. *9*, 67–78 (1989).

[11] Garland, M., Heckbert, P. S.: Fast polygonal approximation of terrains and height fields. Technical Report TR CMU-CS-95-181, Carnegie Mellon University, School of Computer Science, 1995.
[12] Graham, R. L.: An efficient algorithm for determining the convex hull of a finite planar set. Information Proc. Lett. *1*, 132–133 (1972).
[13] Hamann, B.: A data reduction scheme for triangulated surfaces. Comput. Aided Geom. Des. *11*, 197–214 (1994).
[14] Hamann, B., Chen, J. L.: Data point selection for piece-wise linear curve approximation. Comput. Aided Geom. Des. *11*, 289–301 (1994).
[15] Hamann, B., Chen, J. L.: Data point selection for piecewise trilinear approximation. Comput. Aided Geom. Des. *11*, 477–489 (1994).
[16] Hoppe, H.: Progressive meshes. In: Proceedings of SIGGRAPH 1996 (Rushmeier, H., ed.), pp. 99–108. New York: ACM Press 1996.
[17] Kreylos, O., Hamann, B.: On simulated annealing and the construction of linear spline approximations for scattered data. In: Proceedings EUROGRAPHICS-IEEE TCCG Symposium an Visualization, Data Visualization '99 (Groeller, E., Loeffelman, H., Ribarsky, W., eds.), pp. 189–198. Wien New York: Springer, 1999.
[18] Lounsberg, M.: Multiresolution analysis for surfaces of arbitrary topological shape. Dissertation, Department of Computer Science and Engineering, University of Washington, Seattle, WA, 1994.
[19] Nielson, G. M.: Scattered data modeling. IEEE Comput. Graph. *13*, 60–70 (1993).
[20] Nielson, G. M., Tvedt, J.: Comparing methods of interpolation for scattered volumetric data. In: State of the art in comput graphics (Rogers, D. F., Earnshaw, R. A., eds.), pp. 67–86. New York: Springer, 1993.
[21] Preparata, F. P., Shamos, M. I.: Computational geometry, 3rd ed., New York: Springer 1990.
[22] Scarlatos, L. L., Pavlidis, T.: Hierarchical triangulation using terrain features. In: Proceedings IEEE Conference on Visualization '90 pp. 168–175, 1990.
[23] Schumaker, L. L.: Computing optimal triangulations using simulated annealing. Computer Aided Geom. Des. *10*, 329–345 (1993).
[24] Shepard, D.: A two-dimensional interpolation function for computer mapping of irregularly spaced data. Technical Report TR-15, Harvard Univ., Center for Environmental Design Studies, Cambridge, Cambridge, MA, 1968.
[25] Sobolev, S. L.: The Schwarz algorithm in the theory of elasticity. Sokl. Acad. N. USSR, *4*, 236–238 (1936).
[26] Gieng, T. S., Hamann, B., Joy, K. I., Schussman, G. L., Trotts, I. J.: Constructing hierarchies for triangle meshes. IEEE Trans. on Visualization and Computer Graphics, *4*, 145–161 (1998).
[27] Hamann, B., Jordan, B. W., Wiley, D. A.: On a construction of a hierarchy of best linear spline approximations using repeated bisection. IEEE Trans. Visual. Comput. Graph. *5*, 30–46, 190 (errata), 1999.
[28] Trotts, I. J., Hamann, B., Joy, K. I., Wiley, D. F.: Simplification of tetrahedral meshes. In Proceedings IEEE Conference on Visualization '98 (Ebert, D. S., Hagen, H., Rushmeier, H. E., eds.), pp. 287–295. IEEE Computer Society Press, 1998.

R. Schätzl
H. Hagen
Fachbereich Informatik
Universität Kaiserslautern
D-67653 Kaiserslautern
Germany
e-mails: schaetzl@informatik.uni-kl.de
hagen@informatik.uni-kl.de

J. F. Barnes
Vanderbilt University School of Engineering
Box 1679 STA B
Nashville, TN 37235
USA
e-mail: J.Fritz.Barnes@vanderbilt.edu

B. Hamann
K. I. Joy
Center for Image Processing
and Integrated Computing
Department of Computer Science
University of California
Davis, CA 95616-8562
USA
e-mails: joy@cs.ucdavis.edu
hamann@cs.ucdavis.edu

Computing [Suppl] 14, 323–335 (2001)

Computing

Implicit Surfaces Revisited – *I*-Patches

T. Várady, **P. Benkő** **G. Kós**, Budapest, and **A. Rockwood**, Cambridge, MA

Abstract

Techniques to combine implicit surfaces have been widely used in the context of blending surfaces, but not for making *n*-sided patches. This is mainly due to the lack of proper control for the interior of complex shapes and control of separate branches. The main attraction of implicit formulations is, however, that they represent a *general paradigm* based on distance functions. This property motivates our scheme, wherein classical implicit techniques are mixed with new features. Several examples are given to prove the feasibility of *I*-patches for shape design.

AMS Subject Classifications: 68U07, 65D17.

Key Words: Computer aided design, implicit surfaces, n-sided patches.

1. Introduction

Generating smooth, connecting surfaces between given primary surfaces is one of the central problems of Computer Aided Geometric Design. A significant part of the related literature deals with connecting only two adjacent surfaces – see for example reviews on blending by [22, 24]. Another significant part of the literature investigates general *n*-sided patches – see for example the recent review of [13]. Methods vary (i) in the mathematical equations used, (ii) in the creation of boundaries for the transition surfaces (these are either explicitly specified or are byproducts of the construction applied), (iii) by the degree of smoothness, which is assured between the original and the transition surfaces and finally (iv) by the free shape parameters, with which the shape of the transition surface are controlled. In practice, smoothness means G^1 or G^2 continuity, but often approximating solutions are adequate.

The advantages and disadvantages of using implicit (algebraic) or parametric surface representations are well-known. Implicit surfaces represent half-spaces and it is trivial to decide by simple substitution whether a point lies on the surface or not. However, to generate sequences of points lying on an implicit surface can be computationally demanding and for higher degree implicit surfaces singularities and self-intersections may occur. Parametric surfaces are bounded portions; while it is simple to generate points on the surface, it is hard to decide whether a point lies off the surface or not. The control points of parametric surfaces directly determine the shape of the surface, however, the coefficients of implicit surfaces do not typically have intuitive meaning.

Current CAD/CAM systems use implicit surfaces for the common engineering surfaces, such as planes, natural quadrics and tori. Generally, the parametric representation is used to define geometrically complex free-form shapes and to approximate various transition surfaces, such as rolling ball blends.

Several implicit solutions have been published for blending two surfaces. Here the primary surfaces are given in implicit form and the blend surface is also described by an implicit equation, i.e. the surface is given as the locus of all points $\mathbf{x}$, for which $P(\mathbf{x}) = 0$. The classical concept of Liming [11] was improved and extended in many various ways, see [8, 9] and solutions by Hoffmann and Hopcroft [5, 6], Middleditch et al. [14] and Rockwood et al. [15, 16], where special combinations of the primary implicit functions lead to the final surface equation. A common feature of the above methods is that the boundaries of the blends – in other words the *trimlines*, where the original primary surfaces need to be trimmed back – are indirectly determined. If two primary surfaces $P_1 = 0$ and $P_2 = 0$ need to be blended, the trimlines will be computed as the intersection curves between the surfaces $P_1 = 0$ and $P_2 = r_2$ or $P_1 = r_1$ and $P_2 = 0$, respectively. Although advantageous in certain situations, this is obviously a strong limitation when more general boundary configurations are needed.

In another group of implicit surface methods the boundaries are explicitly given in the form of intersection curves. For each primary surface P_i there is an associated *bounding surface* B_i (or in other words a cutting surface), which locates the patch boundary on P_i. The final blend surface provides a smooth connection to the primary surfaces across these intersection curves. (The term rail curve is also frequently used.) This solution was suggested by Zhang [25], Warren [23], and later for functional splines in [3, 4, 10]. Implicit patches in Bézier form were also investigated, amongst others, by Sederberg [17] and by Bajaj and Ihm [1].

In summary, it seems that implicit methods have been successfully applied to blend two surfaces, but they have not been extensively used for generating implicit n-sided patches connecting a given closed loop of boundaries. Many of the previously mentioned methods fail, when we want to extend them for three or more surfaces, and the functional spline method also has certain practical limitations, as will be shown later. The problems are partly explained by convexity constraints and by the high degree of the algebraic surfaces obtained, which may result in undesirable singularities. The appearance of these surfaces is often unpredictable due to oscillations and branching, the latter causes unwanted folding back on the primary surface.

As indicated before, the purpose of the current paper is to bring implicit formulations back to light for defining complex n-sided surfaces. It will be shown that by improving and extending former methods, natural shapes can be generated in a relatively simple manner. Our investigations started with the analysis of parametric n-sided patches pointing out that, while the boundaries and cross-derivative functions of parametric surfaces can be defined in a straightforward manner, to connect them smoothly and specify the interior is difficult – see various solutions to fill in n-sided holes [13]. The reasons behind this are complex – the boundary functions themselves are not sufficient to determine an ideal, overall transition surface, and often additional internal structures, such as subdividing curves need to be defined – the question is how. The parameterization associated with each boundary curve also causes problems, local parameters are artificial quantities and their overall assignment is difficult.

Our current view on blending is based on the following general principle: take a primary surface and a bounded curve segment on it; the effect of this surface to the n-sided patch dominates in the vicinity of this boundary, but as we get closer to the other boundaries, it must gradually vanish. This immediately suggests the use of some 'natural' distance measure associated with each primary function in such a way that some combination leads to a good transition surface.

We describe the so-called I-patch formulation, which makes it possible to obtain smooth, user controllable shapes. Here we deal only with connecting simple implicit surfaces. However, the method is valid for any type of surface for which a good distance measure can be defined. The primary application of I-patches we anticipate is free-form shape design and/or vertex blending in solid modelers. I-patches can be sampled and approximated by standard surface representations such as NURBS, and thus converted for practical use in CAD/CAM systems.

The outline of the paper is the following. After presenting the basic formulation of I-patches, its basic features are analyzed. Next I-patches and functional splines are compared briefly. Several simple examples illustrate how I-patches work. Open questions and future research issues conclude the paper.

2. The *I*-Patch Formula

For simplicity's sake let us first investigate the three-sided I-patch. Three primary surfaces and three bounding surfaces are given, which are denoted by P_i and B_i, respectively. (Note: capital letters always denote implicit surfaces, ordinary small letters stand for constants.) The I-patch is given in the following form:

$$I = w_1P_1(B_2B_3)^2 + w_2P_2(B_3B_1)^2 + w_3P_3(B_1B_2)^2 - w_c(B_1B_2B_3)^2 = 0.$$

1. The I-patch interpolates the three boundary curves. Consider the first one, for which $P_1 = 0$ and $B_1 = 0$. Note that all four terms in the equation will be zero, consequently all points of the intersection curve of P_1 and B_1 also satisfy the I-patch equation.

2. The I-patch guarantees first order continuity to the primary surfaces. The gradient vector of the I-patch is parallel to that of the related primary surface in any point of the $P_1 \cap B_1$ boundary curve. Rewriting I as

$$I = GP_1 + HB_1^2$$

the partial derivative of I is

$$\frac{\partial I}{\partial x} = G'P_1 + GP_1' + 2HB_1'B_1 + H'B_1^2.$$

For any point of the first boundary curve, the first, third and fourth terms will fall out, and the three components of the gradient of I will be equal to those of P_1 multiplied by the scalar function G evaluated at the given point of the boundary. This fits the theory given by Warren [23].

Note: the exponent of the bounding functions is 2 in the above formulation, however, by raising it to 3 or more, it is assures higher degree continuity to the primary surfaces. Fractional degree can also be used to adjust the interior of the shape for finer control.

3. As noted earlier, the 'effect' of P_1 will disappear as we get closer to the second and third boundaries; then the first term becomes almost zero due to the fact that, the squared boundary functions B_2 and B_3 become zero, and the other remaining terms will dominate.

4. It is best to use truncated bounding surfaces B_i^+, after carefully setting their signs. In this way we define the I-patch only for points where $B(\mathbf{x}) \geq 0$ and we can get rid of various undesirable branches of the surface. Further operations, for example rendering, also become simpler.

5. For each primary function we can also assign a positive weight w_i, which makes it possible to adjust the fullness of the patch in an asymmetric way. As can be seen, there is a fourth, *correction term* added, multiplied by a scalar value w_c, which is also a free shape parameter. The correction term obviously interpolates the three boundary curves. It can be used to prevent the I-patch from passing through the intersection point of P_1, P_2 and P_3, which is undesirable in certain situations. It also makes it possible to control the interior of the patch.

There are two ways of interactively setting the above shape parameters. Either the user explicitly sets the weights w_i and w_c, or he defines a characteristic point Q to be interpolated by the patch. The individual weights can be all set to 1 or to arbitrary positive values. In both cases, after substituting the Q_x, Q_y, Q_z coordinates into the equation of the I-patch, w_c can be expressed directly.

6. One of the crucial issues with implicit surfaces is the distance measure. In former approaches the composite surfaces were thought to need a low algebraic degree, this is why mostly the algebraic distance, obtained by substitution, was used. For example, Hoffmann and Hopcroft in [5] created quartic blends between quadric surfaces. For I-patches, unconsidered algebraic distances will often lead to unacceptable shapes. Since we consider the I-patches not as a final CAD representation, but rather as a procedural representation, we can apply different distance measures, which assure more natural transitions.

A well-known way of normalising distances is to divide by the absolute value of the gradient of the surface equation, see for example [18] amongst others. If in the equation of the I-patch we use $P_i^N = P_i/|\nabla P_i|$ instead of the original P_i, well-controlled shapes result. Of course, special care is required to avoid singularities. This normalization gives a very good approximation of the Euclidean distance close to the surface, a first order approximation. If we have a polynomial function of degree n, the distance will vary close to linear, as the ratio of a degree n polynomial divided by a degree $n-1$ polynomial.

For planes, natural quadrics and tori, it is straightforward to compute the exact Euclidean distance instead of the algebraic or normalised algebraic distance. For example, in the case of a cylinder, instead of the original $P_i = x^2 + y^2 - r^2$, it is much better to use $P^E = \sqrt{x^2 + y^2} - r$. Here the signs of the terms must be considered carefully, depending on the location of the boundary segments.

Note, distance measure can be associated with parametric surfaces as well, see for example, the solution suggested in [19]. This demonstrates that implicit techniques, and the I-patch formulation are not restricted to algebraic primary surfaces, but may include parametrics as well.

7. If two of the bounding functions happen to be identical, the patch equation degenerates to zero. To avoid this, assume that there are n primary functions denoted by $P_1, \ldots, P_n$ and m different bounding functions $B_1, \ldots, B_m$, where $n \geq m$. Define an index function $\beta(i)$, which selects the index of the corresponding bounding function for P_i. Then the general equation of the I-patch can be given as follows:

$$I = \sum_{i=1}^{n} w_i P_i^X \frac{\prod_{j=1}^{m} \left(B_j^X\right)^d}{\left(B_{\beta(i)}^X\right)^d} - w_c \prod_{j=1}^{m} \left(B_j^X\right)^d.$$

Superscript X indicates that one should use not only the algebraic distances, but the normalised N or the Euclidean E distances, as explained in point 6. The quantity d denotes the degree of continuity $+1$, i.e. for G^1 it is 2, for G^2 3, as noted in point 2. The use of truncated bounding functions is also recommended (see paragraph 4).

3. Evaluation

Assuming that the bounding functions and the weights are properly chosen I-patches represent a special surface class, for which well-behaved transition surfaces can be generated. It is akin to functional splines (see [3, 4, 10]), given as

$$F = (1-\lambda)\prod_{i=1}^{n} P_i + \lambda \prod_{i=1}^{m} B_i^2, \lambda \in [0, 1]$$

In fact, more sophisticated distance functions help to improve functional splines. At the same time, it was found that several shapes, which were defined by I-patches, cannot be realized by functional splines. This is explained by various

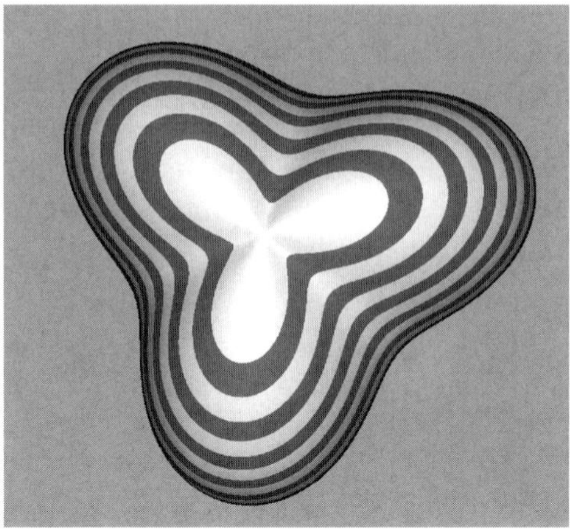

Figure 1. One-sided I-patch

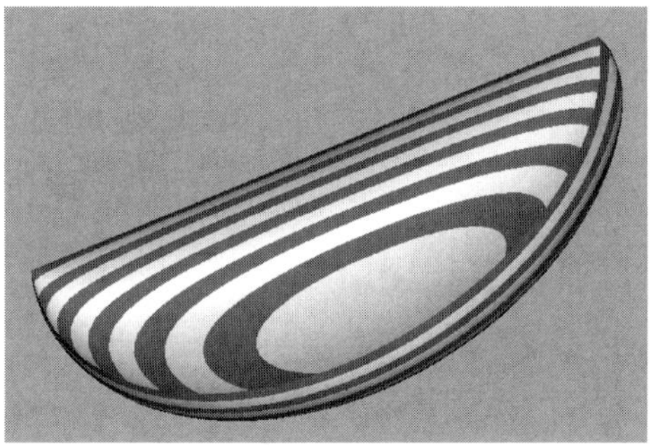

Figure 2. Two-sided I-patch

convexity constraints and a missing feature, which I-patches have: the individual terms of the primary surfaces are separated. While the three-sided I-patch will interpolate the $P_1 \cap B_1$ curve, but not the $P_1 \cap B_2$ and $P_1 \cap B_3$ curves, functional splines will interpolate the latter two as well, which is undesirable in many cases. Another advantage of I-patches is that it is possible to assign fullness weights to the individual components.

To compare I-patches and 'genuine' n-sided parametric patches such as the approaches in [2, 12, 20] is quite difficult – see the review in [13]. Here a few remarks follows related to 'composite' n-sided patches, which are created as a collection of four-sided patches. Boundedness and the control point representation are attractive features from geometric point of view, but for the definition of these types of parametric patches, it is necessary to define a proper midpoint and appropriate subdividing curves, which connect the midpoints of the boundaries and the midpoint of the surface. Moreover, for internal smoothness several constraints need to be added such as compatibility of twists. In the case of I-patches, the interior is wholly defined by a single formula, no need for extra terms and internally the patch is infinitely smooth. To assure G^2 or higher degree

Figure 3. *I*-patch representing a suitcase corner

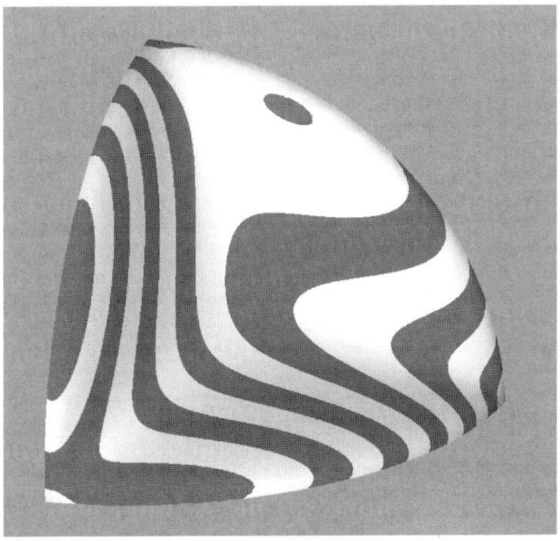

Figure 4. Suitcase corner with modified midpoint

continuity to the primary surfaces is also easy, unlike parametric constructions. Finally, standard polynomial patches cannot handle incompatible cross-derivative functions. Though Gregory twists [8] can overcome this situation, they are described by much more complex equations with parametric singularities, and they are not standard. *I*-patches can handle certain singularities. For example, they can connect two faces with different normal vectors at a common corner point – see, for example, Figs. 5 and 6 later.

4. Implementation and Examples

To make experiments with *I*-patches, an interactive test program was developed (LINUX, C++, VTK graphic package). This helped to test various distance

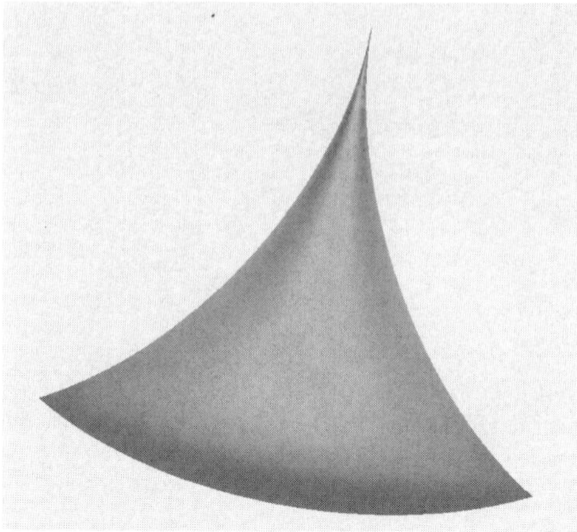

Figure 5. Triangular I-patch with one singularity

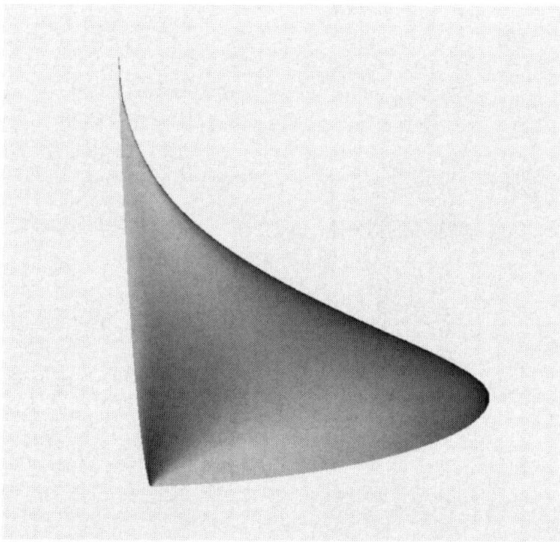

Figure 6. Triangular I-patch with three singularities

functions, to assign various weights to the primary surfaces and to make comparisons between the I-patches and the functional splines. To render I-patches is not an easy task. The following pictures were rendered by a special 'moving front' triangulator, which adaptively marches from the outside loop of the patch boundaries inwards until the whole area is evenly covered by triangles – Figs. 11, 12 and 13.

Example 1: a one-sided patch. It is quite straightforward to formulate a one-sided patch using the I-patch scheme:

$$w_1 P_1 - w_c B_1^2 = 0$$

For example, the smooth termination of a closed, translational object, such as a bar defined by sweeping an implicit profile, is shown in Fig. 1.

Figure 7. Four-sided I-patch – default fullness

Figure 8. Four-sided I-patch – fullness adjusted I

Figure 9. Four-sided I-patch – fullness adjusted II

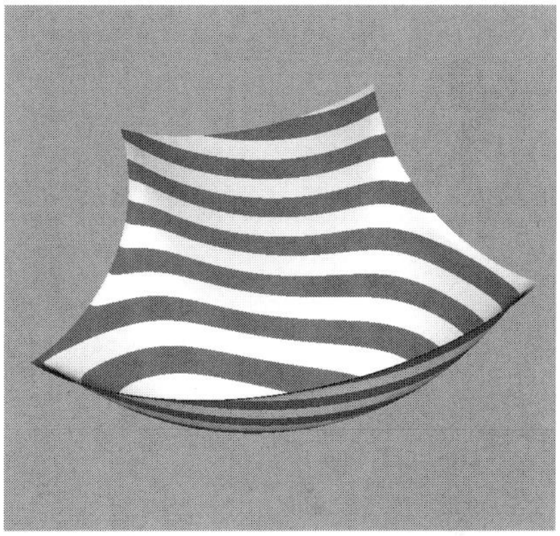

Figure 10. Four-sided I-patch – fullness locally adjusted III

Example 2: a two-sided patch. To formulate two-sided patches is also straightforward. A half-cylinder and a plane is terminated by the patch shown in Fig. 2. At the corners there are singular points.

Example 3: the suitcase corner. The classical suitcase corner configuration is shown in Fig. 3. If required, the interior of the patch can be adjusted by specifying an internal surface point. A particular example is shown in Fig. 4.

Example 4: three-sided singular cases. As explained before, singularities may occur at the corner points. For example, the connecting surface between two horizontal quarter cylinders lying on the $z = 0$ plane will have contradicting cross derivative functions at the point $(0, 0, 1)$. The patch in Fig. 5 illustrates that this sort of singularity does not destroy the shape of the patch; a natural transition is created.

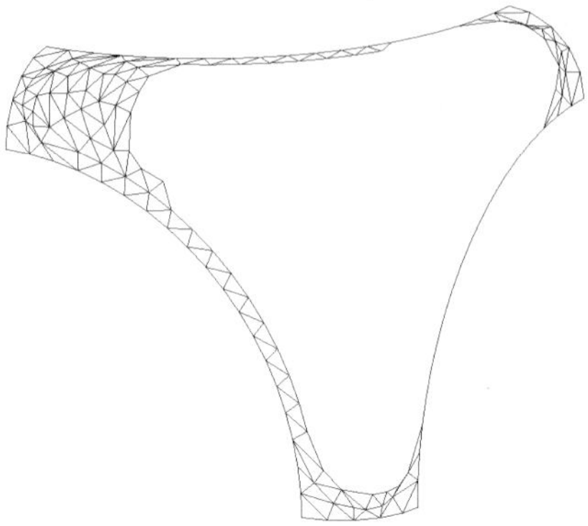

Figure 11. Growing triangulation/1

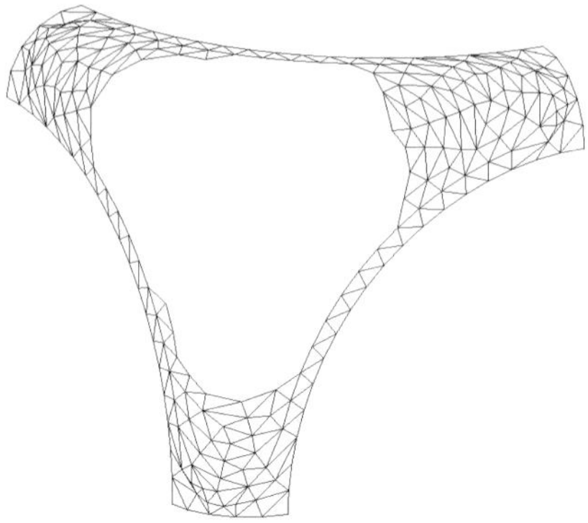

Figure 12. Growing triangulation/2

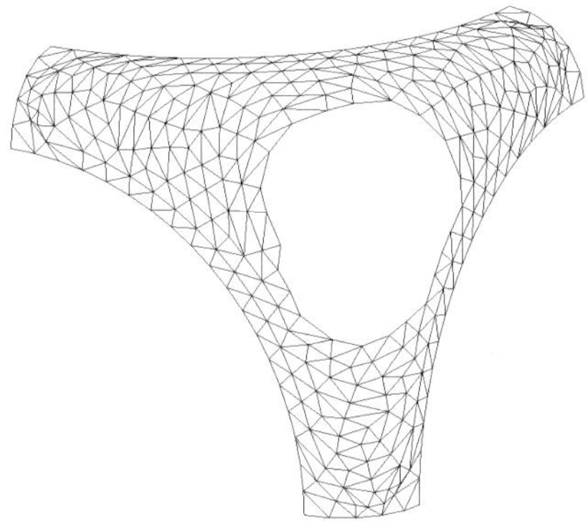

Figure 13. Growing triangulation/3

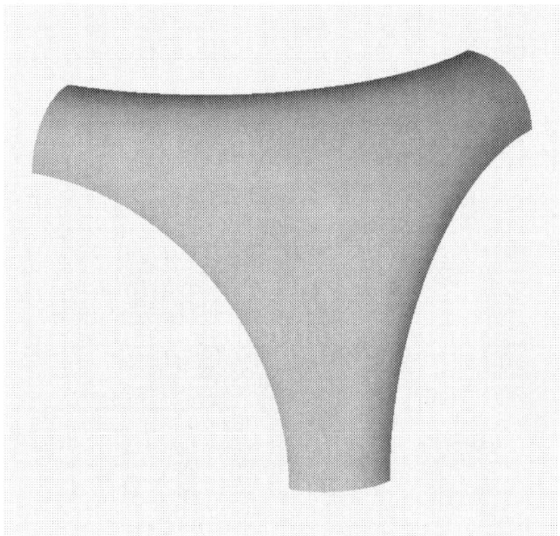

Figure 14. Setback-type vertex blend

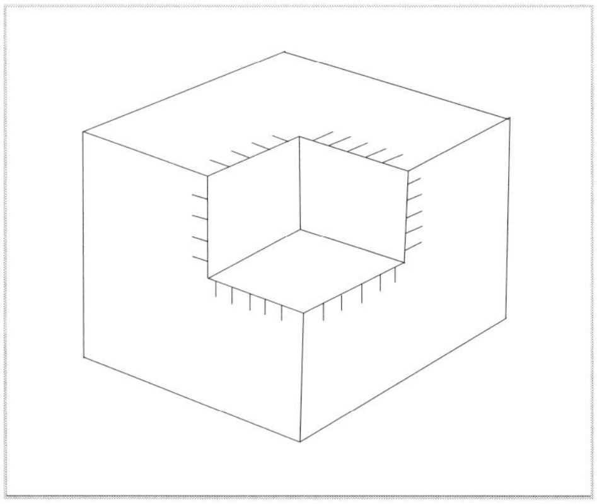

Figure 15. Six sided face using two-cubes

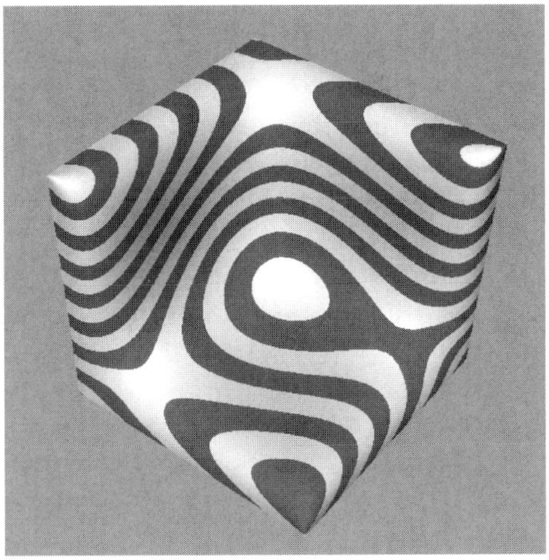

Figure 16. Six-sided *I*-patch with slicing, midpoint = (0.3, 0.3, 0.3)

Figure 17. Six-sided *I*-patch with slicing, midpoint = (0.7, 0.7, 0.7)

In Fig. 6, in addition to the two horizontal cylinders, not the $z = 0$ plane, but a third vertical cylinder represents the third primary surface. All three corners are singular, but the *I*-patch created represents a natural transition.

Example 5: a torus like shape. Figure 7 illustrates a torus like shape created by connecting two small horizontal cylinders, one larger vertical cylinder and a plane for the bottom face. The *I*-patch joins the primary surfaces smoothly and approximates the mathematical torus.

Example 6: adjusting fullness locally. It may be necessary to assign different weights to individual surface components. To illustrate this, the previous piece of torus is taken with weights assigned 1(left cylinder):1(plane):1(right cylinder):1(vertical cylinder), see Fig. 7. In the next three figures exaggerated weights were applied. A large weight was assigned to the left and right cylindrical surfaces in Fig. 8 – (20:1:20:1). A large weight was assigned to the planar surface in Fig. 9 – (1:10:1:1). Finally, a large weight was assigned to the vertical cylinder, resulting in a strange shape in Fig. 10 – (1:1:1:25).

Example 7: setback vertex blending. *I*-patches are well suited to generate setback type vertex blends (e.g. [21]). Figure 14 shows three mutually orthogonal cylindrical edges, which are connected by a six-sided *I*-patch.

Example 8: Six-sided *I*-patches. Imagine that a unit cube is subtracted from one twice as large. The closest corner of the small cube is identical to the closest corner of the large cube, all faces set parallel. The missing cube represents a six-sided face set within the large cube, which is smoothly interpolated by *I*-patches (see Fig. 15). The *I*-patch is everywhere tangential to the L-shaped faces of the large cube. In Figs. 16 and 17 the midpoints were chosen in a different way.

5. Conclusion

The basic concepts of the *I*-patch occur previously in various contexts. Our form of implicit patches, however, have not been described and demonstrated earlier, perhaps due to the percieved difficulties of higher degree implicit functions, which may have deterred other authors. Our salient contribution is that by modifying the former implicit formulations – non-algebraic distance functions, weights, correction term, truncation – it has been shown, that implicit techniques can be used intuitively for complex free-form shape definition. We are at the beginning of this research and there are many open questions. These include a thorough analysis of the shapes obtained, how to more fully avoid self-intersections and undesirable branching and how to set the most appropriate bounding functions, which obviously influence the actual shape. The automatic setting of the scalar weights also requires further analysis.

The *I*-patch approach invites us to rethink methods for generating transition surfaces. The results we have obtained indicate considerable promise in this invitation.

Acknowledgement

This research was supported by the US-Hungarian Joint Science and Technology Fund, No. 396 and by the National Science Foundation of the Hungarian Academy of Sciences (OTKA 26203).

References

[1] Bajaj, C., Ihm, I.: C^1 Smoothing of polyhedra with implicit algebraic splines. Comput. Graphics *11*, 61–91 (1992).

[2] Charrot, P., Gregory, J. A.: A pentagonal surface patch for computer aided design. Comput. Aided Geom. Des. *1*, 87–94 (1984).
[3] Hartmann, E.: Blending implicit surfaces with functional splines. Comput. Aided Des. *22*, 500–506 (1990).
[4] Hartmann, E.: On the convexity of functional splines. Comput. Aided Geom. Des. *10*, 127–142 (1993).
[5] Hoffmann, C. M., Hopcroft, J.: Quadratic blending surfaces. Comput. Aided Des. *18*, 301–306 (1986).
[6] Hoffmann, C. M., Hopcroft, J.: The potential method for blending surfaces and corners. In: Geometric modelling, algorithms and new trends (Farin, G., ed.), pp. 347–365. Philadelphia: SIAM, 1987.
[7] Holmström, L.: Piecewise quadratic blending of implicitly defined surfaces. Comput. Aided Geom. Des. *4*, 171–189 (1987).
[8] Hoschek, J., Lasser, D.: Fundamentals of computer aided geometric design. Wellesley: A. K. Peters, 1993.
[9] Bloomenthal, J. (ed.): Introduction to implicit surfaces. San Francisco: Morgan Kaufman, 1997.
[10] Li, J., Hoschek, J., Hartmann, E.: G^1 functional splines for interpolation and approximation of curves, surfaces and solids. Comput. Aided Geom. Des. *7*, 209–220 (1990).
[11] Liming, R. A.: Practical analytical geometry with applications to aircraft. New York: Macmillan, 1944.
[12] Loop, C., DeRose, T. D.: Generalised B-spline surfaces of arbitrary topological type. SIGGRAPH'90, 347–356 (1990).
[13] Malraison, P.: A bibliography for *n*-sided surfaces. In: The mathematics of surfaces VIII (Cripps, R., ed.), pp. 419–430. Information Geometers, 1998.
[14] Middleditch, A. E., Sears, K. H.: Blend surfaces for set theoretic volume modelling systems. SIGGRAPH'85. Comput. Graphics *19*, 161–170 (1985).
[15] Rockwood, A. P., Owen, J.: Blending surfaces in solid modelling. In: Geometric modelling, algorithms and new trends (Farin, G., ed.), pp. 367–384. Philadelphia: SIAM, 1987.
[16] Rockwood, A. P.: The displacement method for implicit blending surfaces in solid models. ACM Trans. Graphics *8*, 279–297 (1989).
[17] Sederberg, T.: Piecewise algebraic surface patches. Comput. Aided Geom. Des. *2*, 53–59 (1985).
[18] Taubin, G.: Estimation of planar curves, surfaces and nonplanar space curves defined by implicit equations with applications to edge and range image segmentation. IEEE PAMI *13*, 1115–1138 (1991).
[19] Vaishnav, H., Rockwood, A. P.: Blending parametric objects by implicit techniques. In: 2nd Symposium on Solid Modeling and Applications (Rossignac, J., Turner, J., Allen, G., eds.), pp. 165–168. ACM SIGGRAPH (1993).
[20] Várady, T.: Overlap patches: a new scheme for interpolating curve networks with *n*-sided regions. Comput. Aided Geom. Des. *1*, 7–27 (1991).
[21] Várady, T., Rockwood, A.: A geometric construction for setback vertex blending. Comput. Aided Des. *29*, 413–425 (1997).
[22] Vida, J., Martin, R. R., Várady, T.: A survey of blending methods that use parametric patches. Comput. Aided Des. *26*, 341–365 (1994).
[23] Warren, J.: Blending algebraic surfaces. ACM Trans. Graphics *8*, 263–278 (1989).
[24] Woodwark, J. R.: Blends in geometric modelling. In: The mathematics of surfaces II (Martin, R. R., ed.), pp. 255–297. Oxford University Press: OUP, 1987.
[25] Zhang, D.: CSG Solid modelling and automatic NC machining of blend surfaces. PhD Dissertation, University of Bath, 1986.

T. Várady
P. Benkő
G. Kós
Computer and Automation Research Institute
Hungarian Academy of Sciences
Budapest, Hungary
e-mail: varady@sztaki.hu

A. Rockwood
Mitsubishi Electric Research Labs
Cambridge, MA
e-mail: rockwood@merl.com

Computing [Suppl] 14, 337–351 (2001)

Computing

Radial Basis Functions, Discrete Differences, and Bell-Shaped Bases

J. Warren and **H. Weimer**, Houston, TX

Abstract

In this paper, we introduce the notion of a normalized radial basis function. In the univariate case, taking these basis functions in combinations determined by certain discrete differences leads to the B-spline basis. In the bivariate case, these combinations lead to a generalization of the B-spline basis to the surface case. Subdivision rules for the resulting basis functions can easily be derived.

AMS Subject Classifications: 65D07, 65D17, 15A90, 39A12, 68R99.

Key Words: Splines, stable basis, radial basis, modeling, subdivision.

1. Polynomial Splines

In the early days of engineering design, before the advent of computer aided tools, designers used to draft smooth curves using a simple yet efficient device. A thin strip of metal or wood, called a spline, was attached to the drafting board using pegs. The designer then allowed the strip to slide freely along the pegs into a relaxed configuration. Once the spline had settled, the designer simply followed the shape of the spline with a pen to draw a smooth curve that goes through the points fixed by the pegs.

Looking at the spline more closely, we observe that its use actually invokes a simple form of energy minimization. Allowing the spline to relax while still passing through the fixed pegs yields a shape that has a minimal bending energy configuration. The spline slides into a minimally bending shape – which naturally leads to a smooth curve.

In fact, we notice that the pegs are quite crucial for the spline to be useful at all. Allowing the tool to achieve its relaxed configuration without attaching it to the drafting table at some number of points simply straightens out the shape. As a result, all curves drawn using a spline without pegs are straight. Splines provide the basis for most of the computer aided modeling tools used in practice today.

Mathematically, a spline is described using a function $p[x]$ in one parameter x. The values $p[x]$ simply trace out the shape of the spline as we vary the parameter x.

Requiring the spline to pass through some number of pegs on the drafting table can be captured very concisely. We simply use a set of points p to represent the location of the pegs, providing one entry per attachment point on the drafting table. For the spline to pass through the pegs we have to require that the mathematical model $p[x]$ passes through the points p.

One more difficulty remains to be addressed: We have to find the actual parameter values x, called knots, for which the function p should pass through the respective points in p. A very pragmatic solution is to simply use the integers starting from zero, requiring $p[x]$ to pass through the ith entry of p as $x = \mathrm{i}$,

$$p[\mathrm{i}] = (\mathrm{p})_{\mathrm{i}}. \tag{1}$$

Our next task is to capture the energy optimality of the spline that was achieved by allowing the physical tool to slide along the pegs into a relaxed configuration. The first derivative of the function, $p^{(1)}[x]$, represents the tangent of the curve p at parameter x. The second derivative of the function, $p^{(2)}[x]$, measures how much the tangents of p change at x. In other words $p^{(2)}[x]$ measures how much p bends at x.

Thus, to model the effect of allowing the spline to settle into its minimum energy configuration, the function $p[x]$ is determined such that

$$\varepsilon[p] = \int_0^{\mathrm{n}} (p^{(2)}[x])^2 dx \tag{2}$$

is minimal (while $p[x]$ passes through the prescribed points p according to Eq. (1)). Functions that minimize the functional ε from Eq. (2) while satisfying relation 1 are called natural cubic splines.

In effect, the functional $\varepsilon[p]$ measures the total bending of the function $p[x]$ on the parameter interval $[0, \mathrm{n}]$. ε acts by taking the second derivative of the function $p[x]$, squaring it to yield a positive number, and then integrating to obtain a single scalar value that concisely and quantitatively characterizes the shape of $p[x]$.

The cubic B-spline basis is a particularly interesting basis for solutions to this problem. B-spline basis functions $\mathscr{F}$ of degree m satisfy the differential equation

$$\Delta^{\mathrm{m}}[\mathscr{F}[x]] = 0$$

everywhere except at the integer knots. Here Δ denotes the second derivative operator. For a more detailed introduction to this topic see [3], pp. 75.

In the first half of this paper, we show that two particularly important bases for these functions, the radial basis and the B-spline basis, are intimately related. In the second half of this paper we extend our derivations to the surface case yielding a new and interesting characterization of an important class of minimal energy surfaces.

1.1. Discrete Differences

The key to linking the radial and B-spline bases is a discrete version of the differential operator Δ^m. The discrete version of this operator is a sequence of coefficients that approximates the action of Δ^m at the integer knots.

Functional analysis defines the second derivative $\Delta p[x]$ of a function $p[x]$ as the limit

$$\Delta p[x] = \lim_{t \to 0} \frac{p[x-t] - 2p[x] + p[x+t]}{t^2}.$$

Thus, due to the definition of the derivative, any possible sequence of values for t is guaranteed to converge to the second derivative of $p[x]$, as long as we can guarantee that $t \to 0$. Consequently, we can pick a particularly nice sequence of values for t. Substituting $t = \frac{1}{2^k}$ leads to

$$\Delta p[x] = \lim_{k \to \infty} 4^k \left(p\left[x - \frac{1}{2^k}\right] - 2p[x] + p\left[x + \frac{1}{2^k}\right] \right). \tag{3}$$

Therefore, in terms of generating functions, the coefficients of the approximation of Δ^m are simply the coefficients of the Laurent polynomial $d[x]$ defined by

$$d[x] = \left(\frac{1-x}{x^{1/2}} \right)^{2m}.$$

Here the factor $x^{1/2}$ simply centers the coefficients around the origin. For example, if $m = 2$, then the discrete difference operator is $(1, -4, 6, -4, 1)$ with the coefficient 6 being associated with x^0. As a shorthand, we denote the coefficient of $d[x]$ associated with x^i by d[i]. Similarly, we denote the coefficient of $d[x^2]$ associated with x^i by dd[i].

1.2. Normalized Radial Basis Functions

One approach to generating polynomial splines is to express them as a linear combination of radial basis functions. In the univariate case, the radial basis functions are integer translates of a single fundamental radial function $\psi[x]$

$$\psi[x] = \frac{|x|^{2m-1}}{2(2m-1)!}$$

where $|x|$ denotes the absolute value of x. Note that $\Delta^m \psi[x]$ is zero everywhere except at the origin. At the origin, $\Delta^m \psi[x]$ is a delta function. The main point of this definition is the choice of the normalizing constant $\frac{1}{2(2m-1)!}$. This constant forces the integral

$$\int_{-\infty}^{\infty} \Delta^m \psi[x]\, dx$$

to be exactly one. To compute this integral, we observe that $\Delta^{\mathrm{m}}\psi[x]$ is zero outside the interval $[-1, 1]$, thus

$$\begin{aligned}\int_{-\infty}^{\infty} \Delta^{\mathrm{m}}\psi[x]\,dx &= \int_{-1}^{1} \Delta^{m}\psi[x]\,dx \\ &= \Delta^{\mathrm{m}-1}\psi^{(1)}[1] - \Delta^{m-1}\psi^{(1)}[-1] \\ &= 1.\end{aligned}$$

Here, $\psi^{(1)}$ denotes the first derivative of the function ψ. Note that the radial basis function $\psi[x]$ satisfies a particularly simple scaling relationship with its dilate $\psi[2x]$ due to its definition as powers of x,

$$\psi[x] = \frac{1}{2^{2\mathrm{m}-1}}\psi[2x]. \tag{4}$$

1.3. The B-Spline Basis

The B-spline basis functions can be expressed as a simple combination of normalized radial basis functions. The exact combination corresponds to the discrete differences of order 2m,

$$\phi[x] = \sum_{\mathrm{i}=-\mathrm{m}}^{\mathrm{m}} \mathrm{d}[\mathrm{i}]\psi[x-\mathrm{i}]. \tag{5}$$

We next show that $\phi[x]$ is the B-spline basis function of order 2m. By construction, $\psi[x]$ is a polynomial of degree $2\mathrm{m}-1$ everywhere except at the origin. Since the mask d[i] annihilates polynomials of degree $2\mathrm{m}-1$, $\phi[x]$ is supported exactly on the interval $[-\mathrm{m}, \mathrm{m}]$. Given that $\phi[x]$ is a piecewise polynomial with $2\mathrm{m}-2$ continuous derivatives, $\phi[x]$ must be a scalar multiple of the standard B-spline basis function.

To complete the proof, we show that the functions $\phi[x-\mathrm{i}]$ form a partition of unity,

$$\sum_{\mathrm{i}=-\infty}^{\infty} \phi[x-\mathrm{i}] = 1, \tag{6}$$

and therefore, are exactly the B-spline basis functions. The key is to analyze the behavior of the expression $\sum_{\mathrm{i}=-\infty}^{\infty} \phi[2^{\mathrm{k}}x-\mathrm{i}]$ as $\mathrm{k}\to\infty$. Applying the definition of $\phi[2^{\mathrm{k}}x-\mathrm{i}]$ and Eq. (4), we note that

$$\sum_{\mathrm{i}=-\infty}^{\infty} \phi[2^{\mathrm{k}}x-\mathrm{i}] = \frac{1}{2^{\mathrm{k}}}\sum_{\mathrm{i}=-\infty}^{\infty}\sum_{\mathrm{j}=-\mathrm{m}}^{\mathrm{m}} (2^{2\mathrm{m}})^{\mathrm{k}}\mathrm{d}[\mathrm{j}]\psi\left[x-\frac{\mathrm{i}}{2^{\mathrm{k}}}-\frac{\mathrm{j}}{2^{\mathrm{k}}}\right].$$

The mask $(2^{2\mathrm{m}})^{\mathrm{k}}\mathrm{d}[\mathrm{j}]$ acts as a discrete approximation to the differential operator Δ^{m} on $\frac{1}{2^{\mathrm{k}}}\mathbb{Z}$. Overall, the right hand side of this equation represents a discrete

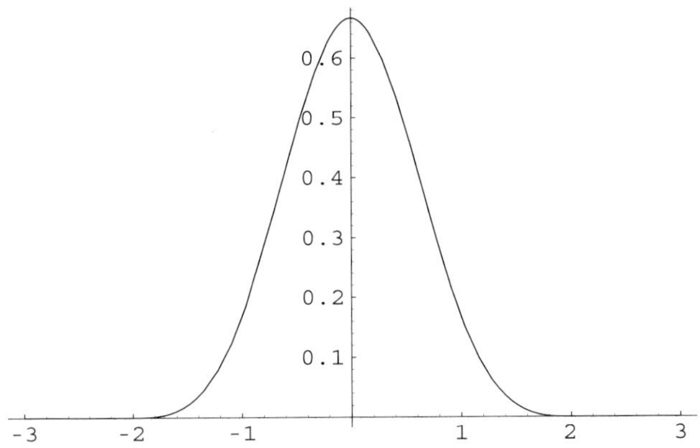

Figure 1. The cubic B-spline basis function $\phi[x]$ defined as a linear combination of radial basis functions $\psi[x]$

approximation to the continuous expression $\int_{-\infty}^{\infty} \Delta^{m}\psi[x]\,dx$ taken over the knot sequence $\frac{1}{2^k}\mathbb{Z}$. Since $\int_{-\infty}^{\infty} \Delta^{m}\psi[x]\,dx$ is one by construction, the residual error

$$\left\| 1 - \sum_{i=-\infty}^{\infty} \phi[2^k x - i] \right\|_{\infty}$$

decreases to zero as $k \to \infty$. However, since this error is independent of k, the error must be zero and Eq. (6) holds.

As a first example, Fig. 1 shows a plot of the cubic B-spline basis function $\phi[x]$. An expansion of $\phi[x]$ in terms of the radial basis functions $\psi[x-i]$ is given by

$$\begin{aligned}\phi[x] &= \psi[x-2] - 4\psi[x-1] + 6\psi[x] - 4\psi[x+1] + \psi[x+2] \\ &= \tfrac{1}{12}|x-2|^3 - \tfrac{1}{3}|x-1|^3 + |x|^3 - \tfrac{1}{3}|x+1|^3 + \tfrac{1}{12}|x+2|^3.\end{aligned}$$

The correctness of our definition also follows from the standard definition of B-splines using divided differences and $(x_+)^{2m-1}$, see [4] pp. 118 for more details. The function $(x_+)^{2m-1}$ can be viewed as a one-sided version of the radial basis function $\psi[x]$. In fact, the definition of B-splines given above also holds for irregularly spaced knot sequences. The key to generalizing this definition is defining appropriate discrete differences. One satisfactory definition is to use the standard divided differences associated with an irregular knot sequence and to normalize them by a factor of $(2m-1)!$ times the size of the support for each basis function.

1.4. Subdivision for the B-Spline Basis

One important property of B-splines is that the B-spline basis function $\phi[x]$ defined on the coarse knot sequence $\mathbb{Z}$ can be expressed in terms of its translates and dilates, $\phi[2x-i]$, on the fine knot sequence $\frac{1}{2}\mathbb{Z}$. The key to deriving this subdivision relation is the scaling relation of Eq. (4),

$$\psi[x] = \frac{1}{2^{2\mathrm{m}-1}}\psi[2x].$$

Taking translates $\psi[x - \mathrm{i}]$ and multiplying by d[i] yields the expanded relation

$$\sum_{\mathrm{i}=-\mathrm{m}}^{\mathrm{m}} \mathrm{d}[\mathrm{i}]\psi[x - \mathrm{i}] = \frac{1}{2^{2\mathrm{m}-1}} \sum_{\mathrm{i}=-\mathrm{m}}^{\mathrm{m}} \mathrm{d}[\mathrm{i}]\psi[2x - 2\mathrm{i}] = \frac{1}{2^{2\mathrm{m}-1}} \sum_{\mathrm{i}=-2\mathrm{m}}^{2\mathrm{m}} \mathrm{dd}[\mathrm{i}]\psi[2x - \mathrm{i}]$$

where dd[i] denotes the coefficient of the generating function $d[x^2]$ associated with x^{i}. The left hand side of Eq. (6) is exactly $\phi(x)$. The right hand side of Eq. (6) can be expressed in terms of a linear combination of fine basis functions $\phi[2x - \mathrm{i}]$. If we denote the corresponding coefficients by s[i], then

$$\phi[x] = \sum_{\mathrm{i}=-\mathrm{m}}^{\mathrm{m}} \mathrm{s}[\mathrm{i}]\phi[2x - \mathrm{i}]$$

where s[i] are the coefficients of the generating function $s[\mathrm{x}]$ of the form

$$2^{2\mathrm{m}-1}s[\mathrm{x}] = \frac{d[\mathrm{x}^2]}{d[\mathrm{x}]} = \left(\frac{1 - \mathrm{x}^2}{1 - \mathrm{x}}\right)^{2\mathrm{m}} = (1 + \mathrm{x})^{2\mathrm{m}}.$$

For example, if $\mathrm{m} = 3$, then $s[\mathrm{x}]$ has coefficients $\left(\frac{1}{8}, \frac{4}{8}, \frac{6}{8}, \frac{4}{8}, \frac{1}{8}\right)$. As a shorthand, we let s[i] denote the coefficient of $s[\mathrm{x}]$ associated with x^{i}.

Using simple linear algebra one can easily verify that the B-spline basis functions satisfy the subdivision formula

$$\phi[x] = \sum_{\mathrm{i}=-\mathrm{m}}^{\mathrm{m}} \mathrm{s}[\mathrm{i}]\phi[2x - \mathrm{i}].$$

As a final note, the subdivision mask s[i] for splines of order 2m can be expressed as the mth discrete convolution of the subdivision mask for splines of order 2. This factorization implies that the B-spline basis functions of order 2m can be expressed as the mth continuous convolution of the B-spline basis functions of order 2 with itself.

2. Poly-Harmonic Splines

Polynomial splines can be generalized to the bivariate case in many different ways. [1] considers the following generalization of the univariate functional for polynomial splines to the bivariate case:

$$\varepsilon[\mathscr{F}] = \int_{-\infty}^{\infty}\int_{-\infty}^{\infty}\left(\sum_{\mathrm{i}=0}^{\mathrm{m}}\binom{\mathrm{m}}{\mathrm{i}}\mathscr{F}^{(\mathrm{i},\mathrm{m}-\mathrm{i})}[x, y]^2\right)dx\,dy \tag{7}$$

As in the univariate case, the function $\mathscr{F}[x, y]$ is constrained to interpolate known values at $(x, y) \in \mathbb{Z}^2$. If $\Delta\mathscr{F}$ is the functional that computes the sum of the second derivatives of $\mathscr{F}[x, y]$ with respect to x and y,

$$\Delta\mathscr{F}[x, y] = \mathscr{F}^{(0,2)}[x, y] + \mathscr{F}^{(2,0)}[x, y]$$

then those $\mathscr{F}[x, y]$ that minimize Eq. (7) satisfy the partial differential equation,

$$\Delta^{\mathrm{m}}\mathscr{F} = 0, \tag{8}$$

everywhere except at the data points, where it is a delta function. Here $\Delta^{\mathrm{m}}\mathscr{F}$ denotes Δ applied to $\mathscr{F}$ m times. Again, [2] and [3] gives a more complete introduction to this topic.

If $\mathrm{m} = 1$, then this differential equation is simply Laplace's equation, also called the harmonic equation, applied to $\mathscr{F}$,

$$\mathscr{F}^{(2,0)}[x, y] + \mathscr{F}^{(0,2)}[x, y] = 0$$

Laplace's equation describes a variety of physical phenomena such as electromagnetism, heat conduction and simple fluid flow.

If $\mathrm{m} = 2$, then Eq. (8) is the biharmonic equation,

$$\mathscr{F}^{(4,0)}[x, y] + 2\mathscr{F}^{(2,2)}[x, y] + \mathscr{F}^{(0,4)}[x, y] = 0.$$

Functions satisfying the biharmonic equation are often referred to as thin-plate splines since the biharmonic equation models the behavior of a thin plate of metal.

In the second half of this paper, we show that taking linear combinations of normalized radial basis functions defines a bell-shaped basis for surfaces, very similar to the univariate B-spline basis. This bell-shaped basis shares many of the important properties of the B-spline basis such as forming a partition of unity and possessing a simple subdivision formula.

2.1. Discrete Differences

The key to defining the bell-shaped basis is a discrete version of the differential operator Δ^{m}. As in the univariate case, the discrete version of this operator is simply a sequence of coefficients that approximate the action of Δ^{m} at the integer knots, $\mathbb{Z}^2$. Recall that for $\mathrm{m} = 1$, $\Delta\mathscr{F} = \mathscr{F}^{(2,0)}[x, y] + \mathscr{F}^{(0,2)}[x, y]$. Therefore, the discrete bivariate mask can be written as the sum of two discrete univariate masks

$$\begin{pmatrix} 0 & 0 & 0 \\ 1 & -2 & 1 \\ 0 & 0 & 0 \end{pmatrix} + \begin{pmatrix} 0 & 1 & 0 \\ 0 & -2 & 0 \\ 0 & 1 & 0 \end{pmatrix} = \begin{pmatrix} 0 & 1 & 0 \\ 1 & -4 & 1 \\ 0 & 1 & 0 \end{pmatrix}$$

This mask can be expressed as a generating function in x and y via

$$(1 \quad x \quad x^2)\begin{pmatrix}0 & 1 & 0\\ 1 & -4 & 1\\ 0 & 1 & 0\end{pmatrix}\begin{pmatrix}1\\ y\\ y^2\end{pmatrix} = -4xy + x + y + x^2y + xy^2.$$

Higher order masks can be generated by simply taking the coefficients of the Laurent polynomial $d[x, y]$ where

$$d[x, y] = \left(\frac{-4xy + x + y + x^2y + xy^2}{xy}\right)^m.$$

Again, the action of the factor $\frac{1}{xy}$ consists in centering the coefficients of $d[x, y]$ around the origin. As a short hand we again denote the coefficient of $d[x, y]$ associated with $x^i y^j$ by d[i, j] (where i and j range from $-m$ to m). Similarly, the coefficient of $d[x^2, y^2]$ associated with $x^i y^j$ is denoted by dd[i, j]. For example, d[i, j] for $m = 2$ represents the coefficient mask

$$\begin{pmatrix}0 & 0 & 1 & 0 & 0\\ 0 & 2 & -8 & 2 & 0\\ 1 & -8 & 20 & -8 & 1\\ 0 & 2 & -8 & 2 & 0\\ 0 & 0 & 1 & 0 & 0\end{pmatrix}.$$

2.2. Normalized Radial Basis Functions

Just as in the univariate case, functions $\mathcal{F}[x,y]$ that minimize Eq. (7) can be written as a linear combination of translates of a single fundamental radial basis function [1], [2]. This function has the form

$$\psi[x,y] = \frac{(x^2+y^2)^{m-1}\mathrm{Log}[x^2+y^2]}{2^{2m}((m-1)!)^2\pi}.$$

Note that $\Delta^m\psi[x,y]$ is a delta function centered at the origin, i.e. $\Delta^m\psi[x,y]$ tends to $\pm\infty$ as (x,y) approaches the origin. The key distinction here is that $\psi[x,y]$ is normalized such that this delta has unit integral,

$$\int_{-\infty}^{\infty}\int_{-\infty}^{\infty}\Delta^m\psi[x,y]\,dx\,dy = 1. \tag{9}$$

We prove this fact by induction on m. First, we restrict the integral of Eq. (9) to the unit disc. This restriction does not affect the integral since $\Delta^m\psi[x,y]$ is zero outside of the unit disc. For the base case $m = 2$ we can apply Green's theorem rewriting this integral as

$$\int_{|v|=1} \frac{\partial \Delta^{m-1} \psi[dv]}{\partial v} \, dv.$$

where v is an outward unit normal to the unit disc. Since the integrand remains unchanged as v varies, the value of this integral is exactly

$$2\pi \Delta^{m-1} \psi^{(1,0)}[1, 0].$$

The constant $\frac{1}{2^{2m}((m-1)!)^2 \pi}$ in the definition of $\psi[x, y]$ normalizes this expression to be exactly one. Finally, the inductive step, $\int_{-\infty}^{\infty} \int_{-\infty}^{\infty} \Delta^{m+1} \psi[x, y] \, dx \, dy = 1$, follows by simple algebraic manipulations.

As in the univariate case, the radial basis function $\psi[x, y]$ shares a scaling relation with its dilate $\psi[2x, 2y]$. For $m = 1$, this relation is

$$\psi[x - i, y - j] - \psi[2x - 2i, 2y - 2j] = -\frac{\mathrm{Log}[2]}{2\pi}. \tag{10}$$

More generally, the functions $2^{2m-2} \psi[x - i, y - j]$ and $\psi[2x - 2i, 2y - 2j]$ differ by a constant multiple of $(i^2 + j^2 - 2ix + x^2 - 2jy + y^2)^{m-1}$. Again, this fact follows from simple algebraic manipulations.

Many important physical problems are modeled by functions of this class. For example, poly-harmonic splines of order $m = 1$ model the behavior of an elastic membrane as well as the pressure potential of a perfect fluid; poly-harmonic splines of order $m = 2$ model the behavior of an elastic plate.

2.3. The Bell-Shaped Basis

In the univariate case, the translates of the radial basis functions $\psi[x]$ defined the B-spline basis function $\phi[x]$. In the bivariate case, we follow the same approach. The bell-shaped basis function $\phi[x, y]$ is defined as

$$\phi[x, y] = \sum_{i=-m}^{m} \sum_{j=-m}^{m} d[i, j] \psi[x - i, y - j].$$

One interpretation of this definition of $\phi[x, y]$ is that the coefficients $d[i, j]$ act as a discrete version of Δ^m applied to $\psi[x, y]$. Since $\Delta^m \psi[x, y]$ was the unit delta centered at the origin, $\phi[x, y]$ is a smooth bump function centered at the origin. Figure 2 depicts the bell-shaped basis functions $\phi[x, y]$ for $m = 1, 2$.

Note that the radial basis function $\psi[x, y]$ is unbounded at $(x, y) = (0, 0)$. Consequently, for $m = 1$, the bell-shaped basis function $\phi[x, y]$ is unbounded at $(x, y) = (0, 0), (1, 0), (0, 1), (-1, 0), (0, -1)$. In Fig. 2, the unbounded parts of the graph were truncated to allow plotting.

Partition of unity: The translates of the bell-shaped basis functions $\phi[x - i, y - j]$ form a basis for the poly-harmonic splines. At first, this fact might seem count-

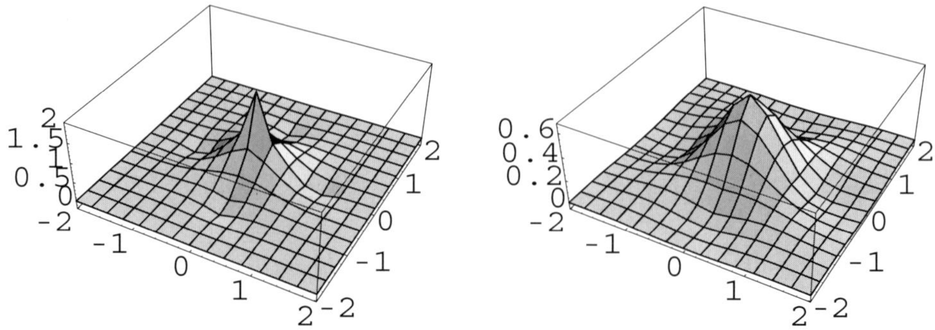

Figure 2. Bell-shaped basis functions $\phi[x, y]$ for m = 1 (left) and m = 2 (right)

erintuitive since the poly-harmonic splines of order m have polynomial precision of order m. However, due to their normalization, the bell-shaped basis functions also have polynomial precision. In this paper, we will prove that they have constant precision,

$$\sum_{i=-\infty}^{\infty}\sum_{j=-\infty}^{\infty}\phi[x-i, y-j] = 1. \tag{11}$$

As in the univariate case, the key idea is to analyze the behavior of the expressions

$$\sum_{i=-\infty}^{\infty}\sum_{j=-\infty}^{\infty}\phi[2^k x-i, 2^k y-j]$$

as $k \to \infty$. Substituting the definition of $\phi[2^k x - i, 2^k y - j]$ and Eq. (10) into this expression yields

$$\frac{1}{2^{2k}}\sum_{i=-\infty}^{\infty}\sum_{j=-\infty}^{\infty}\left(\sum_{u=-m}^{m}\sum_{v=-m}^{m}(2^{2m})^k d[u, v]\psi\left[x-\frac{i}{2^k}-\frac{u}{2^k}, y-\frac{j}{2^k}-\frac{v}{2^k}\right]\right).$$

By definition, this expression is a discrete approximation of the continuous integral $\int_{-\infty}^{\infty}\int_{-\infty}^{\infty}\Delta^m\psi[x, y]\,dx\,dy$ taken on the uniform knot sequence $\frac{1}{2^k}\mathbb{Z}^2$. By construction, this integral is normalized to be one. Therefore, the residual error

$$\left\|1-\sum_{i=-\infty}^{\infty}\sum_{j=-\infty}^{\infty}\phi[2^k x-i, 2^k y-j]\right\|_\infty$$

converges to zero. However, since the value of this expression is independent of k, the error must be zero for all k and Eq. (11) holds.

Localization of the bell-shaped basis: As noted before, the basis function $\phi[x, y]$ has a bump-like shape due to its definition in terms of radial basis functions and discrete differences. In fact, it is possible to show that this basis function has very rapid decay. To facilitate this proof, we convert to polar coordinates using

$$x = r\text{Cos}[\theta],\ \ y = r\text{Sin}[\theta].$$

For m = 1, the basis function $\phi[x, y]$ has a simple expression in polar coordinates

$$\phi[r, \theta] = \frac{1}{4\pi}\text{Log}\left[1 + \frac{1}{r^8} - \frac{2\text{Cos}[4\theta]}{r^4}\right]. \tag{12}$$

This expression can be derived in two steps. First, we convert $\phi[x, y]$ to polar coordinates using the substitutions for x and y listed above and simplify the resulting expression.

$$\begin{aligned}\phi[r\text{Cos}[\theta], r\text{Sin}[\theta]] = \frac{1}{4\pi}(&-4\text{Log}[r^2] + \text{Log}[1 + r^2 - 2r\text{Cos}[\theta]]\\ &+ \text{Log}[1 + r^2 + 2r\text{Cos}[\theta]]\\ &+ \text{Log}[1 + r^2 - 2r\text{Sin}[\theta]]\\ &+ \text{Log}[1 + r^2 + 2r\text{Sin}[\theta]])\end{aligned}$$

Next, we apply the two laws of logarithms, Log[a] + Log[b] = Log[ab] and aLog[b] = Log[b^{a}] to simplify further. (Note that we clear the leading constant of $\frac{1}{4\pi}$)

$$\phi[r\text{Cos}[\theta], r\text{Sin}[\theta]] = \frac{1}{4\pi}\text{Log}\left[1 + \frac{1}{r^8} - \frac{2\text{Cos}[4\theta]}{r^4}\right].$$

So, by Eq. (12), $\phi[r, \theta]$ decays at a rate of $O(r^{-4})$ as $r \to \infty$. For $m > 1, \phi[r, \theta]$ exhibits even higher rates of decay. This observation follows due to the fact that higher order basis functions can be defined via convolution.

Given that the bell-shaped basis functions are highly localized, we conjecture that for $m > 1$, the integer translates $\phi[x - \text{i}, y - \text{j}]$ form a stable basis for the space of poly-harmonic splines. In the case of m = 2, the stability of the bell-shaped basis has been previously studied in [2]. There, the authors proposed preconditioning the interpolation matrix for the radial basis by a discrete version of Δ^{m}. This preconditioning simply amounts to a change into the bell-shaped basis.

2.4. Subdivision for the Bell-Shaped Basis

In the univariate case, the B-spline basis function $\phi[x]$ could be expressed as a linear combination of translates and dilates $\phi[2x - \text{i}]$. These functions corresponded to B-splines with knots at the half-integers, $\frac{1}{2}\mathbb{Z}$. A similar construction holds in the bivariate case. We next express the bell-shaped basis function $\phi[x, y]$

as a linear combination of its translates and dilates, $\phi[2x - \mathrm{i}, 2y - \mathrm{j}]$, the bell-shaped basis functions for poly-harmonic splines over the knot set $\frac{1}{2}\mathbb{Z}^2$.

Derivation of the subdivision mask: The key to deriving the subdivision mask is to recall the scaling relationship shared by $\psi[x, y]$ and its dilate $\psi[2x, 2y]$ from Eq. (10),

$$\psi[x - \mathrm{i}, y - \mathrm{j}] - \psi[2(x - \mathrm{i}), 2(y - \mathrm{j})] = -\frac{\mathrm{Log}[2]}{2\pi}.$$

Since the discrete difference mask d[i, j] annihilates constants, these constant factors cancel in the definition of $\phi[x, y]$. For m = 1 we can easily verify that

$$\sum_{\mathrm{i}=-\mathrm{m}}^{\mathrm{m}} \sum_{\mathrm{j}=-\mathrm{m}}^{\mathrm{m}} \mathrm{d}[\mathrm{i}, \mathrm{j}]\psi[x - \mathrm{i}, y - \mathrm{j}] = \sum_{\mathrm{i}=-\mathrm{m}}^{\mathrm{m}} \sum_{\mathrm{j}=-\mathrm{m}}^{\mathrm{m}} \mathrm{d}[\mathrm{i}, \mathrm{j}]\psi[2x - 2\mathrm{i}, 2y - 2\mathrm{j}].$$

For m > 1, recall that the functions $2^{2\mathrm{m}-2}\psi[x - \mathrm{i}, y - \mathrm{j}]$ and $\psi[2x - 2\mathrm{i}, 2y - 2\mathrm{j}]$ differ by a constant multiple of $(\mathrm{i}^2 + \mathrm{j}^2 - 2\mathrm{i}x + x^2 - 2\mathrm{j}y + y^2)^{\mathrm{m}-1}$. Since the difference mask d[i, j] annihilates polynomials of degree 2m − 2, a similar relation holds for higher order m. For example, for m = 2

$$\sum_{\mathrm{i}=-\mathrm{m}}^{\mathrm{m}} \sum_{\mathrm{j}=-\mathrm{m}}^{\mathrm{m}} \mathrm{d}[\mathrm{i}, \mathrm{j}]\psi[x - \mathrm{i}, y - \mathrm{j}] = \frac{1}{2^{2\mathrm{m}-2}} \sum_{\mathrm{i}=-\mathrm{m}}^{\mathrm{m}} \sum_{\mathrm{j}=-\mathrm{m}}^{\mathrm{m}} \mathrm{d}[\mathrm{i}, \mathrm{j}]\psi[2x - 2\mathrm{i}, 2y - 2\mathrm{j}].$$

The left hand side of this relation is exactly the definition of $\phi[x, y]$. If we let dd[i] denote the coefficients of the generating function $d[\mathrm{x}^2, \mathrm{y}^2]$ associated with $\mathrm{x}^\mathrm{i}\mathrm{y}^\mathrm{j}$, then

$$\phi[x, y] = \frac{1}{2^{2\mathrm{m}-2}} \sum_{\mathrm{i}=-2\mathrm{m}}^{2\mathrm{m}} \sum_{\mathrm{j}=-2\mathrm{m}}^{2\mathrm{m}} \mathrm{dd}[\mathrm{i}, \mathrm{j}]\psi[2x - \mathrm{i}, 2y - \mathrm{j}]. \tag{13}$$

The right hand side of Eq. (13) can now be expressed in terms of a linear combination of fine basis functions $\phi[2x - \mathrm{i}, 2y - \mathrm{j}]$,

$$\phi[x, y] = \sum_{\mathrm{i}=-\infty}^{\infty} \sum_{\mathrm{j}=-\infty}^{\infty} \mathrm{s}[\mathrm{i}, \mathrm{j}]\phi[2x - \mathrm{i}, 2y - 2\mathrm{j}].$$

This subdivision mask s[i, j] corresponds to the coefficient of the generating function

$$2^{2\mathrm{m}-2}s[\mathrm{x}, \mathrm{y}] = \frac{d[\mathrm{x}^2, \mathrm{y}^2]}{d[\mathrm{x}, \mathrm{y}]} = \left(\frac{-4\mathrm{x}^2\mathrm{y}^2 + \mathrm{x}^2 + \mathrm{y}^2 + \mathrm{x}^4\mathrm{y}^2 + \mathrm{x}^2\mathrm{y}^4}{-4\mathrm{xy} + \mathrm{x} + \mathrm{y} + \mathrm{x}^2\mathrm{y} + \mathrm{xy}^2}\right)^{\mathrm{m}}. \tag{14}$$

Computation of the subdivision mask coefficients: Unfortunately, $d[\mathrm{x}, \mathrm{y}]$ does not exactly divide $d[\mathrm{x}^2, \mathrm{y}^2]$. This fact is expected since $\phi[x, y]$ has infinite support.

However, *s*[x, y] does exist as a Laurent polynomial. This polynomial corresponds to the expansion of *s*[x, y] as a bi-infinite power series centered at the origin that is convergent at $(x, y) = (1, 1)$. To compute this series, we focus on the case of $m = 1$ since higher order *s*[x, y] are simply powers of *s*[x, y] for $m = 1$.

At first glance, one might doubt whether this series actually exists since $d[1, 1]$ is zero. However, if we for example expand both $d[x, y]$ and $d[x^2, y^2]$ at (1, 1) for $m = 1$, then

$$d[x, y] = -x^2(y-1) + y^2 - xy^2, \text{ and}$$

$$d[x^2, y^2] = -4x^3(y-1)^2 + x^4(y-1)^2 + (y-2)^2y^2 - 2x(y-2)^2y^2 + x^2(2-2y+y^2)^2.$$

The low order terms of $d[x, y]$ and $d[x^2, y^2]$ are $x^2 + y^2$ and $4x^2 + 4y^2$, respectively. Thus, *s*[x, y] converges to 4 as (x, y) approaches (1, 1). Using simple linear algebra we compute a finite power series approximation to $\frac{-4x^2y^2+x^2+y^2+x^4y^2+x^2y^4}{-4xy+x+y+x^2y+xy^2}$ of a given size and use the coefficients of this mask as an approximation of the subdivision scheme. Based on our arguments above, the coefficients of this power series rapidly converge to zero as we increase the support.

Figure 3 shows a plot of the coefficients of such a 5×5 approximation. Note the similarity of this plot to the plot of $\phi[x, y]$ for $m = 1$, see left half of Fig. 2.

Examples: At this point we can use the finitely supported power series approximations as subdivision masks *s*[x, y].

One nice property of capturing the subdivision mask as a generating function is that the effects of $n + 1$ steps of subdivision can be captured as a product of the generation functions

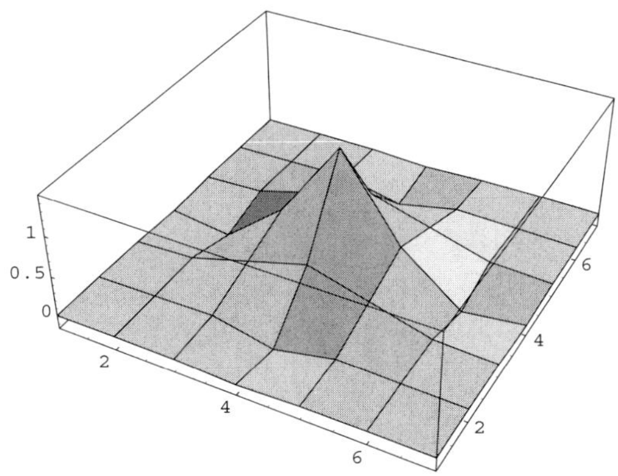

Figure 3. Local approximation of the subdivision mask *s*[x, y] of support 5×5, $m = 1$

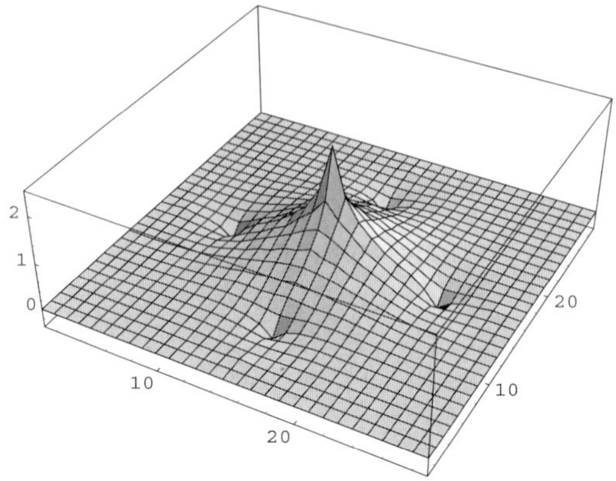

Figure 4. Three rounds of local subdivision for the modeling of the bell-shaped basis for m = 1

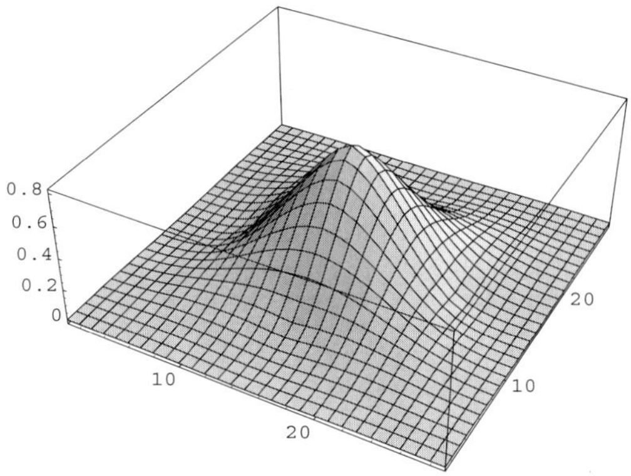

Figure 5. Three rounds of local subdivision for the modeling of the bell-shaped basis for m = 2

$$\prod_{i=0}^{n} s[x^{2^i}, y^{2^i}].$$

As a first example, Fig. 4 shows the results of three rounds of subdivision for the basis function $\phi[x, y]$ for m = 1. Note that the subdivision scheme is converging to $\phi[x, y]$ everywhere except at points in $\mathbb{Z}^2$.

Figure 5 shows a plot of $\phi[x, y]$ after three rounds of subdivision for m = 2.

Due to the factorization of Eq. (14), the bell-shaped basis functions of order m can be expressed as m continuous convolutions of the bell-shaped basis functions of order 1 with itself.

In fact, the corresponding subdivision scheme has the property that it diverges (very slowly) at the integer grid points (just as the analytic basis does) and converges everywhere else. Thus, the graphs of the basis function produced by subdivision always appear to be bounded for a small (say < 10) number of rounds of subdivision. Since poly-harmonic basis functions (i.e. $m > 1$) can be expressed in terms of the $m = 1$ harmonic basis function through convolution, we felt that the case of $m = 1$ was worth directly addressing.

3. Conclusions

In this paper we exposed the link between radial basis functions and the B-spline basis for piecewise polynomial splines. Taking the same approach in two dimensions, we can define a surface basis, called bell-shaped basis, for poly-harmonic splines which behaves much like the B-spline basis for curves. Subdivision schemes for these bases follow naturally and provide for the efficient implementation of these schemes.

To conclude, we note that bell-shaped bases can also be defined for irregularly spaced sets of knots. The key problem is to generalize the discrete differences used in defining $\phi[x, y]$. One possibility is to use the energy matrices arising from the variational approach of [5] as discrete approximations to Δ^m. We intend to address this problem in a future paper.

Acknowledgements

This work was supported in part under NSF grant number CCR-9732344. The authors would like to thank the anonymous reviewers for their helpful, constructive criticism.

References

[1] Duchon, J.: Splines minimizing rotation invariant semi-norms in Sobolev spaces. In: Constructive theory of functions of several variables (Keller, M., ed.), pp. 85–100. Berlin Heidelberg New York: Springer.

[2] Dyn, N., Levin, D., Rippa, S.: Numerical procedures for surface fitting of scattered data by radial functions. SIAM J. Stat. Comput. 7, 639–659 (1986).

[3] Hoschek, J., Lasser, D.: Fundamentals of computer aided geometric design. Wellesley: A. K. Peters, 1989.

[4] Schumaker, L.: Spline functions. New York: J. Wiley, 1981.

[5] Warren, J., Weimer, H.: Variational subdivision for natural cubic splines. In: Approximation Theory IX, Vol. 2, (Chui, C.K., Schumaker, L.L., eds.), pp. 345–352., Vanderbilt University Press, 1988.

J. Warren
H. Weimer
Department of Computer Science
Rice University
P.O. Box 1892
Houston, TX 77251-1892
USA
e-mails: {jwarren, henrik}@rice.edu